Construction Failure

Wiley Series of Practical Construction Guides

M.D. Morris, P.E., Series Editor

William G. Rapp

CONSTRUCTION OF STRUCTURAL STEEL BUILDING FRAMES

Jacob Feld

CONSTRUCTION FAILURE

Construction Failure

Jacob Feld
Consulting Engineer

John Wiley & Sons, Inc.
New York London Sydney Toronto

2 3 4 5 6 7 8 9 10

Library of Congress Catalog Card Number: 68–30908
SBN 471 25700 1
Printed in the United States of America

This book is dedicated to my two grandsons

Michael E. and *Andrew S. Marrus*

with the hope that they will continue their early interest in the construction field as a life work of fulfillment and satisfaction.

Series Preface

The construction industry in the United States and other advanced nations continues to grow at a phenomenal rate. In the United States alone construction in the near future will exceed ninety billion dollars a year. With the population explosion and continued demand for new building of all kinds, the need will be for more professional practitioners.

In the past, before science and technology seriously affected the concepts, approaches, methods, and financing of structures, most practitioners developed their know-how by direct experience in the field. Now that the construction industry has become more complex there is a clear need for a more professional approach to new tools for learning and practice.

This series is intended to provide the construction practitioner with up-to-date guides which cover theory, design, and practice to help him approach his problems with more confidence. These books should be useful to all people working in construction: engineers, architects, specification experts, materials and equipment manufacturers, project superintendents, and all who contribute to the construction or engineering firm's success.

Although these books will offer a fuller explanation of the practical problems which face the construction industry, they will also serve the professional educator and student.

M. D. MORRIS, P.E.

Preface

In some 40 years of consulting civil engineering practice and, to a smaller extent in the previous 10 years of work in engineering design and construction, I have seen many instances of construction failure, both collapse and distress, which seemed inexcusable. In return for a half century of satisfying and profitable occupation I offer this collection of experiences to the construction industry, not as a castigation or indictment but rather as a warning in a series of lessons to be learned from unfortunate or ill-advised procedures which led to these failures. The building of structures is the contribution made by this industry to the advancement of the living standards of humanity. It can be done better, however, with less grief and fewer troubles. A review of what has not worked so well as it might have can help to provide the necessary improvements in technique and control in every facet of the construction industry.

It would be impossible to give the names of the hundreds of people associated with the investigations and reports of these incidents. Even a list of references became too voluminous for inclusion in the book. Thankful credit is due to Professor Seymour S. Howard, Jr., of Pratt Institute, Brooklyn, New York, for his critical comments on the final text and for the preparation of the diagrams used as illustrations. The horrendous typing job of the several preliminary texts was performed by Mrs. Esther Whitton. The final version was produced by Ruth Wales of Typing Unlimited.

JACOB FELD

New York, New York
June 1968

Contents

1

Introduction; Concept Error

Introduction, Definition of Failure, Historical Notes, Cause of Failure, Design and Drawing Cause of Failure, Production Deficiency.

1.1 Introduction

"Of making many books there is no end," said Koheleth in Ecclesiastes 12:12, so a new book should have justification for its existence to avoid overpopulation. This volume is a compilation of experiences, that of the author and of many others. Experience is one of the greatest assets of a professional in any field of endeavor and equally valuable to the creative person. Construction involves many facets of training, the preparation of designs based on concepts of the planner, delineation of such information into the universal language of drawings and models, the preparation of materials and component assemblies, and finally the production of the finished product. It is the production of goods, both software and hardware, that improves the standard of living of a people. The climate for a better and fuller life of a community is set by such production. Ideas have never improved the lot of people; only through production can that be attained. And the construction industry, from ancient days, has been in the forefront of such progress.

The various and separate phases of the construction industry increase their experience and capability by learning from their mistakes, but many of these could be avoided if each knew the mistakes made by others. In the hope that this compilation will tend towards that end, this

book is thus presented to the construction industry as one man's contribution. Because many failures are not explained and often not even publicly reported, no complete coverage of the subject is claimed or even attempted. Lessons can be learned from the material presented here; many more lessons could be made available from the unlimited experience of others. Hopefully this work will loosen up valuable information in the files and records of private investigation for the benefit of the entire industry.

1.2 Definition of Failure

If we define failure as observed collapse, there are few failures. But if nonconformity with design expectations is defined as failure, and this is the more logical and honest approach, and one takes the trouble to measure the shape, position, and condition of completed structures, there are many failures—far more than the list of incidents that are covered by news media, both technical and public. This statement is more applicable to the complicated space framings than to the simple or pin-connected structures. Unwanted displacements, unexplainable deformations, are often found and it is questioned whether they are failures or normal (but unexpected) strains or merely "incidents," using a foreign term to describe the unexpected results.

Whether failure or not, there usually follows long and expensive argument and litigation, during which the experts are pumped by their clients' lawyers and crossexamined at great length by the opponents' counsel in an attempt to pin down the "proximate" cause of the failure or incident—as if the legal fiction that responsibility stems from a single and only cause makes it possible to determine such cause by either observation or deduction. Sometimes there is a single simple explanation for a failure, but usually it is a combination of conditions, mistakes, oversights, misunderstandings, ignorance, and incompetence, even dishonest performance, but not a single item by itself that can be picked as the sole and only cause of failure. Yet each in a way can be said to be the responsible straw that broke the camel's back, and if one straw had not been added, the camel's back would not be broken. The dimension between success and failure in any construction operation is that of the thinnest hair.

The construction industry may be aided by a book of this nature to produce and perform more successful designs, from the lessons describing errors in judgment, in design, in detail, in control, and in performance. This type of book is not easy to write since the author is not a

news reporter interested in what the public has been conditioned to accept as "news" with little emphasis on truth, nor is the morbid account of accidents and mistakes a pleasurable task. For many years he has felt that one of the worst indictments of his own profession is in the statement; "Doctors bury their mistakes, architects cover them by ivy, engineers write long reports which never see the light of day." To include the rest of the anticipated readers, one should add "and contractors call their lawyers and notify their insurance carriers."

Failures are defined here as a behavior not in agreement with the expected conditions of stability, or as lacking freedom from necessary repair, or as noncompliance with the desired use and occupancy of the completed structure. Failures occur in all types of structure, small and large, low and tall, minimal and monumental, whether framed or wall bearing, whether with timber, steel, or concrete as the basic supporting material.

Construction is taken as the sum total of project concept, choice of materials, structural design, production of materials, erection of the components, and even final cleanup and equipment installation. All factors involve two related requirements: sufficiency and necessity. Sufficiency provides safety, not only from collapse but also from undue deterioration. Necessity is a measure of economy, very important in this industry, but should only be considered after sufficiency has been satisfied.

Continuous pressure for greater economy, from private financial competition as well as from public demand that budgets be met even if unreasonable, both in design and in construction, has often resulted in safety being reduced below the minimum sufficiency. Failure of part or even serious collapse of a structure usually comes during construction, when the latent uncalculated space frame strength is not yet available. The boundary between stability and unstability, between sufficiency and failure, is a thin line. Ignorance of the boundary is no excuse when a failure occurs.

Much more emphasis is needed in technical education to teach what not to do and when to say "no," rather than to give the impression to the untutored and inexperienced mind that blind compliance with minimal code provisions and reports of committees, signed by all members after several years of disagreement but with ultimate consent to the strongest willed minority, is a guarantee of sufficiency. Perhaps every technical and trade organization should award annual prizes for complete descriptions and analyses of failures, so that everyone can learn from that experience and avoid similar trouble in his own practice.

There is sufficient professional precedence for reporting incidents of

nonsuccessful nature. The medical journals describe in detail the diagnosis and treatment that did not result in a successful cure. The legal tomes give in detail the arguments on both sides of a trial, in which exactly half the litigants did not win. Scientific literature is full of reports of detailed and lengthy research ending in blind alleys with no positive results, which become warnings of which directions are nonproductive of results. The construction industry has a duty to record for the use and profit of the present and of the future workers in the industry, the experiences which have not been successful with as much information as can be gathered to explain in detail wherein the fixed and uncompromising laws of nature were violated and insufficient resistance was provided against internal and external forces.

1.3 Historical Notes

The science or art of archeology is dependent to a large extent upon the uncovered debris of engineering failures. If the code of Hammurabi (about 2000 B.C.) is typical of ancient regulations, few engineer builders had the opportunity to learn from their own mistakes. The five basic rules covering construction failures are reproduced in original, transliteration, and translation in Fig. 1.1. Whether these rules stopped all failures is not known, but they certainly must have been a deterrent to shoddy construction practice and eliminated the possibility of repetitive malpractice. It is of passing interest to note that the symbol for "builder" is the framework of a house, consisting of a main center post with four corner posts all connected and braced by roof beams, and the center post is the symbol for a man.

Some warnings in the Hebraic literature against building on shifting sand may stem from experiences encountered when the Hebrews were the construction battalions in Egypt. History and fable report instances of summary execution in ancient civilizations of the designer-builders of successful monuments as a guarantee that competing or better constructions would not be performed to outrank the architectural gem. If these stories are based on fact, there would have been an incentive to perform poorly and imperfectly.

That entire ancient city complexes collapsed from normal aging, although probably also from enemy attack, is found in the deep excavations at city sites with definite separate layers of rubble. In the exploratory trenches dug at the Walls of Jericho Major Tulloch, then retired chief engineer of the Allenby invasion of Palestine in World War I,

A. If a builder build a house for a man and do not make its construction firm and the house which he has built collapse and cause the death of the owner of the house – that builder shall be put to death.

B. If it cause the death of the son of the owner of the house – they shall put to death a son of that builder.

C. If it cause the death of a slave of the owner of the house – he shall give to the owner of the house a slave of equal value.

D. If it destroy property, he shall restore whatever it destroyed, and because he did not make the house which he built firm and it collapsed, he shall rebuild the house which collapsed at his own expense.

E. If a builder build a house for a man and do not make its construction meet the requirements and a wall fall in, that builder shall strengthen the wall at his own expense.

Translated by R.F. Harper.
"Code of Hammurabi" p. 83 – seq.

Jacob Feld 1922.

Fig. 1.1 Code of Hammurabi (2200 B.C.).

uncovered at least six distinct collapsed walls before coming to the wall which Joshua felled. Reportedly that collapse was the first failure from sonic forces; there are many now being investigated in the fields of supersonic aviation and high-velocity missiles. Tulloch however claimed proof that the Jericho wall fell from the undermining of the foundation stones, operations being carried on by sappers while the defenders were distracted by the blowing of horns.

In Greco-Roman times the construction industry was in the hands of trained slave artisans and successful work was rewarded by gifts of substance as well as of freedom. With no competitive financial or time schedule to worry the builder, work was well done and with apparent great success, as is seen in the many examples still existing. Similarly, in the Medieval period time was no object and the great successful religious and governmental structures have lasted for centuries. One wonders whether our modern works will exist to comparable age.

In the common law developed in England, as is found in the fifteenth century court records from the reign of Henry IV, the rule was, "if a carpenter undertake to build a house and does it ill, an action will lie against him." Of course, by "ill," is meant "not well," and if all work were done well, there would be no record of failure. So historically the burden rests on the construction industry to see that work is done "not ill" and each partner in the industry, from the architect who conceives the project to the foreman who directs the performance of a part thereof, must lend every effort to avoid and eliminate every possible cause of "ill" structures.

The Napoleonic code has become the basis of the common law wherever the original settlers were French. In many ways there is a greater responsibility placed on the designer and the professional in charge of the work as agent of the owner, to safeguard the investment and guarantee proper and adequate performance, than is expected under English legal procedure.

In recent years some books and many articles have appeared to warn of the danger and to plead for open discussion and information of actual incidents. There seems to be a smaller restriction to such open discussion in the fields of foundations, earth structures, and bridges than in the work generally classified as superstructures and buildings.

In the period from 1870 to 1900 construction became a large industry in the United States, chiefly influenced by the expansion of the transportation system with its necessary bridges, storage facilities, and industrial plants. Comparatively little is recorded of failures in buildings, most of which were heavy masonry wall bearing structures with good timber floors. Bridge spans soon exceeded the capability of timber trusses and a

great competition in the sale of iron spans resulted in many failures. These were spectacular and were almost daily newspaper headlines. Even foreign technical journals commented on the great number of unsuccessful designs of bridges in the United States. Engineering magazines from 1875 to 1895 were as full of reports of railroad accidents and bridge failures as today's daily newspaper reports of automobile traffic accidents. Even as late as 1905, the weekly news summary in *Engineering Record* described the most serious railroad wreck of the week, usually tied in with a bridge failure. The *Railway Gazette* in 1895 published a discouraging summary of iron bridge failures resulting from railway traffic, listing 502 cases in the period from 1878 to 1895 inclusive, collected by C. F. Stowell, and noting that the first 251 occurred in 10 years while the second 251 occurred in only eight years. In the years 1888–1891 inclusive there were 162 such accidents. These reports received wide publicity and attention and must have had considerable influence on the designing engineers as well as the bridge salesmen of those days.

Tank failures also made common news. Bryant Mather in a discussion of a paper by the writer (*ASCE Proc.*, **81,** separate 632, 1955) reported 65 tank failures from 1879 to 1953, all similar to the failure of a molasses tank in Boston in May 1919 that put the elevated railroad out of service for several weeks. Incidentally, an identically designed tank for the same use failed in Havana Harbor in 1925.

Every new development in the construction industry is accompanied by a concentration of failures. If they are brought to light and the warnings are heeded, repetition of failures stemming from the same causes can be prevented. When reinforced concrete was being accepted as a satisfactory structural material, economy being then as now one of the important arguments in favor of its use, the editors of the *Engineering News* in the April 9, 1903 issue wrote:

"One of the special advantages usually claimed for concrete work is that it can be safely built by unskilled labor, but in view of some recent accidents which occurred it appears well to point out that this principle is not of universal application. For concrete in large masses such as abutments, foundations, retaining walls, etc. there is no doubt that unskilled labor should be employed, under proper supervision. But for certain other classes of work, girders and floors in concrete buildings, it appears that some degree of skilled labor should be employed, or at least the entire work should be under strict and constant supervision by skilled foremen, architects or engineers. In the carpentry work for forms and falsework especially, there is frequent evidence that the weight of the mass to be supported and the hydrostatic pressure of very wet concrete in columns are not realized by the men who are entrusted with the construction of this part of the work. When concrete has once thoroughly set it will stand very hard service and even overload or abuse, but it should be very strongly impressed upon the men engaged in concrete construction that the wet mass is simply a dead weight to be supported, having absolutely no supporting power in itself. If steel bars, rods, etc. are used, these simply add to the weight of the wet mass. The fall of a concrete floor at Chicago, noted in our issue of December 4, seems to have been due entirely to an ignorant man blindly following out instructions which were probably extremely indefinite. He was told to go in and knock out some of the shoring, and he proceeded to knock out every bent, until the unsupported concrete gave way, and its fall broke through other completed concrete work below.

"Another accident which is said to be due to failure of falsework is recorded elsewhere in this issue. In another case which came under our personal observation a laborer was found knocking away some of the struts and braces under a green concrete floor, simply because (as he told the superintendent when discovered) a carpenter told him to get some lumber. When informed that he stood a good chance of killing himself and other men in the work, as well as wrecking the building, he simply became surly and appeared to think that the superintendent was making a fuss about nothing. The same applies to the concreting gang. The men usually employed for this class of work simply dump the concrete into the forms without discretion, frequently being left without observation by even a foreman for a considerable time. This is especially serious at the junctions of concrete columns and girders. We have heard of instances where forms for girders were filled before the form at the junction with the column was complete, the men simply putting a piece of plank at the end of the form for the girder to make a stop-off and prevent the concrete from running out. If the column is built immediately and the block pulled out there may be little harm done, but there is the liability that the plank may be forgotten, or the concrete of the girden allowed to set before the concrete for the column is deposited. In either of these cases the girder becomes a cantilever, being partially or entirely separated from the column, instead of forming a homogeneous mass with it. There is no doubt that a great deal of concrete work is built by men who are really not competent to undertake it, and that much more work is done without sufficiently strict and continuous expert supervision to ensure the best and safest results. In view of the enormous increase in the use of concrete, and in the variety of purposes for which it is used, it is well for engineers to bear these facts in mind."

This editorial should be required reading for everyone in this industry at least once a year.

In 1918 the American Railway Engineering Association published an editorial article on "Study of Failures of Concrete Structures" with the subheading "A Compilation of Failed Concrete Structures and Lessons to be Drawn Therefrom." The study covers a period of 25 years and classifies the causes under the following headings.

1. Improper design.
2. Poor materials or poor workmanship.
3. Premature loading or removal of forms before complete setting.
4. Subsidence of foundations, fire, etc. The final conclusion given is, the following:

"The one thing which these failures conclusively point to is that all good concrete construction should be subjected to rigid inspection. It should be insisted upon that the Inspector shall force the Contractor to follow out the specifications to the most minute details. He must see that the materials used are proper and are properly mixed and deposited, also that the forms are sufficiently strong and that they are not removed until after the concrete has set. It is believed that only by this kind of inspection is it possible to guard against the failure of concrete structures."

So the 1903 editorial advice still held in 1918, and the lesson was not learned and is still not learned.

In 1924 Edward Godfrey, a structural engineer consultant well-known for his frank and tireless criticism of improper design techniques, published a book on *Engineering Failures and their Lessons*, consisting in a large part of discussions and letters contributed in the years 1910–1923. Included is a copy of a letter to W. A. Slater published in *Concrete*, Detroit, February 1921 in which are summarized the reports of reinforced concrete structures failures from 1900 to 1920. Of the 24 examples cited, most are explained as resulting from frost, insufficient shear resistance, lack of overlap of reinforcement at beam to column connections, and excessive sag.

Every phase of civil engineering is covered in the 20 chapters, the variety of references and the scope of coverage indicate the large amount of effort to present a whole picture. The concrete examples are somewhat clouded by the two fixed ideas of Godfrey in reinforced concrete design, that stirrups are of no value as shear resistance and that hoop-tied column bars are a detriment to a concrete column.

In 1952 Henry Lossier published a small book on *La Pathologie du*

Beton Arme and cites many examples of failures, mostly in concrete frames and special structures, with conclusions similar to those of the 1918 AREA paper plus warnings that errors in the choice of framing and in design details are most important factors causing failures. A revised and expanded edition of this book, published in 1955 under the title *La Pathologie et Therapeutique du Beton Arme* was translated into English and published as *Technical Translation 1008* by the National Research Council of Canada in 1962. This work covers reports on a great variety of structural and foundation failures both in design and in execution. There are some interesting examples of false reasoning in the design of concrete structures and on the misdeeds of certain engineering experts. Applicable to all problems is the outline of steps to be taken in investigating an accident happening to a structure.

Under the title *Engineering Structural Failures,* Rolt Hammond published a book in 1956 to cover the causes and results of failures in modern structures of various types. In an interesting forward Sir Bruce White quotes Robert Stevenson, when President of the British Institution of Civil Engineers in 1856 in summing up a paper, that

"... he hoped that all the casualties and accidents, which had occurred during their progress, would be noticed, in revising the Paper; for nothing was so instructive to the younger Members of the Profession, as records of accidents in large works, and of the means employed in repairing the damage. A faithful account of those accidents, and of the means by which the consequences were met, was really more valuable than a description of the most successful works. The older Engineers derived their most useful store of experience, from the observations of those casualties which had occurred to their own and to other works, and it was most important, that they should be faithfully recorded in the archives of the Institution."

Hammond covers in a very broad way the spectacular collapses in the fields of earthworks, dams, maritime structures, buildings and bridges, underground structures, vibration problems and welded structures. Although the author is British, only the chapters on Earthworks and Tunnels cover troubles in English work, with only passing references if any to local items in the other fields of work. Some fields such as reinforced concrete are completely ignored. The overall picture does not permit any instructive or usable recommendations in the general field of construction, since only spectacular examples are cited, and the last chapter on "Lessons of Failures" is a description of laboratory and full scale stress and strain testing procedures, exclusively as performed in Britain.

C. Szechy in 1957 published in book form a selection of foundation failures in Hungary, with a description of causes and remedial measures adopted. To obtain greater circulation, this Hungarian book was translated into English with examples added from the literature of other countries. In all 72 foundation failures are described, 44 from the original text and separated under design, execution and external influences as causes. The investigations are described in great detail in *Foundation Failures,* Concrete Publications Ltd., London, 1961. This book also appeared in a Russian translation.

Following the same format, Szechy added many more examples and expanded the work to 228 pages printed in German as *Grundungsschaden* in 1964 and as *Alapozasi Hibak* in Hungary. The German work is published by Bauverlag GMBH, Wiesbaden.

Dealing chiefly with maintenance problems and repair are books by S. Champion on *Failure and Repair of Concrete Structures,* 1962, and by S.M. Johnson on *Deterioration, Maintenance and Repair of Structures,* 1965.

The author has contributed several articles on general and restricted parts of the failure problem in the technical press here and abroad and also written the American Concrete Institute Monograph No. 1 on *Lessons from Failures of Concrete Structures,* 1964 and the section on "Concrete Failures" for the *American Jurisprudence Proof of Facts,* annotated, **19,** Bancroft-Whitney, San Francisco, 1967, a guide book for the legal profession interested in such matters.

1.4 Cause of Failure

What makes a structure misbehave? Why do some fail? Following a lecture on the design and construction of the New York Coliseum, scheduled a week after the formwork failure at that project, this author was asked as the first question, "What made the floor fall?" The answer, at the spur of the moment, was: "Gravity." Such is probably the universally true answer to any state of affairs where insufficient resistance is provided to resist vertical fall, since gravity always acts. However where a vertical assembly of components is stable and failure still occurs, either some outside influence has degraded one of the chain of vertical resistances (supports) or else a horizontal force has laterally displaced one of the components so that the chain of vertical resistances has been broken.

Many failures can be avoided, as is illustrated by case histories in the succeeding chapters. Avoidance requires the implementation of certain

skills and the continuous control of performance to avoid omission of a necessary factor. Where construction booms, the sudden large increase in volume of work outruns the availability of trained personnel to exercise the necessary control, and failures become more common.

Such an apparently uncontrollable rash of structural failures hit England in 1966. The London *Construction News* of November 10, 1966 asked "When is it going to stop?" and listed the sequence of seven major failures in the past 12 months. The attempt to correlate available information to determine an underlying common reason for the failures did not lead anywhere. The variety of failures and of the types of structures involved seemed to indicate that every kind of construction was vulnerable. The list included the Ferry-bridge, three concrete cooling towers that collapsed, the stoppage of work at the Bedford County Hall where the foundations and structure were found unsound, the collapse of the steel roof of the Ford Research Center at Basildon, the partial collapse and further cracking of precast concrete beams at the M.4 Coldra-Crick viaduct, the reinforced concrete silo collapse at Whitehaven, the collapse of the partially completed steel frame for the Zoology Building at Aberdeen University, and the successive rejection of welded steel frame for the carrying trusses at the Fenchurch Street Building in London. The conclusion was that close examination is needed for the system and method used for checking both design and construction, and also of the allocation of responsibility.

The Sunday Times of November 13, 1966 took up the story and pointed out that the city of London has had a freedom from building collapse over a century or more, unparallelled in the world. The Great London Council area maintains 28 district surveyors, all qualified engineers, who have the status of commissioners and are responsible for the correctness of calculations and the safety of design and construction of all buildings constructed therein. No such control exists in most of Britain, although the suggestion is weakened by the admission that the designing engineer of the Bedford County Hall was also the engineer responsible for checking major designs for the county. There is no legal registration of engineers or architects in Britain.

Almost a year earlier H. L. Childe, editor of *The Builder* wrote an article headed "Why Do Structures Fail?" Complaining that little of the data collected in inquiries and inquests reaches the public, he notes that "It is important that the causes of failures be known so that they can be avoided in future. This is particularly so at a time when many new ideas and methods and materials are being used." Childe cites several failures where the causes have not been disclosed. The failure of the concrete cooling towers was a mystery if the force of the wind was below the

probable maximum, the effect of wind was calculated by the most modern method then available and accepted by all the leading concerns designing such towers, and the materials and workmanship were generally satisfactory. Another example was the failure of a multistory precast concrete building exactly identical with the first building which was erected successfully. A new church collapsed after it was completed and the explanation was that it was a novel design, but it appeared to be similar to many other structures built in recent years. In the autumn of 1965 the Minister of Housing and Local Government was scheduled to open a large new housing community, but the ceremony was cancelled because the houses were so badly built with bulging walls and fittings that "came away from the walls at a touch." In one of the new towns scores of flat roofs blew off new houses but no one reported why or how they were fixed. A case of complete information was the bulging brick wall of a new home, awarded a certificate of good quality by the National House Builders' Registration Council, where it was determined that the mortar comprised one part of cement to 14 parts of sand because cement had been diverted illegally by the foreman. Another is the failure of welded joints in the steel frames of the Fenchurch Street building and the substitution of prestressed concrete girders. The M1 Motorway is continuously under repair and restricted use; is the use of a flexible base, the first time in England for such a large contract, the real trouble? The new Rochester bridge had an unexplained slip of one abutment. The Sevenoaks bypass was built at a site where local people advised no structure would be stable, and was abandoned after completion when the piers tipped 5 to 10 deg with a few weeks of use. A year later a second location was again abandoned. The writer completes the record with a plea for some unbiased group to publicize the necessary lessons required to stop the series of failures which waste so much public money.

Similar public requests for information and control of the ills of construction have appeared at several conferences, reported in the technical press, in the United States. "Failures: Too Often, Too Similar," a discussion by three engineers, Boyd G. Anderson, Frederick S. Merritt, and Jacob Feld, before a meeting of the New York Chapter of the American Institute of Architects in 1960, summarized causes in the following sequence of statements and questions:

1. A frequent faulty design practice is changing the design without the knowledge of the original designer.
2. Another faulty design practice leading to failures is poor drafting and insufficient checking of drawings.

3. Thorough checking for errors is an added design expense, but worth every cent it costs.

4. Too often, design contracts do not include provisions for supervision of construction.

5. Should the engineer alone be responsible for the integrity of a completed structure?

6. Design oversight includes a number of errors that recur frequently:

   Inadequate attention to thermal effects.
   Insufficient bearing by framing on walls and wall or lintel resultant failure.
   Short reinforcing steel and insufficient or badly placed laps.
   Shear and flexure reinforcing omitted from beams.

7. Faulty construction practices cover sins of commission as well as omission, such as:

   Too many temporary holes in the framing.
   Completed frame erection without permanent connections or installation of floor systems.
   Faulty bracing.

8. "Adequate" temporary bracing sometimes is not enough.

Recognition of the possibility of failure has caused a number of engineers to study the likely causes and sound warnings to the industry. Technical committees have been active recently in evaluation of the true strength of existing structures and in the study of the possibility of prewarning that a structure is degrading and may require a greater restriction on permissive load capacity or even should be demolished. To date the architectural, contracting and vendor segments of the industry have not contributed to this necessary activity, an insurance against public demand that legal restrictions be enacted to force compliance with the basic laws of nature. Gravity always acts to cause an unsupported component to fall; lateral forces can only be neutralized by lateral resistances; all inanimate objects tend to disintegrate.

Failure from tropical environment deterioration of military material instigated the formation in 1945 of the Prevention of Deterioration Center, as a branch endeavor of the National Academy of Sciences-National Research Council in Washington. The Center's responsibilities until 1960 were in the field of "natural" environment material deterioration and its prevention, providing the Armed Services with advisory, consulting, and general information services in the detection and cure of such losses. Emphasis is on the effects of moderate heat, sunlight, wind,

sand, dust, moisture in all forms, oxygen, ozone, salts, acids, alkalies, mildews and rot organisms, bacteria, insects, and marine organisms. The Center's functions were enlarged in June 1960 to include the "induced" environment effects on materials and equipment in the space program.

Collection of difficulties in maintaining use of structures located in areas of adverse weather, was the basis of a series of reports by Palmer W. Roberts, Captain, CEC, U.S. Navy on the effect of such conditions on design and construction. Adverse weather areas of the world are classified as (a) wet hot, (b) dry hot, (c) wet cold and, (d) dry cold. Each environment requires special precautions to avoid early failure of new facilities. The destructive elements in the "wet hot" tropical area are the combination of high temperature, high humidity, heavy rainfall, drenching dews, and radiation. The requirements include protection against galvanic corrosion of all metals, pressure treatment of all lumber, continuous sealing of shrinkage cracks in all masonry, control of fungus and algae on all exposed surfaces, and the use of materials only if stable under large and variable moisture content. The chief troubles in "dry hot" areas comes from the large daily temperature fluctuations, high-temperature effects on plastic sealants and water-tight joints, as well as on concrete during the setting period. In the "wet cold" areas, permafrost in the ground and drifting snows are the chief problems. In the "dry cold" arctic regions, disintegration is at a minimum. In none of these areas is the normal design, acceptable for temperate zones, a safe or a sufficient solution for comparable construction problems.

A realistic approach to protection against failure, both total loss and deterioration of structural integrity is the maintenance program for structures of the Port of New York Authority. Basically it is similar to the enforced inspection and maintenance schedules of aviation equipment. The axioms for the program are the three "facts of life" about structural maintenance listed by Kavanagh and Johnson in *Civil Engineering*, page 52, May 1965.

"1. There is hardly a structure in existence in which there are not defects that could have been prevented by better engineering and more careful construction. In the main, these defects are not serious, but they exist and inevitably shorten the service life of the structure.

"2. The same problems consistently recur in similar structures under similar conditions of exposure. This recurrence appears to be the result of an inadvertent but repeated use of unsuitable details and/or practices in design and construction. It indicates that design engineers, in general, are not aware that such details and practices proved unsatisfactory.

"3. Defects are seldom detected before they become so serious that repair must be accomplished on a rush basis. The result, at best, is that management's planning and maintenance budgets are upset and, at worst, that the facility is put out of service while repairs are being carried out."

The Authority defines structural integrity as that complete unimpaired state of a building or structure which, when achieved and maintained, assures that the structure or its major components will not collapse or fail during construction and for the period of its planned life. This state of being results from the integration of sound engineering with the proper application of construction materials, equipment, and methods. The practical and proper interpretation of the term "structural integrity" requires that reasonable and knowledgeable men focus their attention on the likelihood of failure of the structure or its major components with consequent danger to the people using it.

The Authority's Operations Standards Division has prepared routine inspection instructions in great detail, covering every type of structure and every component part, analogous to a thorough medical examination. First started with steel bridges and concrete piers, the program is extended to vehicular and railroad tunnels, aircraft hangars, airport facilities including bridges carrying taxiways and pile-supported runways, marine piers, terminals and warehouses, bus terminals, and elevated parking structures. Implementation of such a program can go a long way in preventing the cause of construction failures.

From the experience of the Pierose Building Maintenance Company, Los Angeles, a firm which employs some 3600 persons, a computerized record of maintenance work provides a list of common errors in detail which should be avoided. The cost of such maintenance must be balanced against the original cost of substitute materials or facilities. Some of the suggested revisions to their clients for future buildings to avoid the high maintenance costs have been to avoid exterior louvers set too close to the building face so that the window washer must resort to extension equipment, flush window sills and frames that permit water to run over the face of the buildings and stain the exterior, vertical venetian blinds requiring two men to wash a window since one man must hold the blind away from the glass, and the provision of lights where scaffolding is necessary to change a bulb, an operation that may cost $30 each time. The total list could well become a check list of what not to do in future planning of buildings.

1.5 Design and Drawing Cause of Failure

Few modern projects fail because of error in concept, although historical record of such failures in ancient work indicates that many projects were abandoned during construction, when it became apparent that the concept was not feasible. In the Halley lecture at Oxford University J. R. Pierce on May 15, 1963, described this world of science and technology as the "World of natural law and of the understanding and application of natural law. Some things are possible, some are beyond possibility. At a given time we can in part distinguish between the possible and the impossible. Of those things which are possible, some we can this very day achieve by an effort commensurate with their value. Some we can have, but at far too dear a price. Some, though possible in principle, are beyond our present grasp, however hard we may strive."

The story of the U.S. Navy "Big Dish" project is the record of a failure due to error not in design or in construction, but in concepts. With the rapid development of radio-wave reception and radar measurements in the field of astronomy coupled with similar techniques for control of motion, location, and communication with objects in space, the scientists demanded instruments of unprecedented mass in accurately controlled motion with surfaces always of fixed shape. The surface for receiving and sending radio waves is only usable if the mathematical shape of the dish is maintained at an accuracy within about one-tenth of the length of wave received. And since the aperture or face of the dish must change position continually and along two axes, gravitational distortion and the varying thermal exposure must not modify the shape beyond the permissible tolerance. When dealing with meter-long waves, the several inches of tolerance will not seriously modify the normal structural design procedures. When, as is now common, radiation of wave lengths in the few centimeter, or even millimeter, dimension is to be measured, a new concept of structure must be the criterion of design. Failure to recognize this can only result in the failure of the project; if not physically, certainly economically.

In 1948 the Naval Research Laboratory suggested the construction of a large radio telescope for purely scientific work. By 1956 this idea developed into an approval of a budget of $20 million by Congress for the construction of a 600-ft diameter—over 7 acres in area—paraboloid shell, mounted to provide full coverage of the sky above the horizon. By the end of 1957 feasibility studies reported that such a structure could

Fig. 1.2 Model of Navy "Big Dish."

be put together, but at a cost of $52.2 million. Soon thereafter the Navy decided to combine military functions with the instrument's scientific capabilities and the cost would then be $79 million (Fig. 1.2). These estimates of cost were subsequently changed to $126 million and then to over $200 million. In September 1961, however, some years after construction had begun and with designs still not fully crystallized, Congress set a ceiling of $135 million on the project. In 1962 with expenditures of $63 million and a number of contractors' claims for other payments, work was stopped and the project was abandoned.

For various and sundry reasons it was decided that construction would proceed concurrently with design, and foundation contracts were let while the design on the superstructure was hardly started. Soon it was determined that a complex electronic control system, not an impossible device, was required to maintain the dish shape automatically. Any increase in rigidity to prevent deflection of the reflector would be self-defeating since the added weight of steel would in itself cause more deflection. While a new design team worked on the apparently endless problem, structural steel contracts were let and the foundation with its circular carrying rails had been built. The new design indicated a

Fig. 1.3 Progress at termination.

superstructure weight of 36,000 tons, more than double the estimated weight used for the design of the foundations and the rails. In the final committee investigation, it was also disclosed that the original estimate of 30 persons to man the completed instrument had been modified to a complement of 1146, with housing, offices, laboratories, and services not yet provided.

At the closure of the work, there were two 400-ft-high erection towers, completely equipped work shops and administration building, 17,000-cy concrete ring foundation, a 550-ton pintle bearing in place and a 116-ton shaft at the site ready for erection, with a total of 623 tons of steel erected. All 7 acres of aluminum skin were in fabrication or already delivered. The turntable trucks and wheels at the site would have required doubling for the new loads (Fig. 1.3).

Expecting the completed installation to become a tourist attraction, West Virginia planned motels and services to be financed by the Area Redevelopment Administration, and a grant of $1.4 million was approved three weeks before the "Big Dish" project was canceled. On top of it all the Controller General's report stated as justification for termination of the project "Our belief is based upon reports prepared in 1960 and earlier by scientists within and outside the government who reviewed specific problem areas and indicated serious doubt that the instrument, if completed, would have the desired capabilities."

Deficiency in the basic design of a structure, such as amount of reinforcement at points of maximum moment, or incorrect dimensions of concrete or steel sections to provide sufficient resistances for normal loading, is an extremely rare cause of failure. One case that was caught just before placing concrete in forms for roof beams spanning 64 ft over a school auditorium in Yonkers, N.Y., (1925) was where about 4 in.^{2} of rods had been called for and placed while the design should have required 40 in.^{2}. At the time girders of such size were unusual and the error was picked up by an inquisitive young engineer in the contractor's staff. The girders would have undoubtedly failed if the error had not been corrected.

In 1961 a young construction engineer again picked up an evident error at the almost completed upper level addition to the Birmingham, Ala., Legion Field Stadium. A vertical extension in structural steel to provide 8632 seats was being rushed for the Alabama-Tennessee game of October 21, 1961. The second deck consisted of sloping plate girders on two columns with cantilevers fore and aft. Service and exit mezzanine space was added in the center half only and the load of the mezzanine was suspended from the girders. During a final inspection of the steel work a junior employee of the contractor asked the designing engineer why the girders were all alike, whether with or without the hanging load (Fig. 1.4). Very little checking soon revealed that the computations had omitted the hangar reactions. The designing engineers immediately notified the city that the new deck was not safe for the intended load and ordered additional angles welded to both chords to make up the deficiency. Outside consultants were retained by the city to confirm the sufficiency of the corrective work and the upper stand was not used for several months. The action of the professionals in safeguarding the public was most commendable.

For simplicity of detailing and assembly of reinforcement it is sometimes expedient to use the same reinforcement for a roof slab as in the typical floor of a multistory building. This may explain why in a three-story school the main reinforcement of the floors was incorrectly copied from the design notes and the corresponding roof steel was shown on the drawings and so installed. Although this was only one of several non-compliances discovered in the investigation, it may have had considerable influence on the large slab deflections which required some major repair and strengthening.

Multistory parking garages are marginal economical structures and every possible reduction in construction cost is necessary. Continuity of

Fig. 1.4 Deck addition on grandstand.

steel beams and girders is commonly used to save weight. Many designs are based on an arbitrary assumption of the location of the points of inflection, where bolted connections are detailed. The varying live-load distribution combined with an inaccurate choice of the joint location results in cracks in the floor slab with expensive claims from water leakage damage to stored cars. A more careful analysis of the design can eliminate such distress and trouble, with little if any added costs.

A six-story steel frame for the new Business Administration School at Georgia State College was erected in 1966 when the architect notified the owner that an additional $2000 was required to pay for adding beams at the elevator penthouse level to support machinery and to frame escalator openings, for some added reinforcing steel and changes to the metal deck forms. The college engaged an outside consultant to review the request and check the structural plans. The report cited a number of computed weaknesses in the existing frame based on the requirements of the AISC structural steel code, especially in some half of the columns in the first and second stories. One major difference in design approach was the assumption that the masonry walls prevented sidesway and therefore much more slender columns could be used. Overstress under full load was also indicated in roof girders, floor beams, and foundations.

Fig. 1.5 Failure of expanded steel joist by pullout of end member.

Among the corrective work necessary was the removal of concrete fireproofing from 34 columns so that cover plates could be welded to the flanges.

Completely inadequate designs prepared by laymen (sometimes even by churchmen) in frontier days and locations collapsed regularly, but such occurrences are rare in present day recognition of the necessity and availability of proper talent. Yet in 1964 a completed concrete church roof collapsed during Sunday mass in Rijo, Mexico, killing 55 persons and injuring 63. The roof had been designed by the brother of the parish priest and to complete the building in time for the Christmas holidays, with insufficient money available, the required bars had been changed to ½ in. throughout.

There have been a number of part failures of roofs over stores, factories, and warehouses in the past few years where a similar steel joist detail is at fault (Fig. 1.5). In every case the actual load on the roof, even with allowance for maximum rain or snow accumulation, has been less than half of the design load. Included in this group of failures is the rear portion of a distribution warehouse in 1967 at Edison, N.J., an almost identical area in a warehouse in 1966 at Raleigh, N.C., 3700 ft^2 of roof over a department store in 1967 at Blackstone, Mass., where plugged

leaders were blamed but the maximum possible water accumulation was 8 or 9 in. with a 30 psf design load, and incidents in the metropolitan New York area. The steel joists, spaced some 5 ft apart, carry light deck roofs of either metal or precast concrete plank. The joists are made of double angle chords with ¾-in. rod webbing formed by a continuous 60-deg bending of a single rod over most of the span. To make up the necessary full length of the joist, a separate piece of ¾ in. rod is filled in at each end as a sloping end member and welded, more or less, within the double angle chords. The bottom of these rods is always found pulled out, since apparently the weld does not develop the strength of the rod and in some joists, the rod as a sloping bottom chord member is far too weak. In failures at considerable distances from each other the investigation proved that the joists came from the same fabricator, in one case almost 1000 miles away.

Large structures are vulnerable even if only a single detail is of insufficient strength. The Pan American Terminal at the Kennedy Airport in New York has a cable-suspended steel roof deck. An elliptical roof is carried by 32 radial steel girders which sit on columns and at the same point carry masts for the cable suspenders. The bearings under the girders were double rockers with malleable iron casting equalizers. Due to some defects in the castings, one equalizer failed under the erected load-dropping part of the roof 1 in. before the rest of the structure came to rest on the supports. All 32 bearings were replaced with steel weldments, with more certain strength of the metal.

Drafting and detailing errors usually result in localized failure which is field corrected, but sometimes at great cost and delay. An exception was the collapse of a 24-ft high reinforced-concrete retaining wall in 1950, the rear wall of an automobile showroom in Manhassett, N.Y., caused by a poor drafting job (Fig. 1.6). The design called for 1¼-in. square bars vertically and the draftsman had placed the "1" of the "1¼" on a dimension line. Actually ¼-in. smooth wires had been supplied and installed. Total collapse started as soon as the wall was backfilled. It is a strange coincidence that no one questioned the rods that were delivered and installed. The design had been checked by an outside engineer upon request of the local building inspector before the permit for construction was issued and the design had been approved. Shop drawings of the reinforcement were prepared, with the detailer reading the plan as requiring ¼-in. rods, but there was no record of any checking or approval; yet these shop details were used in the actual construction as a correct guide. The print from which the detailer worked could easily be read to call for ¼-in. rods.

Another detailing error, of structural steel, caused by poor drafting

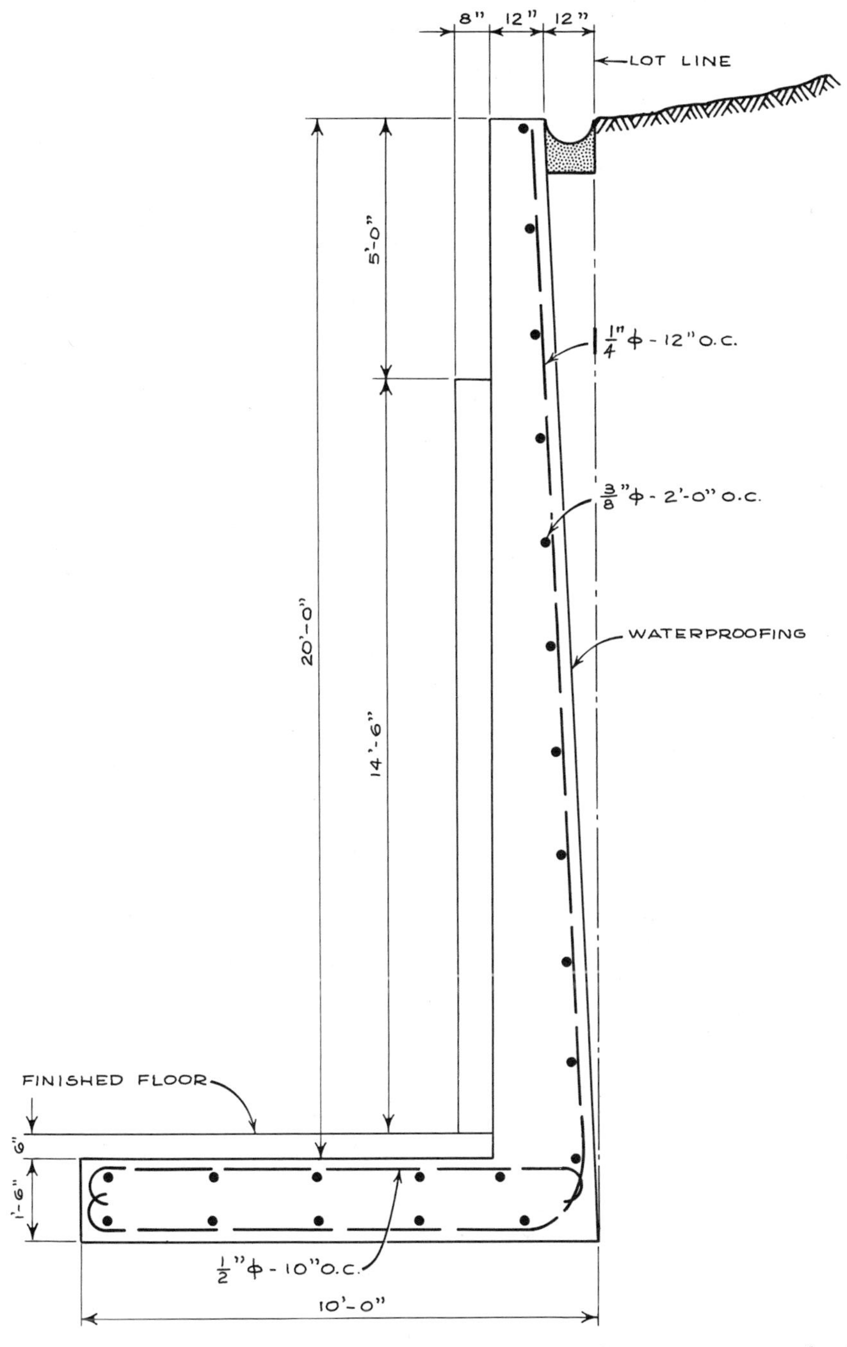

Fig. 1.6 Drawing of retaining wall that failed because of misinterpretation of rod size.

resulted in all the corner stones of a limestone faced courthouse in Queens, N.Y., rotating out of position under the thermal difference from sun-exposure on the adjacent faces. The contract drawings called for a steel angle lintel continuous at each face extending to 2 in. from each corner. Where angles met at a corner, the vertical leg of one was cut out to clear the 5-in. stone thickness. Under each lintel angle was a lead joint to relieve any vertical compression on the stones covered with a 2-in. depth of caulked joint. The structural steel details followed the requirements and the correct dimension of 1 ft–2½ in. was shown from the corner column center to the edge of the steel angles. The drafting was poorly done and the material list noted on the sheets, for the lintel angle lengths, took this dimension as 12½ in. As a result every lintel at the corners on all floors was short by 2 in. and the lead pads were similarly short. The corner stones had cement joints which permitted some compression in the stack of units. With temperature change on one face, the corner stones rotated and opened up the vertical joints. The distress was not stopped until the cement joints were raked out and the stones reset on a flexible joint.

An even more dramatic failure caused by a drafting error was the collapse of 4200 ft^2 of roof over the cafeteria of a Junior High School in Charlotte, N.C. The roof had stood for some four years until an accumulation of 4 in. of snow and ice during a storm in January 1968. The roof was framed with open web steel joists supported on intermediate line of girders. Two of the columns under the girders had been omitted when the plans were finalized to incorporate fireproofing requested by the insurance division during state review. It is inconceivable how such an omission was not detected in the checking by various agencies or how the steel could be erected without the necessary number of supports. The drafting error was publicly admitted by the architects when they checked the plans following the accident, which came after the cafeteria had been in use for over three years.

1.6 Production Deficiency

Errors of omission and commission in the performance phase of the construction industry probably are the major cause of failures. Some cases are a combination of minimal design with not too careful performance. A recent example is the collapse of the roof at the Minnesota State Fairgrounds in 1967. The elaborate report of the investigating committee cites a list of deficiencies stemming from poor structural design, field changes in the contract drawings, deviation from plans and

specifications, and sloppy field work. Some 33,000 ft^2 of roof, about two-thirds of the total, collapsed under a roof loading less than the 40 psf live load used in the design. Thermal change in the unheated one-story exhibition hall was the trigger of failure in the 18-month-old building. Columns are set in 30 × 40 ft array with continuous steel girders on the 30-ft spans, and 40-ft long precast prestressed channel slabs sitting on top of the steel girders. No cross bracing was provided between columns or between girders. Bearing of the concrete channels varied between 1⅜ and 2 in. with only a few imbedded shoe plates welded to the girders. The steel had been designed as laterally restrained, but not enough bracing was provided by the channels. Workmanship was deficient in bad field welding, bolts omitted in beam connections, misplaced anchor bolts compensated by burning large holes in base plates, substitution of plug welds for bolts and omission of some anchor bolts. In all not a complimentary record for the design, the construction, and the inspection services at this project.

A similar mess came out in the testimony concerning the collapse of a gymnasium roof just completed in 1963 at the Klondike School in Texas. The frame consisted of four H-columns 22-ft high with two main trusses bolted to the column cap plates. Each truss was fabricated in five sections and by welding and bolting assembled in place into 96 ft long by 8-ft deep units. A series of 25-ft long tapered trusses, 4 in a line, were field bolted and then welded to the main trusses. In the center two tapered units met and were connected by gusset plates with slotted holes and four ⅝-in. bolts. The other pair cantilevered out from the trusses and a short distance beyond continuous glass in the walls. The main trusses had a 2-in. built-in camber most of which did not come out under dead load and the dead level roof design became a depressed pond in the middle of the roof. The structural plans were prepared by the architect's draftsmen from design sketches of the engineer, who signed but did not check the final drawings. No one checked the shop drawings and no one checked the suitability of the field welding, although everything was formally approved. Butt-welded splice plates were not bevelled, and one such weld in the bottom of a main truss was isolated as the cause of failure. The welds at the top of the tapered trusses to the main trusses were omitted to provide a "breathing roof," with no formal advice to the designers. The connection was made by the field erection bolts. The flexible roof ponded water which caused more deflection until failure occurred, and then everyone sued everyone.

Structural frame redesigns to save material are quite common in speculative building but it is strange to find an example in a public

building. In 1931 the Yonkers, N.Y., City Health Center was contracted and the four-story steel frame was redesigned. The erected steel was so bad, with poor details, that the entire frame had to be demolished.

Some noteworthy examples before 1945 in the category of contractor misdeed are the collapse of the 12-story Darlington Apartments in New York (1904), where 10 stories of steel were erected and eight floor levels concreted, because of a missing column footing; the failure of the Home Theatre in Chicago (1912) where 48-ft spans consisted of 24 I's with top and bottom plates, made up of two pieces field-spliced in both web and flange plates with unit stress of 44,000 psi since the flanges were not spliced; the collapse of the frame for the theatre in South St. Paul, Minn. (1914), when 25 tons of steel were loaded on the roof at one corner; the collapse of the Majestic Dance Hall in Salt Lake City, Utah (1916), where 110-ft wood trusses with no lateral bracing and green lumber that shrank in two years so that the connections were broken; the Knickerbocker Theatre collapse in Washington, D.C. (1922), with roof trusses on high tile masonry walls where the balcony had been extended 18 in. and fell because of poor framing connections pulling the walls and roof down; the garage structure at Dallas, Tex. (1928), originally designed for 11 stories with four stories built in 1925, and when it was decided to add 14 more floors, the columns were found on the edge of caissons and the caissons were not carried down to rock, so that a great deal of reconstruction of the original work was found necessary; the case of the eight-story office building in Buffalo, N.Y. (1930), where a concrete mixer was set on the roof and materials stocked to a load of 350 psf around it, with the result that the mixer fell through the bar joist and concrete floor down to the cellar; and the gymnasium roof in Detroit, Mich. (1937), where 60-ft trusses fell because the bottom chord splice did not have the angles spliced out, only the cover plate.

The variety of the causes of failure in these older cases would seem to cover the field, but some more recent incidents added new examples. Diversion of cement to the black market so reduced the concrete mix that at least two buildings in Brazil completely collapsed even before occupancy, the 10-story apartment buildings in Rio on the Copacabana (1946) and a 12-story apartment in Sao Paulo which was completed and failed as the roof tank was being filled. In the investigation of the latter (1948) it was determined by analysis of the rubbish and confirmed by testimony that an average of 4½ bags of cement of the seven sent to the job for each cubic meter of concrete had been diverted and bags of lime and dust substituted.

A concrete school building in Salt Lake City, Utah, failed in 1947, at

Fig. 1.7 Failure of guyed mast pavilion frame.

22 years of age, and only then was it discovered that the members were smaller than called for and the concrete included considerable wood left in the forms.

As a final example is the collapse of the New Jersey pavilion framework at the New York World's Fair in 1963. The various unit exhibit areas had individual roofs suspended by cables attached to 12 steel pipe booms. These booms were in sets of three, sloping outward from four common footings and linked together by cables near the tips. The cables had been carefully detailed and were all prestretched and fabricated to close tolerance. All steel work was fabricated under close inspection and each member was marked as satisfactory by the inspector. During the intricate carefully planned erection procedure, the end of one hangar cable did not engage the connection to its roof. To make the connection, the cable was pulled by a hoist until the attached boom crippled and the entire frame fell to the ground (Fig. 1.7). Repairs and replacements were rushed to avoid missing the opening date and all computations and cable length carefully checked and found correct. In the second erection sequence, very carefully done under close supervision, the same cable was short again. This time, the 80-ft boom was also checked and the connection for the cable was found 10 in. too high, which explained the former collapse. Two other booms had connections out of position but by only small amounts.

Competence in every phase of the production segment of the construction industry is needed, if failures are to be avoided. Succeeding chapters will describe similar shortcomings under the various kinds of work and use of basic materials.

2

Foundations of Structures

Undermining Safe Support; Load Transfer Failure; Lateral Movement; Unequal Support; Dragdown and Heave; Design Error; Construction Error; Flotation and Water Change; Vibration Effect; Earthquake Effect.

The necessity for a proper foundation support for any structure is an accepted axiom in engineering. All loads must be so transferred to the underlying soils that the resulting settlements can be tolerated by the structure and so that stability will be maintained over the life of the structure. The introduction of a foundation into the soil mass causes a new set of physical conditions. The later addition of loadings during and after construction, and the construction of foundations on adjacent areas may modify these conditions and affect an existing satisfactory support capacity.

Foundation design must be a combination of scientific deduction with the art developed from records of successful and nonsuccessful examples. Foundation construction is a developed technique of maintaining stable conditions during the operational shock period when minimum support is provided until the final design is incorporated into the ground. The general rule is that the unforeseen and the unexpected is always encountered so that secondary lines of defense must be provided. Careful analysis of the effects from the most unfavorable conditions must be the guide for proper design and construction methods. Disregard of a single factor which will degrade soil supporting or resisting strength, for

long time periods in the case of designs, and for shorter time periods in the case of construction, almost always result in failure, either in the form of distress and distortion, or even complete collapse of the structure.

2.1 Undermining of Safe Support

The necessity for careful and thorough soil investigation before design of a foundation or earth structure is too often not carried beyond the actual confines of the new project. The foundation conditions, and sometimes even the existence, of adjacent existing structures is often disregarded, with tragic results as shown by later examples. Excavation adjacent to and below existing wall or column footings reduces the support value and such evident error has been often committed, such as at Wheaton, Ill., in 1926 where excavation some 2 to 3 ft below the adjacent wall footing in stiff blue clay caused a one-foot settlement, or the excavation for a deep cellar near a five-story wall of a residential building in New York in 1960 caused a total collapse of the 90-year-old building.

Again in New York, prior to the excavation for a three-level garage below street grade for a multistory apartment, an adjacent five-story wall-bearing building was underpinned to rock level. During the excavation of the bedrock, using controlled blasting because of the adjacent buildings along two lot lines, a major slip of the rock carrying the underpinning removed all foundation support under the rear part of the adjacent building. Part of the wall fell and part hung to the stable portion with the masonry corner remaining plumb. The cost of providing a new foundation exceeded the value of the building. None of the borings within the project limits, indicated such a loose condition within the bedrock. Seamy rock must be provided with temporary shoring until covered and braced by the permanent construction. In some soils the excavation need not be carried even down to adjacent footing level and yet cause trouble. The soft clay excavation for the Lafayette Hotel in Buffalo, N.Y., in 1924, caused the adjacent wall footings to settle 6 to 8 in. and push out the same amount, even before the old footings were exposed. Similar lateral clay squeeze during the extension of the cellar for the Meyers Department Store in Albany, N.Y., in 1905, caused partial wall collapse when only 5 ft of the old wall footing was exposed.

The Ford Theater Building in Washington, D.C., after its conversion to an office building used by the War Department was involved in an undermining failure in 1893. An excavation in the cellar under the piers of the front wall caused some 40 ft of the wall to collapse below the third floor,

hurling occupants and equipment into the cellar. In 1964 a foundation for a two-story building in Ottawa, Canada, was being placed in a trench in clay soil about 18 in. deeper than the footing of an adjacent building. Part of the new wall was concreted, but the remainder of the trench was left exposed to the weather and after two days, the neighboring wall collapsed as the clay below its footing softened. Alongside a three-story government building in Brussels, Belgium, an uncontrolled excavation was noted in 1962 as causing hollow sounds in the sidewalk. As the decision was being reached to evacuate the building, it collapsed, causing 10 casualties and 20 injured or missing.

Overloading stable foundations can cause failure as was proven empirically in Cairo, Egypt, in 1964. A two-story concrete building built in 1948 with 5-in. concrete walls was raised successively to three, four, and five stories. It was then 35 ft^2 in plan and 60 ft high with 15 dwelling units. The original foundations overloaded the soil to the point of failure and the apartment building collapsed suddenly taking 31 lives.

Even rock excavation below structures resting on rock but with unfavorable sloping seams will cause loss of support and failure of the wall. In one such case in New York, the rest of the building was reinforced by cable ties introduced for all footings and anchored to the rock at the opposite side of the building. The undercutting of the sloping rock sheets disturbed the continuity of support when the rock sheets moved slightly to reach a new position of stability.

Excavation for drains and pipe changes along existing buildings being considered minor works usually disregard effect of undermining, as happened in 1965 at a new shopping center in Atlanta, Ga., where the footing was undermined by a sewer excavation. The sewer excavation went 15 ft below the footing and permitted a shear failure in the firm sandy micaceous silts and dropped the column together with 60-ft prestressed double T roof beams. Similarly, excavation for a fuel oil tank, 10 ft deep and 2 ft from a wall footing but 7 ft deeper, in Jamestown, N.Y., caused the collapse of 35 ft of wall of the Hotel Samuels, after a heavy rain saturated the exposure of the fine soil.

Excavation for caissons of a new skyscraper bank building in New York in 1934 caused sufficient loss of adjacent soil while penetrating layers of varved silt which tend to flow readily, to settle the spread footings of the adjacent building on Wall Street. Prior construction of two other buildings each had the same effect. In the Chicago clays a tunnel excavation in the street caused the 16-story Unity building to lean 6 in. in 1900, which in 10 years increased to 27 in., requiring expensive foundation underpinning and correction of the column connections, to avoid collapse.

In many countries coal mining removes natural support sometimes with catastrophic results. As early as 1923 the partial failure of the Spring Creek bridge near Springfield, Ill., was diagnosed as being caused by coal mining. Suddenly, the center pier and one abutment of the two-span concrete bridge settled from 8 to 16 in., even though all foundations were founded on rock when the bridge was constructed in 1921. Inspection of a coal mine 130 ft below the surface showed that the roofs over seven rooms directly below the bridge had collapsed and filled the 5 ft of excavation height. The bridge was jacked to grade and underpinned with brick masonry as an emergency measure with permanent reconstruction postponed until all subsequent consolidation of the bedrock had been completed.

In 1963 the Zeigler-Royalton High School in southern Illinois was closed when the building sank 4 in. without more damage than falling plaster. The settlement was blamed on the collapse of an abandoned coal mine. Similar problems in the coal mining area of Nottinghamshire, England, were taken care of by a design of an articulated steel frame which can adjust itself to ground movements, which were expected to be in the range of a 4 in. per horizontal stretch of 100 ft with several inches of differential settlement. Columns are pin connected at the base and cross braced with spring-mounted diagonals. Wall panels are ship-lapped and all partitions are connected with flexible joints.

Formation of sink holes in the dolomites of the Transvaal, South Africa, have caused complete loss of foundation support under entire buildings. One such hole was measured as 300 ft diameter and 120 ft deep. In 1962 a complete crushing plant at the Westdriefontein Mine, about 50 miles west of Johannesburg, dropped without warning into an unknown cavity. The frequency of sink hole formations increased greatly after 1960, when underground mine pumping caused a substantial lowering of the water table—in places by as much as 400 ft. In the design of the concrete viaduct for the Johannesburg freeway roads, special provisions were installed for jacking to correct for foundation movements. Some signal devices are also planned for remote warning of overstress resulting from distortions. Adjustments are available for a 2 in. change of grade in 100 ft, and a longitudinal movement of 3¾ in. in the same distance in any span.

2.2 *Load Transfer Failure*

A stable frame whether designed to be rigid or merely obtaining rigidity from the assembly of walls, frame, floors, and partitions, adjusts its load

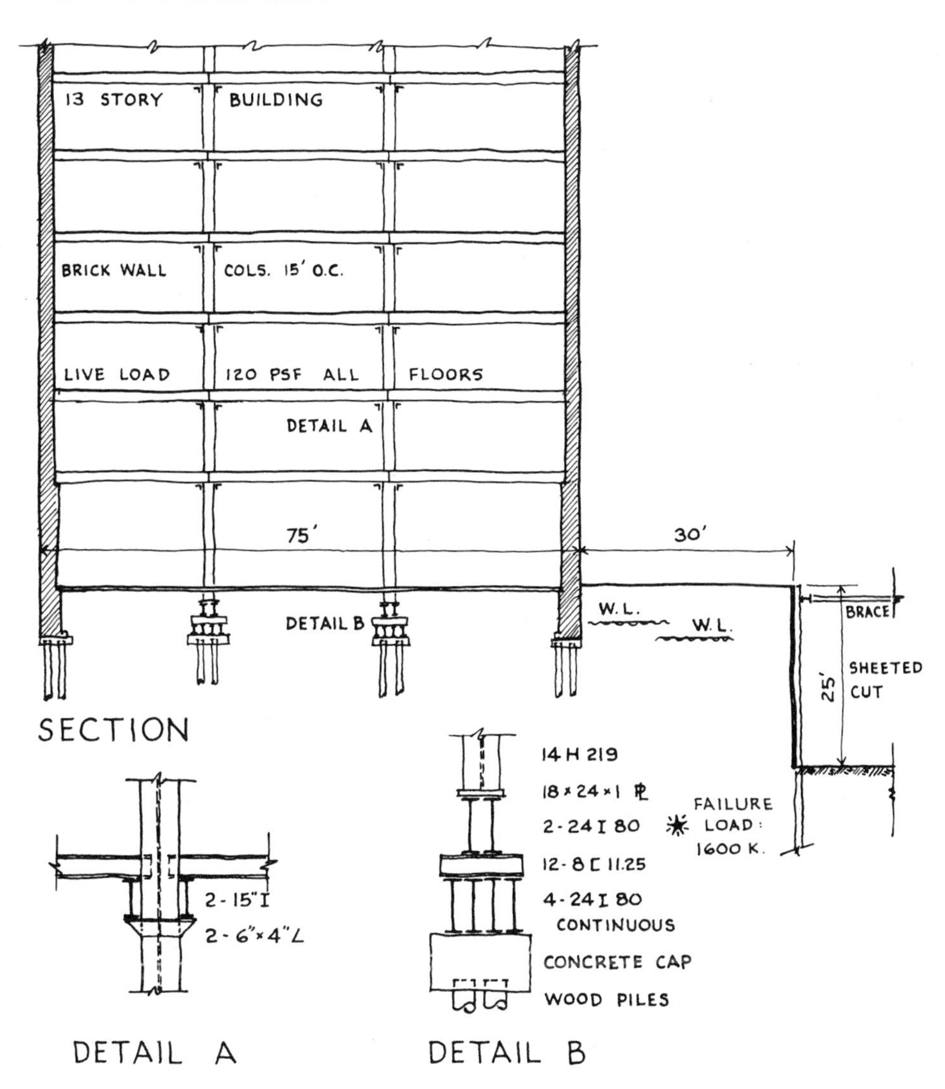

Fig. 2.1 Drawing of building frame and load transfer caused by settlement of exterior pile support.

to compensate for differential foundation settlement. When the state of stability is disturbed by partial support loss, the reactions are transferred within the available supports. In a 13-story commercial building in New York, supported by four continuous pile caps and timber piles, pumping operations from a neighboring excavation for a depressed

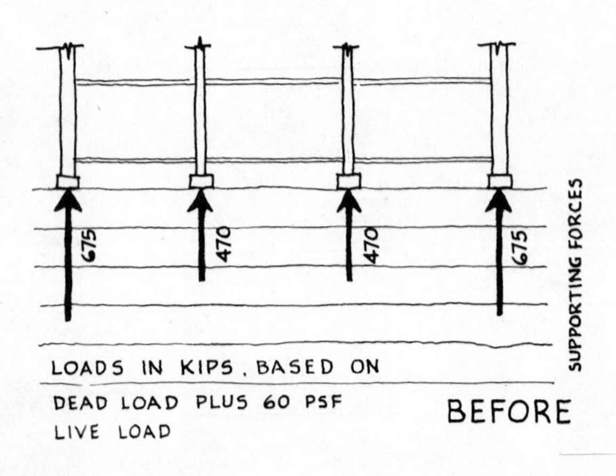

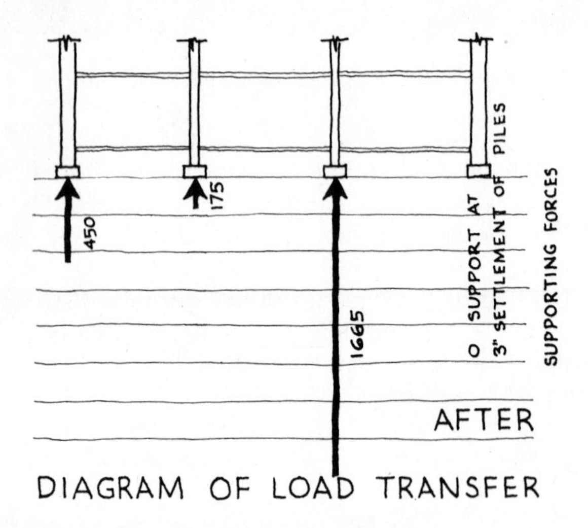

DIAGRAM OF LOAD TRANSFER

roadway lowered the water table and caused one line of piles to settle which slightly pulled that pile cap down (Fig. 2.1). This released the exterior wall support and the load transfer to the interior line of columns overloaded their steel beam grillages; the 24-in. beams collapsed to a 12-in. depth with consequent 12-in. settlement of every floor in the building (Fig. 2.2). The floor beams rested on girders between columns and had only a loose bearing so that they took the new trough shape without

Fig. 2.2 Failure of column grillage beams overloaded by load transfer.

failure. After the entire building was internally shored, new shallow grillages were provided and each floor was releveled. The small settlement of the exterior wall was neutralized and its bearing restored on the existing piles.

When foundation disturbance results in masonry distress, the investigation to localize the trouble should consider that the failure came from a load transfer and the actual foundation failure is not at the same position as the evident distress.

2.3 *Lateral Movement*

It is usually conceded in the foundation construction fraternity that 1 in. of lateral movement in a structure causes much more damage than 1 in. of settlement. Lateral movement develops from the introduction of unbalanced horizontal pressure, usually from the removal of the resisting component, but sometimes from temporary or permanent buildup of the active pressure without providing corresponding added resistance. In the design analysis consideration must be given to possible change in

both active pressure and passive resistance with water level conditions; saturation increases the active and often decreases the passive.

A heavy double-deck concrete viaduct built in 1966 into a clay hillside in Cleveland, Ohio, suffered shear cracks in the partly imbedded columns when lateral pressures built up from saturation. The columns are 50 ft high and 6 ft in diameter set on 6-ft thick caps with 70-ft long piles. The piers had been built prior to the installation of an adjacent embankment. Normal procedure is to construct the embankment so as to consolidate any underlying compressible materials before inserting the structure but the pressure of schedules caused a reversal of the sequence. The amount of consolidation would have been the same but the effect on the piers, much less. The columns were jacketed to an 8-ft diameter and a larger pile cap built on top of the original cap.

Excavation for roads in the Palo Verde Peninsula near Los Angeles started lateral movements which seriously damaged 145 fine residences. Dredging of the waterway of the Mississippi River in 1962 removed the lateral support of the grain elevator foundation at Jackson and seriously damaged the bins.

Normal scour of the river together with a high-water table level unbalanced the abutment of the Peace River Bridge on the Alcan Highway and the slide of the bridge abutment into the river required the complete reconstruction of the structure in 1957. Failure began as a longitudinal compression distortion of the bottom chord of the stiffening truss at the rocker arm. The Peace River Bridge was built in 1943 as a necessary link of the Alcan Highway, using a suspension design to permit more rapid completion. Actually, it was open to traffic nine months after the decision on bridge type. Foundations were set on the shale considered strong enough for the imposed loadings. There were a series of necessary corrective measures in the 15-year life of the bridge. The cables vibrated in the wind and this was stopped by adding cable clamps to hold the strands together. Some broken wires were repaired in 1947, but the oscillation of the span was tolerated as an infrequent condition. Also in 1947 the shale under the upstream end of the North Tower pier was scoured out and the concrete was underpinned to undisturbed rock. Ten years later a deep seated slip in the rock, possibly triggered by some industrial construction on the north bank, caused the anchorage movement.

Tilting of a bridge pier at the Grand Coulee Bridge in 1935, a 950-ft cantilever, was caused by a slip within a 20–30 ft overburden layer of gravel with lenses of fine saturated silts. The pier was founded on granite and tilted 9 in. exactly in line with the bridge axis. Cable holdbacks and jacks inserted in the closure of the steel truss provided

enough resistance to stop further movement, and permanent changes in the expansion joint details were made to compensate for the slope of the pier support.

Lateral flow of the soil under a 24-story office building in Sao Paulo, Brazil, in 1943 caused total collapse. A similar subsurface slide at Altena, Germany, in 1960 under an eight-story office building was stopped by inserting 1¾ in. rods in 150 holes drilled 20 ft deep. At Topeka, Kan., the resisting berm for a rear wall construction had some filled-in wells that permitted soil loss and the wall collapsed.

Several instances of collapse are reported when adjacent buildings are demolished leaving an open cellar lower than the foundations of the remaining building. Although stability, with probably a much lower than expected factor of safety, exists with adjacent buildings intact, aging of the soil imposes extra loads on the lower cellar walls which can resist the horizontal pressure as long as they are loaded, but are insufficient when the upper stores are removed. Such an example is the complete failure of a four-story wall in Richmond, Va., in 1922, 10 years old, when the adjacent building left a cellar wall as its contribution to stability. In 1957 the 11-story residence building in Rio de Janeiro collapsed completely after excavation adjoining the wall removed the lateral support of its building piers. Underpinning operations were not ordered until too late, after the settlements and rotation were noted for almost two years. Even restricted excavations can cause similar failures as was proved by the collapse of the Boston club building in 1925, with the loss of 44 lives, caused by digging seven underpinning pits, 4 ft wide and 7 ft deep below the foundations without any lateral bracing. The result was to provide no lateral support for the pressure of 14 ft of earth.

During the construction of the conduits for the Niagara Power Project in 1959, damage to buildings and equipment was claimed by the Kimberly-Clark Corp. and suit was brought for $2.3 million. The claim was based on an alleged lateral movement of the natural rock resulting from the adjacent excavation. At the same time the city of Niagara Falls claimed $300,000 damage to a 60-in. sewer from the same cause.

There are three recorded identical wall failures from broken drains laid alongside the footing with wash out of the soil during heavy storms. Such vulnerable conditions, especially in clay soils, must be guarded against.

2.4 Unequal Support

All footings will settle when loaded, but not equally unless soil resistances and load distributions are equal. If not differential settlements

will result, with load transfer and tipping of the structure. When the framework is not structurally rigid, the brittle masonry enclosure will crack along the shear surfaces bounding the volume that can be supported by the less yielding support. There are a great number of such structures, with foundations partly on rock and partly on soil, or with piles of unequal bearing value at the same settlements, or with foundations of more than one structural type but resting on soils of unequal compressive support. Correction of these support deficiencies is usually an underpinning of the weaker foundation to equalize support or in some cases the addition of a subsurface enclosure to increase the bearing value of the weaker soil.

An early example of the latter method, in 1919, was the addition of 2600 pinch piles to enclose the foundation of the Portland, Ore., grain elevator placed on 25-ton piles which showed progressive unequal settlement. The added piles were 20–25 ft longer than the original ones.

Aging of the support must be considered as was shown by the settlement of the Baltimore, Md., pumping station founded on large concrete piers carried to rock. Inspection of the bottoms in 1908 showed that some of the rock bearings were no longer on rock but on material of clay consistency. Certain metamorphic rocks, such as schists and shales, are known to lose their solid characteristics when loaded, especially if water saturated, and more rapidly in unclean waters.

There are some classical examples of tilting buildings due to nonuniform foundation conditions. The Tower of Pisa is probably the best-known and researched leaning example and is still in use with or perhaps because of its 5-deg list, as is the Palace of Fine Arts in Mexico City with settlements measured in meters, and still progressing.

The Pisa Tower is 179 ft high and with a wall thickness of 13 ft at the base, is now 14 ft out of plumb, with a rate of increase in lean of one inch in 25 years. The foundation mat, a 60-ft diameter plate with a hole in the center, rests on a fine volcanic silt. The foundation has rotated so that it is 7 ft out of level, with the highest base pressures at the low side. Built in the latter part of the 12th Century to three tiers high, tilting became serious and work stopped. A 60-year cooling-off period elapsed and since the tower was not moving, another story was added and work stopped again when additional tilting was noted. Another 100 years went by, and the tower was completed with some changes in design in an attempt to counterbalance the tilt tendency. In 1932 stabilization of the soil by grout injection proved unsuccessful, and 1000 tons of grout added weight but little stability to the soil. Plans for the stabilization (but without righting) are continuously in the news. A vertical tower would probably remove Pisa as a tourist attraction, but so would a collapsed tower.

In 1906 at Tunis, Tunisia, two warehouses on concrete grid mats tilted

some 25 deg but were righted by loading the high sides. The first building was leveled and became stable after settling 15 ft. The second building had tipped so that the overhang increased to 17 ft in 17 hr. It also leveled, but only after settling 18 ft. In 1914 the Transcona grain elevator in Manitoba, Canada, consisting of 65 concrete bins, 93 ft high on a concrete mat 77 by 175 ft, settled 12 in., then tipped 27 deg when about 85 percent full of grain. The structure weighed 20,000 tons and the grain 22,000 tons. By a fine engineering feat the structure, which was 34 ft out of level, was righted by underpinning with piers to rock support and the structure is still in normal use. A smaller grain elevator, 20 bins 120 ft high on a 52-×-216-ft mat in Fargo, N.D., tilted in 1955 when one year old, but the bins broke up and the structure was wrecked.

A dramatic failure by rotation due to unequal support was the nine-story apartment in Rome that settled at each end and split into two sloping halves. The separation into two buildings with a triangular gap about 3 ft wide at roof level came in 1959 when living space was at a premium and usable space could not be legally kept vacant. So 19 of the 28 apartments remained in use after some stiffening beams were added in the lobby and glass rods were inserted across the break in walls and partitions to warn of added movements.

The Bedford County Hall in England had been constructed partly on rock and partly on clay in 1965, and completion has been delayed. The solution of the unequal settlement was to demolish to the first story, underpin the footing and then reconstruct (Fig. 2.3).

Buildings partly on earth and partly on rock have too often been designed on the assumption that the local building code permissive bearing values are assurance of equal settlement. If such soil conditions exist, a separation joint must be provided and each unit considered an individual structure. These buildings, especially if wall bearing, almost immediately crack when nature provides the omitted joint. Underpinning and repair costs together with the economic loss of delayed or vacated use far exceed the cost of proper separated foundation and structure design. Some instances have been investigated where soil shrinkage due to local dewatering triggered the differential settlement even where the original construction indicated temporary sufficient design.

Foundations on soils with large water level fluctuation are also vulnerable. At Corning, N.Y., a hospital addition in 1949 on loose gravel soon showed greater settlement of the exterior walls than at the interior columns and the differential movements caused considerable distress in the interior partitions and finish surfaces.

New York City had two multistory building collapses in the early days

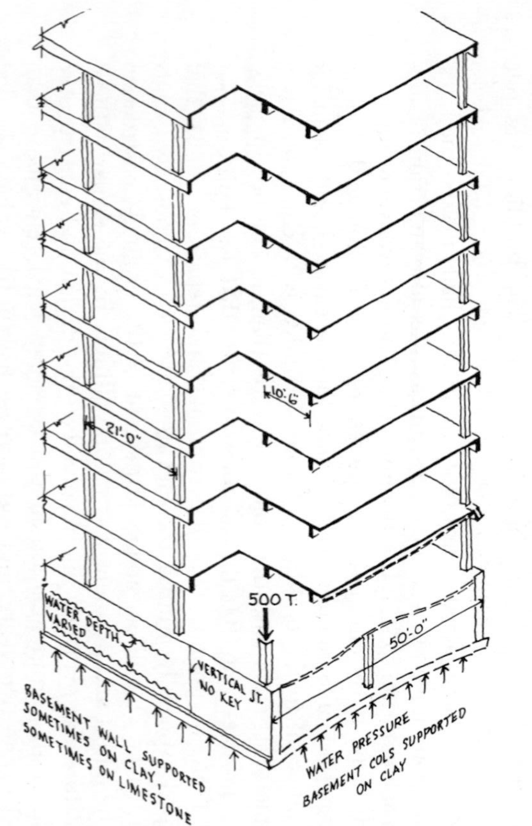

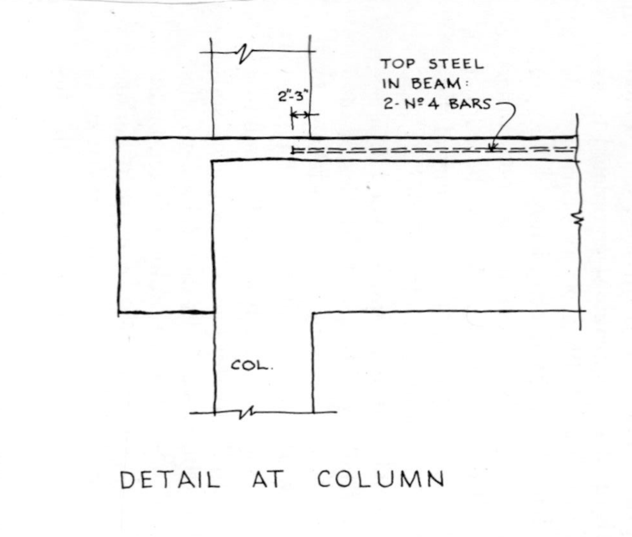

Fig. 2.3 Drawing of failure pattern from improper foundation design of British public building. (Prepared from information in *Architects' Journal*, March 9, 1966.)

of such construction due to improper foundations. In 1895 the Ireland building at Fifth Avenue and Third Street was almost completed to its eight-story height as a fireproof iron frame with masonry bearing walls when almost half of the 100 ft length collapsed completely, killing 16 workers. An interior column was set on a 10-ft^2 footing with one corner above a backfilled brick-lined well only 18 in. deep below the normal footing bottom. The collapse was triggered by the shearing of this footing, resting on very fine wet sand, and the column with its base punctured through the break and penetrated into the sand over 3 ft.

The collapse of the Darlington Apartment House, 13 stories of steel frame and concrete floors with cast iron columns in all wall lines, in 1904 caused 25 deaths among the workers. The collapse was total and came without any warning. Foundations at the front of the building were on rock, which shelved off rapidly to the rear where clayey soils covered some natural springs. Since this design was then considered the most modern type of structure and many similar multiresidence buildings were planned and occupied, the investigation of cause of failure was in great detail. Although the cast-iron columns and connections to the beams were admittedly vulnerable to distortions, it was the inequality of foundation support that caused the collapse.

A 300-ft-high, 24-story concrete office building in Sao Paulo, Brazil, was found to be over 2 ft out of plumb when the elevators were being installed in 1941. One corner leaned over the street 25.5 in. and over an excavation for a neighboring building, 23.6 in. The building was pile supported, but it was assumed that lenses of fine wet sand had been punctured by the adjacent excavation and its flow had affected the piles. Cement grout injection was unsuccessful; as was also injection of aluminum salt solution to stabilize the fine sand. The soil was then frozen to a depth of 60 ft by circulating brine through 160 double-walled pipes over a period of 8 months. Holes were then excavated, 4-ft diameter shafts, adjacent to the building columns and filled with concrete. Using these piers as reactions, the columns were then raised by 40 jacks, ranging in size from 100 to 950 tons each, so that the building was within ¼ in. of plumb. The underpinning and repair of cracks that developed especially around the elevator shafts took almost two years and cost over $1 million, about 50 percent of the original construction cost. The additional payment was made willingly by the owner with no attempt to claim over any one involved in the design and construction. Because of the rapid inflation in the economy legal rents more than doubled in the two years, and the increase in building value exceeded the repair costs.

Fig. 2.4 Drag down of existing wall from settlement of adjacent new building.

2.5 *Dragdown and Heave*

Once a footing is loaded and the bearing soil compresses to provide resistance, rapidly in the case of granular soils and over considerable time period for clays, the structure remains stable since the foundation no longer settles. The stability depends on fringe areas as well as the soil directly below the footing or in the case of piles, on the soil near the pile tip. If the soil below the footings is removed or disturbed, settlement or lateral movement is induced. If the fringe area is removed, the soil reaction pattern must change and even if there remains sufficient resistance without measurable additional settlement, the center of the resistance is changed and a moment is affecting the stability of the support. When the fringe area is loaded by new structure, causing new compression in the affected soil volume, the old area can, by internal shear strength, be required to carry some of the new load. In that case there is an unexpected new settlement of a previously stable building. If the new building is not separated from the existing construction, the soil compression settlement from the new load will cause a partial load transfer to the existing wall, with possible overloading of the previously stable footing (Fig. 2.4). In plastic soils these new settlements often are accom-

panied by upward movements, heaves, at some distance away. Liquid soils cannot change volume, and every new settlement must produce an equal volume heave. Disregard of such elementary considerations is the cause of most of the litigation stemming from foundation construction in urban areas where neighboring existing structures are suddenly placed in danger of serious damage or even complete failure.

Under the New York City Building Code, if new construction requires excavation more than 10 ft in depth adjacent to an existing foundation, all precautions to avoid damage to the existing conditions are the obligation of the new project. If the excavation depth is 10 ft or less, the obligation for protection falls on the owner of the existing building, should his foundations not be below the depth of the new excavation.

To avoid the high cost of underpinning a six-story wall-bearing building occupied as the research center of overseas cable telegraphic transmission, a new multistory office building was designed with spread footings on coarse sand at a depth of exactly 10 ft below street grade. As construction progressed, the adjacent six-story wall settled with considerable wall distortion and even more serious displacement of research facilities on the floors. The measured total settlement was almost exactly ¾ in., the amount to be expected from imposing a load of 4 tons per square foot on the medium compact sands found at subgrade. No separation had been provided between the new and old structures to isolate them as a protection against dragdown. The settlement was legally declared to be in effect an excavation and the bottom of the new work ultimately being below the 10 ft criteria, the liability for the damage was placed on the new building.

On similar sand soil, with the proposed work going lower than 10 ft, an adjacent wall was carefully underpinned by concrete piers carried down to the new excavation depths. After the new walls were constructed, again with no separation which could have been easily provided by a layer of composition board laid against the completed underpinning and the old wall, and the new footings were loaded, the small compression of ½–⅝ in. of the sand pulled the old wall down and caused cracks in the street wall and the interior partitions. The damage was an unwarranted result of omitting a small inexpensive precaution after the large expenditure for underpinning.

In this type of medium dense sand, pile driving is extremely difficult since the sand densifies under the vibration and the shock impulses can shake adjacent structures to the point of serious damage. To avoid the cost of underpinning a neighboring theatre wall and to save the cost of possible damage from pile driving, a project was designed with piles inserted into the sand by jetting a hole for some 15 ft for each pile. The

old adjacent theatre wall cracked up and started to lean outward. In the litigation which followed, the legal decision was to include jetting in the definition of excavation and since it had gone below the 10-ft criteria, the damages were assessed to the new work.

Dragdown from soil shrinkage, especially clay, when ground water tables recede or from desiccation by tree growth, with consequent differential settlements is a phenomenon observed in many countries. The failure of a theatre wall in North London was correlated with the growth of a line of poplar trees in 1942. In Kansas City, Mo., 65 percent of the homes in one residential area were found to be affected by soil desiccation from vegetation growth. The Justice Building at Little Rock, Ark., showed delayed settlement when the ground water was considerably lowered in 1960. After some 15 years of satisfactory behavior, structures at the Selfridge Air Force Base in Michigan suddenly started settlement and tipping from a volume change in the underlying clays as the water table receded aided by tree evaporation. A long time study of Ottawa, Canada, homes which similarly correlated soil desiccation with such damage, has been reported in 1962.

Piles embedded in soil layers that will consolidate from dewatering or from surcharge, are subject to overloading when the greater soil density increases surface friction and greater soil loads hang on the piles. The added load may cause considerable increases in settlement and may even pull the pile out of the pile cap or pull the pile cap free of the column or wall. Several New York public buildings are located on an island in the East River which has some 30 ft of rubbish fill over 50–75 ft of soft silt and clay above a decomposed rock. Normal aging and desiccation aided by tidal action suction of the soft silts has shrunk the fills and dragdown on the piles nearest the shore line has caused serious damage. In one steel-framed building the corner pile cap settled away from the column and a two-in. gap was found under the column base plate. Fortunately, no nuts had been placed on the anchor bolts, otherwise much more damage would have resulted, an unusual condition in which a construction oversight was a help rather than a detriment.

Dewatering of the ground results from many added subsurface constructions. In 1945 a four-story hospital in Buffalo, N.Y., then 20 years old, started to settle when a sewer was constructed 200 ft away; the piles which had been located in a filled creek bed were found to be moving. Holes were drilled in the pile caps and new steel piles were driven to rock support as a corrective measure. The Old Criminal Courts Building, built in New York in 1890, was affected by adjacent subway construction in 1909 and the mat foundation settled, causing the outside wall to drop 4–5 in.

When the four-year-old Justice Building in Little Rock, Ark., showed buckling and cracking of walls, floors, and ceilings in 1960, the investigation proved that foundation design had been governed by a tight budget with a calculated risk taken in not providing footings that would resist the heave of the high-porosity clay soils used for support.

Soil bearing can be neutralized by frost action which will heave the contact surface. In Massena, N.Y., where very cold weather is experienced each winter, a bridge abutment was pushed over by ice formation in the soil and fell in 1959. When at Fredonia, N.Y., the frost from deep-freeze storage froze the soil and heaved the foundations upward 4 in., a system of electrical wire heating was installed in 1953 to maintain soil volume stability. During an extremely cold winter in Chicago, frost penetrated below the underground garage and broke the buried sprinkler line causing an ice buildup which heaved the structure above the street level and sheared a number of the supporting columns.

Certain natural rocks when exposed to air and moisture will expand as the minerals change in composition and in volume. Two such conditions were encountered in 1960. In a strip along the shores of Lake Erie, east of Cleveland, pyrite content in the shales runs to almost 5 percent and the oxidation of the yellowish iron sulfide forms a product of 10 times the original volume. Structures founded on this shale suffer damage from localized heaving. In the Cleveland area several sewers have broken and irregular levels appear in floors and stairways. A newly constructed high school, floors on ground, heaved before completion. The building at Nela Park has shown similar heaving trouble since 1920, with some floor areas raised 1 ft and small footings heaved 4 in. The only cure is an immediate sealing of all rock exposures with a bitumen paint.

An expanding shale was encountered in Denver in 1960 at the concrete reservoir construction. Swelling of the bottom as much as 4 in. indicated that a concrete floor slab would not be satisfactory and the reservoir design was changed to provide 24 in. caissons spaced on 20 × 5 ft grids with barrel-arch concrete floor set up free of the shale, to carry 23 ft of water load. The cost of the reservoir was increased by $500,000 to eliminate the shale heaving effect.

2.6 Design Error

There are practically no recorded failures of a foundation as a structural unit. Exceptions to this rule are the punching failure of a stone footing when a cast iron column with a 36-in.^{2} by 2-in.-thick iron plate broke through the masonry pad causing the collapse of a 50 × 50-ft area of the

Fig. 2.5 Distorted shape of row houses with piles not carried to bearing stratum.

five-story Chicago Club building; the crushing of the foundation which caused failure of the Barentin Viaduct near Havre, France; and the crushing of the lime concrete foundations of the concrete engine shed on the British Metropolitan Railway that caused a complete collapse of the shed. All of these incidents occurred before 1900. In Barletta, Italy, overoptimistic engineers in 1959 placed too much value on the foundations of a one-story garage and to this garage added four stories for apartment use. The structure collapsed and 58 people were killed.

Many pile foundation jobs are designed with insufficient subsurface investigation or with disregard of the true soil conditions, resulting in later expensive corrective work. Timber piles for a row of 12 three-story dwellings in a filled-in marsh area in New York were installed with tip elevations not fully penetrating a peat layer. The walls settled and the buildings tipped, causing complete abandonment in 1966 after only a few years use and total loss to the 12 owners (Figs. 2.5, 2.6). Piles driven for a research laboratory at Holmdel, N.J., in 1960, based on an apparently careful design were noted to be unstable when the pile caps drifted. All piles were then load-tested and redriven where necessary, delaying project completion by a year.

Assumed bearing values on rock can only be relied upon if the rock

Fig. 2.6 Wall cracking from distortion.

exposures confirm the expected strengths. A concrete factory in Cleveland, Ohio, when completed, started to crack at the junction of girders and columns, indicating failure in foundation support. The footings had been designed to bear on shale rock at 13 ton/ft^2, exposures at that depth indicated that the actual soil was a clay with much less value. At Lincoln, Neb., a grain elevator founded on sandstone was noted to have settled 3½ in. in a 283-ft section of the structure. Investigation showed that the sandstone had shrunk in thickness by that amount due to seepage loss of water entrapped within the rock (Fig. 2.7).

Structural design procedures for foundations have sufficient factor of safety; the errors are in the judgment leading to assumptions or deci-

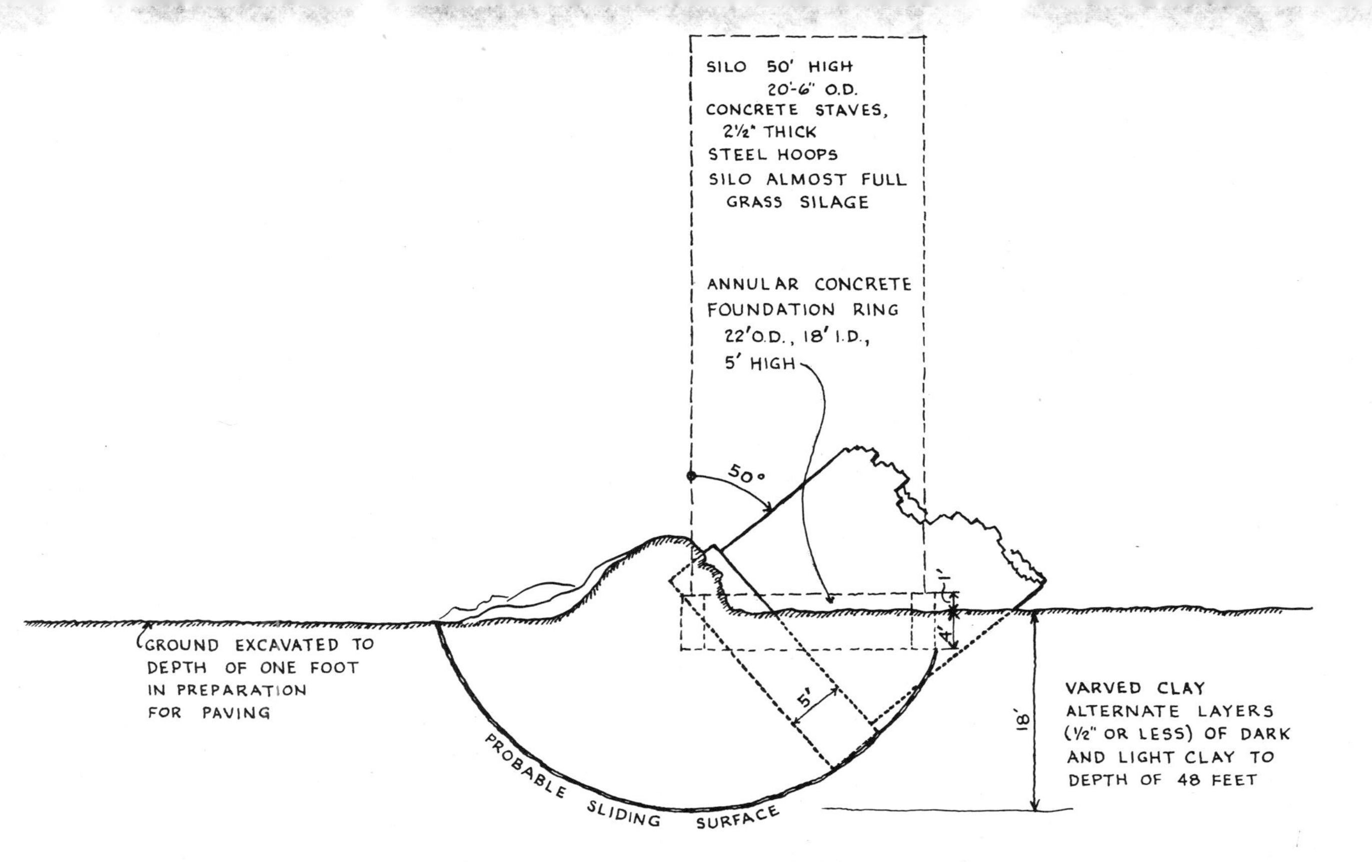

Fig. 2.7 Silo failure from unequal foundation support. (From Research Paper No. 169, Canadian National Research Council; article by W.J. Eden and M. Bozozuk, *Engineering Journal*, September 1962.)

sions which are not complied with by the natural soil. A precast concrete crib wall enclosure of large fuel-oil storage tanks required pile foundations and the designer disregarded the lateral pressure of the sand filling between cribs and tanks. The outer line of timber piles, driven to 15-ton value were then required to support about 72 tons each and the inner line became anchor piles. The cribs failed when the concrete pile cap rotated; new piles were provided.

A common design error is often made, as foolish economy, in one-story industrial buildings located on filled land. Piles are provided to support the walls and roof, but the main floor is placed on more-or-less compacted sand over the fills, rubbish, peats, organic silts, and other compressible layers. As a typical example, a building costing $300,000 was occupied by a lithographic plant with installed equipment valued at over $3,000,000. The color presses require almost perfect dimension stability. Support on a slab resting on filled soils will not provide that stability. It seems more logical to provide pile support for the expensive equipment and let the roof settle. Even in warehouse use, the floor deformations become objectionable. For garage use, the change in drainage pattern with the floor drains always ending up at the high points of the deflected floor, leads to added operation costs which would easily pay for the cost of proper floor support. Economy is not measured by the initial expenditure, but by the total costs of keeping a structure useable.

2.7 *Construction Error*

Construction of a foundation involves two separate sources of trouble. The construction phase requires temporary expedients and protections subject to failure and collapse, often more intricate in nature than the actual foundation work. In such temporary work factors of safety are kept to a minimum, for economy, and expensive design procedures are considered an expensive luxury. Most foundation failures due to construction are in this class and not in the misbehavior of a finished structure. However, the exceptions to the rule are some interesting examples.

The great deal written about responsibility for corrective work to take care of unforeseen conditions never exonerates the builder from not asking for a change in the plans when he finds that conditions do not conform to the design expectations. A collapse of one of the early steel-framed buildings in New York described before, was caused by constructing a column footing partly in natural soil and partly on an abandoned cistern, which of course was not filled with consolidated material of equal-bearing value. In the construction of a five-story garage with column footings on rock bearing, one footing was placed on some loose rock fill in a narrow creek bed crossing the area. By the time that the roof slab was being concreted, this column had settled enough to delete a support for the flat slabs which had not been designed for the double span and the structure collapsed.

In a two-story garden apartment development in Queens, N.Y., three of the buildings crossed a filled-in creek bed and all the buildings were founded on long timber piles. The 20-ton piles were installed under careful control and supervision and capped. Where the creek bed required deeper pile penetration, the caps were placed at a lower depth than the rest of the foundation. Before the construction of a concrete cellar wall which was to act as a grade beam to support the masonry walls, the entire area was graded level and the deeper caps were buried. The grade beam was constructed of uniform depth, partly on the fills above the deeper caps. The completed buildings cracked seriously and the piles were immediately blamed. Exposure at several pile caps with jack-load testing showed these piles in satisfactory condition and fully capable of holding the intended loads. Further investigation uncovered the error and the grade beams were underpinned to the low-pile caps. But the repair and maintenance costs to keep the buildings useable had become so high that the project was abandoned and demolished to be replaced with a high-rise apartment on new pile foundation.

Omission of a pile during construction resulted in a cracked concrete grade beam and the masonry wall was built thereon without correction of the foundation deficiency (Fig. 2.8).

In 1958 concrete cylinders were extruded under pressure to act as piers or piles for the support of a five-story office building near Montreal, Canada. The necking of these cylinders from the intrusion of the compressed surrounding mud when the steel shells were retracted too rapidly, so reduced the strength of the support that the almost completed building, as a rigid mass, tipped 18 in. from the level in the length of 200 ft. The cylinders failed in rapid succession and without warning as the load was transferred from the initial failure at one end of the building. The design was perfectly good; the well-built structure took a sloping position with not a single crack in the concrete superstructure. The owner refused to accept any corrective measures and the building was demolished and rebuilt on new piers. The possibility of mud intrusion was then eliminated by providing permanent thin steel shells within the retracted casings (see Chapter 3).

Fig. 2.8 Broken grade beam covered by brickwork before correction of missing pile.

A somewhat similar incident was found during the enlargement of the LaGuardia Airport in New York in 1961. After the concrete-filled steel-tube piles had been completed, some of the piles had to be cut down to permit installation of added fuel tanks and some piles were found to be not completely filled with concrete. An investigation was started to explore the condition of all the piles. A random sampling of piles was drilled with rock-drills and when found suspect (as indicated by a nonuniform drill resistance and change in dust color) the piles were also cored, and some were extracted to correlate the drilling and coring information with visible exposure after the steel shell was removed. Very good correlation was proven. Enough improper piles were discovered in the random sampling to warrant the decision that the piles were not useable. Two were found with voids over 30 ft in length. All pile caps which had been built were removed and new piles were added to each cluster to carry the full loadings. Since all work had been done by an experienced contractor under continuous supervision, it makes one wonder how many such conditions exist and have not been discovered or even discovered but not reported.

Recently in Chicago, for at least three of the skyscraper developments, the concrete in the deep caissons carried to rock have been found discontinuous. One caisson was so poor that the weight of the first column length, some 30 tons, caused movement. Considerable costs of drilling and coring exploration, grouting, and even replacement have corrected the deficiencies, with consequent delay in project completion. The errors are really those of concrete placement methods, but the foundation is involved so that these cases are here included (see Chapter 3).

One common error in pile driving in plastic soils is the disregard of internal lateral soil displacement. Heaved and buckled or collapsed pile shells can be avoided by thicker shells or a proper sequence of installation with allowance of sufficient time for the induced high-pore pressures to be dissipated. All piles should be driven from one general level, even where this entails more expensive later excavation of part cellars, pits, and trenches. Excavation of lower areas even if sheathed and braced for normal earth pressures will not stop the lateral drift of piles driven adjacent to such excavations. The cost of corrective measures at a hospital in Chicago, where the boiler room area was first excavated, from pile cap redesigns and added piles exceeded the cost of the total original pile installation. This type of failure is such common knowledge, that its repetition is completely inexcusable.

2.8 Flotation and Water Change

Except in well-consolidated granulated soils, a change in water content will modify the dimensions and structure of the supporting soil, whether from flooding or from dewatering. There are two incidents where localized settlements within an existing building resulted from cooling water pumping by the occupant. A 12-story concrete building in Brooklyn, N.Y., supported on spread footings on medium sand suddenly noticed settlement of several interior columns when large scale ground water pumping was installed to provide industrial water. Ground water levels had receded about 30 ft, and the operation was then forced to purchase city water and stop pumping, after several columns had to be underpinned. After about two years of pumping for air conditioning requirements in a large two-story steel-framed factory for television assembly in the Syracuse, N.Y. area, on an apparently good base—thin layers of limestone and shale—but directly above deep layers of salt, the floors and roof dished downward several inches maximum directly around the pump walls. Borings showed that the fast flow of water had dissolved the salt seams and underground settlement resulted. Columns were jacked up and the floors resurfaced, and an outside source of water was obtained for further operation requirements.

Pumping from adjacent construction excavations will affect the stability of spread footings and even cause dragdown on short piles. Such damage often results in expensive litigation and has caused New York City in its construction contracts to require recharging the soil outside of sheathed excavations to maintain original ground water levels (Fig. 2.9). In a long-time job dewatering of the soil around untreated timber piles can start wood rot and pile collapse. The long-time pumping from deep wells for constructing the Brooklyn approach to a subaqueous tunnel in lower Manhattan carried on during the period 1941–1945 to maintain the depressed water table, even with construction closed during the war, dropped the adjacent street by 9 in., affecting all the structures facing the street. Similar pumping for a subway in Brooklyn constructed in fine and medium sands, the excavation enclosed in steel sheeting, required continuous recharging of all the pumped water into the ground on each side of the excavation. There were practically no measured settlements of continuous line of two- and 3-story buildings even with excavations 20 ft below ground water level, and no underpinning of the shallow foundations.

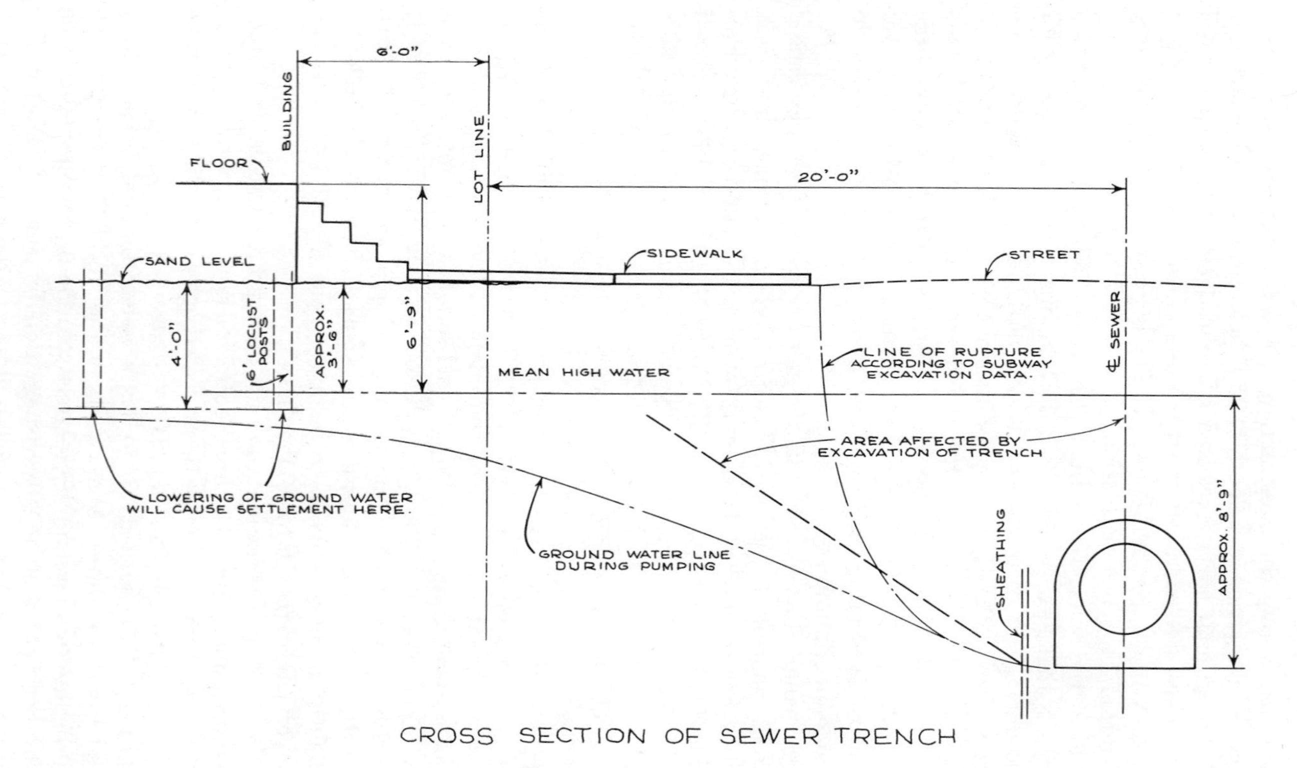

Fig. 2.9 Diagram of dewatering influence.

Without recharging the construction of a sewer near the New Jersey meadows, in silty sands to a depth of 15 ft below water level, which was close to the ground surface, caused consolidations of 8 in. in the soil adjacent to and under an industrial building, fortunately built on a pile foundation with a supported floor slab. The nearest edge of the building was 150 ft from the sewer excavation.

Typical of the suits arising from ground dewatering by construction operations is that filed by Boeing in 1963 against a sewer contractor for $190,000 damage to their Renton, Wash., building. The two-story pile-supported plant suffered settling of the ground floor and cracking of partitions. The trench was sheetpiled but pumping by deep wells lowered the ground water some 30 ft at a distance 1300 ft away from the Boeing building. Another sewer contract, nearer to the building, was then built by excavating in the wet and tremie concreting a base seal to reduce amount of pumping.

Water change from revision in the normal river levels has caused large scale damage to stable structures. When the river level was lowered by construction of dams some 40 miles away and also by a two-year drought, the high school at Elk Point, S. Dak. in 1956, suffered serious settlement with buckled walls and cracks over 4 in. in width, causing evacuation of the building. Regina, Canada, also reported considerable settlements of buildings in 1962 from a recession in general water table.

In Springfield, Mass., two old tenements built along a filled creek bed which was connected to the Connecticut River, following a heavy river flood season, tipped toward each other almost 24 in. The clay under the two adjacent walls, spaced 10 ft apart, had softened and caused the walls to settle. The value of the buildings did not warrant underpinning, so as a protection against further movement steel pipe struts were inserted between the walls at each floor level and some 15 ft apart. The buildings with slightly sloping floors have been used for almost five years and provide very cheap rentals for 20 families.

Clay heave from oversaturation must be expected and in such soils, the structures must either be designed to tolerate upward displacement or else the supporting soil must be protected against flooding. A six-story hospital in Pittsburgh, Pa., in 1908, with walls bearing on hard clay loaded to 2 ton/ft^2 settled up to 12 in. when the soil became fully saturated from rainstorms. Many parts of the world have trouble from heaving bentonitic clays and the perfect solution of a proper foundation design is not yet available. One suggestion for some housing in San Antonio, Tex., was to surround each building with a narrow deep trench

in which is placed a continuous sheet of aluminum foil and the trench filled with tar as a stop against water infiltration to the foundation soils. With no change in water content, the soil volume should be stable. Many articulated footing and slab designs have been suggested as solutions to the expansive soil conditions.

Bentonitic clay foundation supports for homes in Texas, Oklahoma, Colorado, Canada, Mexico, India, Israel, and South Africa and elsewhere have resulted in damage to the structures. More trouble is found in masonry than in wood-frame homes. The cost of a foundation designed to resist all possible heave and shrinkage forces is normally too high and some compromise must be made with an expected repair program in the life of the building.

The lateral pressure of saturated clay can be greater than hydrostatic pressure. In a large housing development on a sloping terrain in Hartford, Conn., the concrete cellar walls were backfilled with the local clay. Before the site had been surfaced and drains provided, a heavy rain caused large flow of water to saturate the backfills on the high side of many buildings and the walls collapsed (Fig. 2.10). After reconstruction of the walls, the builder decided that the high cost of bringing in granular fill was not warranted and again used the local soils. After a second unusual storm caused a repetition of the failures, granular fill was purchased and used. The buildings are now 25 years old and there has been no further failure of the cellar walls.

2.9 Vibration Effect

Earth masses which are not fully consolidated will change volume when exposed to vibration impulses. Vibration source can be construction equipment, especially pile drivers, operating equipment in a completed building, or even traffic on a rough pavement, as well as blasting shock.

Neighboring small residences to a Philadelphia, Pa., bridge construction were seriously shaken from pile hammer impact into the soil, and the contractor was held liable. On a much larger scale, in 1963–1965 the insertion of H-piles by heavy impact hammers and even by vibrating hammers for a tall government building in New York, N.Y., so damaged adjacent buildings, one 19 stories high and resting on a medium dense sand of uniform grain size, that all the adjacent buildings—more than 400 ft of frontage—were condemned as unsafe and had to be demolished. The adjacent street pavement settled up to 16 in. when only a fourth of

Fig. 2.10 Wall failures from liquefied clay backfill.

the piles had been installed, before the vibrating hammers were brought to the job. As a precaution against further street subsidence, a line of steel sheeting was driven and covered with a sand berm on the excavation side. The protection was successful as far as protecting the street against further settlement and damage to the sewer and waterlines buried therein. In a similar building construction in 1960, some 1000 ft away, when the first pile driving indicated serious effect on the adjacent building, work was stopped and the existing footings were underpinned by piles carried to the expected tip elevation of the new building piles. Thereafter pile driving had no effect since the sand layers which were consolidating no longer carried any building load.

At Towanda, N.Y., a four-story factory on sandy clay support suffered local settlement from a vibrating compressor located directly above a footing. To avoid vibration transfer from a battery of air compressors needed for the rock excavation of a subway in the center of New York in 1936, they were set on a concrete box foundation which had 4-in. air gaps at each wall to isolate the support from the adjacent walls and was set on a 6-in. layer of compressed cork above the rock subgrade. Three years of operation did not bring a single complaint from the adjacent buildings, many of them hotels, even with 24-hour use of the compressors.

Blasting operations must be carefully programmed, taking into account the seam structure of the rock, to avoid serious vibration damage to adjoining and adjacent structures. When a parallel tunnel in sedimentary rocks was constructed in the Kittatinny Mountain in Pennsylvania, blasting shock damaged the lining of the operating tunnel. Modern technique in rock removal requires the preparation of an isolation shield made by lines of closely spaced drill holes and presplitting the rock between these holes before any blasting is done for the new excavation. Damage from blasting effects has been so extensive that legal regulations in some areas, as in the State of New Jersey, limit the size of charges to keep vibration transmittal to tolerable limits. These rules can only be of general value, using distance only as a measure of attenuation. When important structures, especially those covered by stone facing or other brittle materials, adjoin a new excavation the type of rock and continuity of seam structures must be carefully studied. In many instances the standard regulations are too restrictive. Similar legal restriction to construction operations can only be avoided by the industry preventing damage to the public as well as to itself by proper control of its work.

2.10 Earthquake Effect

Foundations in earthquake areas must be designed to be tolerant to the shock spectra provided by nature. In the last 10 years there have been more than 20 damaging earthquakes, widely scattered from Alaska to New Zealand. The effect on foundations is usually less than on the superstructures, especially in earthquakes of short duration. Statistically an earthquake duration of 45 sec is a sufficient assumption. Yet the Niigata, Japan, and Anchorage, Alaska, earthquakes of 1964 were of much longer duration, the latter continued with fairly uniform intensity for 330 sec. These long-time vibrations seriously alter the subsoils and a great deal of the damage is the result of foundation failure because of soil bearing collapse.

In the Anchorage earthquake vertical and lateral movements due to the instability of the soil destroyed many buildings, regardless of whether they were properly built or whether they were designed for seismic resistance. In the Turnagain residential area 75 fine houses were made useless by the large consolidation and lateral flow of the underlying clays. Five years earlier the United States Geological Survey warned of the incipient danger from the sensitive clays underlying the sand gravel beds in the Anchorage areas. At four other discontinuous areas, similar soil failures dropped buildings as much as 14 ft vertically and caused total failure of two blocks of small commercial buildings, while similar buildings on the opposite side of the street were unaffected.

Vibration effect at Anchorage, 80 miles from the epicenter of the earthquake, was sufficient to liquefy the saturated fine soils and cause lack of support. Frost depth is from 5 to 8 ft and the weakest soil layers are lower. Similar soil flows occurred at Whittier and at Niigata. Study of the soil characteristics indicates that such failure would not be caused by a single shock, but that at least 1.5–2 min of shaking was necessary. Studies and some full scale tests are going on of methods to reinforce the soil by sand drains, sand anchors, and gravel buttresses.

Foundation design must be coordinated with the expected action of the superstructure. At Anchorage, long narrow dormitory buildings, carefully designed for Zone 3 earthquake forces with wide rattle space expansion joints had common footings for the twin columns at the joint. The distress in the lower floors of the structures were the result of the lack of freedom at foundation level to permit each building section to vibrate independently.

The 1965 earthquake in Mexico City, with the epicenter 250 miles away, caused several 14- to 22-story apartment buildings to lean alarmingly. Even a 42-in. lean of a 22-story building was tolerated by the structural frame when the foundation rotated with the building. Some of the lesser affected buildings were righted by dewatering the soil on the high side and letting the weight of the building right itself. Others required additional piles and basement flotation chambers to aid in stability during future shocks.

Summary

Structural distress from foundation insufficiency at the best causes continuous high-cost maintenance and often results in total collapse. All structures have some tolerance to unequal settlement of the foundation components, but when the support is stressed beyond the elastic limit, ultimate failure is unavoidable without immediate strengthening of the foundation.

The numerical percentage of structures with foundation failures is very small, but every single incident in itself is a serious indictment of the capabilities of the design and construction talent. Foundation failure has a unique property in that adequate adjacent structures are often seriously affected, bringing in serious litigation and expensive liability for damages.

There have been a number of large catastrophic incidents which are well-publicized in technical as well as the general press. Many lessons can be learned from the results of investigation of the less publicized smaller cases of failure. These can set up rules for "what not to do," something that cannot be learned from the most minute study of the theories in soil mechanics or from the glorious record of successful work. Sharing the experience of such information should be of unlimited value as a warning of what pitfalls are to be avoided if the list of foundation failures is not to increase. This warning is of especial value in the design and construction of foundations in congested areas, and in the use and control of the weaker soil conditions.

3

Below Surface Construction

Trenches; Sheeting and Bracing; Piles and Caissons; Tunnels and Sewers; Dams; Rock as an Engineering Material.

In no phase of construction is a keen eye and the skill that comes from experience more important than in the work below surface. Just as in the art of surgery, success does not permit much deviation from continuous and sequential steps in which precaution against a mishap is always at hand. There is considerable merit in the analogy between surgery and below surface construction. Both expose the unknown and must expect to remedy unforeseen conditions and both must prepare to counteract operational shock to the physical conditions, often much weaker than assumed. The recently added expression "systems design" is merely a description of proper technique in the art of surgery or of below surface construction, where every logical step is preconsidered before starting the sequence, with, as in a well-played chess game, alternate protective moves in readiness to correct for a miscalculation. Needless to say, not all work is so planned and an indication of why some were not successful should warn of pitfalls to be avoided. Failures in subsurface work are expensive, in money and in time and too often in loss of life. A generation ago, an accident in a tunnel was an immediate signal for the job to shut down for the day, a sort of protest by the workers that the job control or planning was in error. The higher costs of construction accidents, both from legal and financial considerations, has introduced more and better planning of construction operations, but much still has to be

done to train men to properly plan and execute the work. Such skills are not inherited, neither can they be learned from texts or assigned to the computer. From ancient times, the construction "boss" has been shown as a necessary component, as in the Mesopotamian and Egyptian hieroglyphs. The ancient recognized necessity for such skill is best shown by the assignment to foreign talent by King Solomon of the control of construction for the Temple in Jerusalem. To this very day national pride does not stand in the way when construction talent is needed, and this applies to both the developing and the developed countries.

3.1 Trenches

Digging trenches for defense, for drainage, for waterways, and for the insertion of conduits and foundations has been done for so many centuries that even without any theoretical developments, the unfortunate experience of others should now prevent any failures. Yet the continuous occurrence of trench collapses, often with dramatic news stories of fatal and almost fatal results, is unwarranted and inexcusable. In any kind of soil trenches can be dug which will not harm the workers, the public, or the adjacent existing structures. In this connection it may be of interest to note that the definition of "safety" involves a three-pronged requirement.

1. Freedom from danger or risk.
2. Freedom from injury.
3. Harmlessness which is the state of being without hurt, loss, or liability.

Safety is more than just freedom from failure or collapse. Safety is everybody's business; the author recently pointed out in a letter to a national magazine for the engineering-construction industry that editorials preaching safety for trench excavation are inconsistent with full-page advertising showing evidently unsafe practice in unprotected deep trenches. The reported incidents of open-trench failures are so numerous, with probably at least that number not publicized, that examples are not necessary.

Legal controls resulting from the failures have in many cases imposed unwarranted requirements. The cost of normal sewer construction in the New York area is almost doubled by the requirement that all cuts in excess of 5-ft depth must be sheathed. When the regulation was first enforced, on one job in clean coarse sand, trenches dug with perfectly safe side slopes were required to be sheathed along the slopes, with

claim by the contractor for extra payment. Public control of construction operations will always entail extra costs beyond reason. The construction industry can only avoid such controls by a self-policing of its own work and elimination of unsafe practice.

Typical of the public reaction to failures in trenches is the incorporation of the safety provisions required by the Maryland State Department of Labor and Industry into the Prince Georges County Building Code in 1966. After several cave-ins within a year had taken three lives, the older law making an offense a misdemeanor with a maximum penalty of one year in jail and a $500 fine, became a felony with a maximum penalty of five years in jail and a $5000 fine. Failure to protect the sides of a 7-ft deep storm drain trench which failed three hours after the inspector's order and cost the life of a workman, resulted in the contractors being fined $1000 for not maintaining a safe place of employment. In 1962 a basement excavation in Greenbelt, Md., caved in and killed five workmen, as the unprotected vertical excavation developed cracks in the clay soil and a large block of soil slid out and crashed into wall forms being erected by the men. This was one of the series of accidents which caused public demand for stricter legal control of excavations.

As a defense against the high incidence of trench failures and as an economical solution of the sheeting and bracing requirements, several developments of prefabricated devices are now available. At a site where a wide sloping sided trench could be first dug, Niles Excavating Company of Terre Haute, Ind., made a steel shield with a flat head and two sides 22-ft long with overhead brackets to separate and hold the sides. The shield was lowered into the 11-ft deep trench and pulled forward by a backhoe as a sled to protect the placing of the pipe in 20-ft lengths. Progress averaged over two lengths per hour.

The Torti Equipment Co. of Providence, R.I., make available a shield consisting of two steel rectangular boxes that telescope and are separated by hydraulic rams, in 20-in. steps, after each of which the rear box is pulled forward. Openings in the top of boxes provide for pipe access to the open bottoms. Excavation progresses in front of the shield by backhoe. Progress has been 80 ft per day and the units have worked at 28-ft depth.

Valley Construction Co. of Seattle built a simple shield of two stiffened steel plates for the full height of the trench with welded arch bracing, and pulled it ahead by the backhoe. Morrison-Knudsen developed a trench shield for the 127 miles of sewer built for Boise, Idaho, and also for 14 miles of 8-ft-diameter steel water pipe in Los Angeles. Steel plate 20 ft were braced by a solid plate at the front and beam arch at the rear, using various widths of bracing to conform to pipe size. The

use of the shield required advance cutting of crossing utilities and their repair as the shield passes the obstructions. In spite of some treacherous loose soil no lost time accidents were reported in either job, covering 326,000 man-hours in Los Angeles and 18 months work in Boise.

The question often arises as to what is a trench excavation, especially when an accident has occurred and the legal protection requirements were specifically for trench sides. The Ontario Trench Excavation Protection Act defines a trench as an excavation where the depth exceeds the width (which would include most pits) but did not specify whether the width is to be measured at the top or the bottom of the excavation. In a 1958 trial concerning a cave-in in which six men were killed in an excavation 32 ft deep, 20 ft wide at the bottom, and 55–60 ft wide at the top, the contractor was acquitted as not having violated the Act, but the magistrate recommended an amendment to clarify the intent of the law.

A frequent cause of trouble in trenches is marsh gas seepage in even shallow excavations and sometimes the release of deep gas deposits in shale rocks when the overburden pressure has been removed. Several reports of gas asphyxiation are similar in the serious effect on the rescuers after workers have been felled. Trench gas can easily be purged by air dilution to safe concentrations and indicating instruments for gas presence are readily available and should be required for all work in soils that could contain gas. Boring records often disclose gas pockets and general area incidence is known and should be noted on the subsurface investigation reports.

3.2 *Sheeting and Bracing*

Failures in trench sheeting are becoming less common with the growing awareness by contractors that the minimum timber planks, often hand-mauled into the soil, will not sustain the surcharge pressures of heavy excavation and pipe laying equipment. Hand-excavation methods permitted the addition of cross bracing as lower levels indicated their necessity. More rapid power excavation requires stronger sheeting with planned location of wales and struts. Typical of the failures causing loss of life was the trench collapse in the 30 ft deep sheathed excavation at Varennes, Quebec, in 1956 where tons of damp clay covered the men as the sides crumbled. The top of the vertical plank sheeting, placed with open gaps about the same width as the planks, was several feet below the top of the trench, and all excavated material had been placed on one side of the trench.

Another example of trench sheeting difficulties which delayed comple-

tion of a sewer in an ash-filled marsh above a sloping clay subgrade teaches that economy in construction is not obtained by skimping on sheeting and bracing. A timber-sheeted sewer trench placed parallel to the original beach failed due to unbalanced lateral pressure especially since the excavated soil was placed on the high side. The bottom of the sheeting could not resist the surcharge loads. Steel sheeting was then installed and carefully braced, but some distortion of wales was noticed from the same unbalanced pressures causing the trench to twist. Some diagonal bracing was then installed and the work was satisfactorily completed.

Under the theory that a calculated risk is permissible in heavy construction, as if temporary structures are safe at a standard insufficient for permanent work, sheeting and shoring for subsurface construction is prone to accident and consequent damage. The high cost of repair and delay in progress, even for the small percentage of all such installations which undergo failure, on an insurance basis alone, make it economically advisable to guard against short cuts and protect for the worst possible loading conditions. Description of a few typical cases will show the necessity for some precautions in all such work.

At 7 a.m. on May 21, 1964 the roadway adjacent to a 30-ft deep sheathed excavation in Chevy Chase, Md., suddenly dropped with damage to the buried utility lines. Sheathing consisted of 12 BP53 soldier beams spaced 7 ft 9 in. apart with 4-in. timber lagging placed with 2-in. open gaps. Soldier beams were imbedded 5 ft below subgrade, and were supported by two lines of wales with steel diagonal rakers spaced at every third soldier beam. Soil consisted of 8 ft of filled earth, above 8 ft of silt and sand, above 9 ft of disintegrated rock and sand beneath. Building foundations along the sheeting line were cored caissons to rock level, installed from subgrade. Continuous pumping was necessary to control seepage through the sheeting, from ground water level some 18 ft above subgrade. Some light blasting was used to break up the disintegrated rock. Traffic in the street was very heavy, including trailer trucks. The sheeting was designed for soil pressures and was not sufficient if the soil became saturated, which evidently did occur, either from rain or a leak in the sewer or water mains, after pumping had dehydrated the silt layer and opened up internal seepage channels. The liquid pressures collapsed the sheeting.

A similar 45-ft deep cellar excavation in Montreal, with the soldier beams restrained by prestressed tieback system, anchored into grouted rock sockets, failed as the hand excavation was being completed for foundation concrete. The area of failure of the earth and rock bank was apparently not completely sheathed, being a berm left adjacent to a

Fig. 3.1 Sheeting failure caused by bulldozer hooking a soldier beam.

neighboring underpinned wall. Two closely spaced cave-ins trapped three of the five men working at the base.

In a properly sheathed and braced excavation for a three-tier garage below street level in Washington, D.C., with machine excavation completed, the final earth removal at subgrade was being done by a very small back-hoe scratching the soft disintegrated rock away from the soldier beams. The bucket teeth engaged the flange of a soldier beam and in the attempt to free the contact, pulled the beam inward and caused a local collapse. The soil poured into the excavated area and the added forces crippled the two lower sets of crossbracing, the top frame remaining intact (Fig. 3.1). Street traffic was closed off in one of the

major arteries of travel. Corrective work consisted of a new line of soldier beams surrounding the original sheathing, some 15 ft away with enclosures to tie to undisturbed portions. Fortunately, the small park area adjacent to the point of failure was available for the corrective enclosure.

The purpose of sheeting is not only to protect the enclosed work but also to prevent loss of support of adjacent structures. Movement of the sheeting and especially of the raker supports cannot be entirely avoided, and some method of compensation during the disturbing action of the excavation work, especially blasting and caisson installation, must be available. In the foundation work for the Cleveland Federal Building in 1966 the City filed suit against the contractors for damage resulting from a subsidence of the adjacent street because the shoring did not prevent movement of the sheeting. Similar claims are resulting from the new hotel excavation in the French Quarter in New Orleans. In such vulnerable areas, where sheeting location must be maintained, a system of jacking facilities can and should be provided, so that the required reactions are induced into the supporting members and rapid adjustment to neutralize movement can be done.

Control of movement, sometimes followed by total collapse, requires careful attention to details, even assuming a proper design of the sheathing system. Unwanted surcharge loadings must be prevented, such as unforeseen equipment operation at the edge of the excavation and more often, stocking of steel members and plates at the convenient but hazardous position at the edge of the sheeting. Existing surcharge loadings such as neighboring buildings with shallow foundations must be considered in the design, as well as traffic loading and vibration effect on the disturbed soil or backfill. Ground water control must be positive and continuous, or else the sheathing must be designed for hydrostatic pressure in addition to soil pressure. Loss of soil in back of the sheeting will completely alter the pressure diagram. Reactions imposed on wales change with excavation depth and the concentrations at rakers almost always require welded stiffeners on the webs of economically designed wales. Diagonal rakers impose upward vertical loading on the sheet piles or on the soldier beams. If the soil friction, modified by water and vibration, cannot resist such reactions, the imbedded lower ends must take the full upward pull. Where steel sheeting rests on rock, anchorage may be necessary in the form of grouted rods welded to the sheeting, or in the case of soldier beams either sockets in the rock for grouted imbedment or adding grouted rod anchors welded to the beams. A buckled wale or a bowed brace is a sign of trouble and must not be tolerated. Needless to say, heel blocks for the spur brace reactions must

be stable and large enough to transmit the load to the soil. Disregard of small details generates large failures. Each of the previous items has been noted as the cause of excessive sheathing movements with degradation of the desired harmlessness to adjacent existing structures.

Interior bracing, whether horizontal or sloped, interferes with excavation equipment and usually requires cutouts in cellar walls with incipient leakage in the finished walls. Cable or rod tiebacks to hold the sheathing to the exterior soil or rock are therefore a worthwhile development, eliminating exposed bracing and permitting simple stressing to desired reactions. It should be noted that even for temporary use, tiebacks entering private property forms legal trespass with all its consequences. Even if public or private land is available, the tieback will only serve if the anchorage is reliable. Failure of the entire tiebacked 1:5 sloped earth cut, 35 ft high, with anchors all tested for 10 tons, at the Seattle City Hall construction in 1961, was caused by the internal weakening of the fissured clay. Expanding type anchors had been inserted into 8-in. diameter holes drilled 12–20 ft into the embankment. The surface was covered with wiremesh and gunite and 4 × 12 timbers set in a grid to connect the anchor rods. The slide was 150 ft long and sheared at 15 ft back from the face of the bank.

Sheeting and bracing are seldom installed to the dimensioned tolerances required of permanent framing. To meet a budget limit for a six-story library building in Baltimore, it was decided to place five stories below ground and use the permanent steel bracing as the framing for the lower floors. Soldier beams were first driven and wood lagging placed to a depth of 10 ft below ground level. Sloping berms were then left as the excavation continued to subgrade, clearing space for the interior footings. Niches were dug into the slopes for the wall line footings and the steel framing was then erected for the full height of the building, bracing the soldier beams. To permit concrete encasement of the soldier beams, the wood lagging was placed 2 in. beyond the outer flanges, bearing against two 2-in. pipes welded vertically to the outer face of the beams. Temporary walers were installed above floor levels and reacted against vertical brackets welded to the horizontal floor beams. Site conditions were not favorable to maintaining the cofferdam in an exact position. The outside grades were a full story higher on one side of the building. Subgrade required rock excavation. One corner of the site was adjacent to a building on much shallower footings. Along one side was a concrete service tunnel 10 ft deep and a sanitary sewer 8 ft below that level. Water at the lower levels of the excavation was flowing under constant pressure. The result of all these factors was a lateral drift of the framework, putting columns out of plumb and the

floor beams out of level. Additional concrete encasement was required to provide plumb columns and level floors, somewhat reducing the useable volumes and also reducing the indicated savings of the design.

Subway excavations in built-up areas are normally decked over to provide for street traffic. An economical design is to integrate the side-wall sheeting supports with the decking supports. Both items must therefore remain stable. Normal precautions are to design the elements on a fail-safe basis, where the loss of one support would not stress the carrying beams to failure. Supports are often lost from excavation or blasting effects and sometimes removed for expediency: space needed for equipment operation or steel erection. Two very similar decking failures occurred in 1926 on contracts six miles apart in the construction of the Eighth Avenue subway in New York. Rock blasting loosened large masses of seamy rock at one side of the subway, and the slip caused fall of the earth above, unbalancing the earth pressures. The decking then collapsed into the excavated areas. Failure extended over 100 ft in one job and 200 ft in the other.

Backtied sheetpile bulkheads are stable only as long as proper resistance is provided at the toe of the sheeting and at the anchorage of the ties. Failure occurs when either of these supports gives way. In 1963 a steel sheetpile bulkhead in Santa Cruz, Calif., had a series of failures that illustrate the necessity for each of the stable resistances. Sheetpiling was first installed as Z-27 sheets 34 ft long driven 28 ft into soft silty clay and with a 24 × 18-in. concrete cap. Patented ties of the split-sleeve threaded type were imbedded in the cap and connected to 1⅛ in. high-strength rods, 9 ft apart, which were anchored to a concrete deadman 50 ft away. Water was 6 ft below the top of the sheeting and high water came to the top. Backfill was with dredged material. Some small movements of the wall were noted during the assembly and filling. As a channel was cut in the basin enclosed by the sheeting and water level dropped 4 ft, movements increased to serious magnitude. To stop the movement, the fill behind the sheeting was excavated, but the deadman wall then slipped into the excavation and the sheeting moved outward, up to 13 ft as a maximum. Reconstruction was with longer sheetpiling, heavier tierods and bigger deadman, all to produce the necessary resistances from the saturated silt.

A steel sheetpile bulkhead along the Arthur Kill in Staten Island, driven through soft varved silts to a hard sloping rock surface and tied to paired batter anchor piles, also backfilled with dredged soil, served well for several years but suddenly moved outward some 20 in. with noted heaves in the vicinity of the anchor piles. Investigation revealed that the construction of a new road on a fill being placed on the marsh

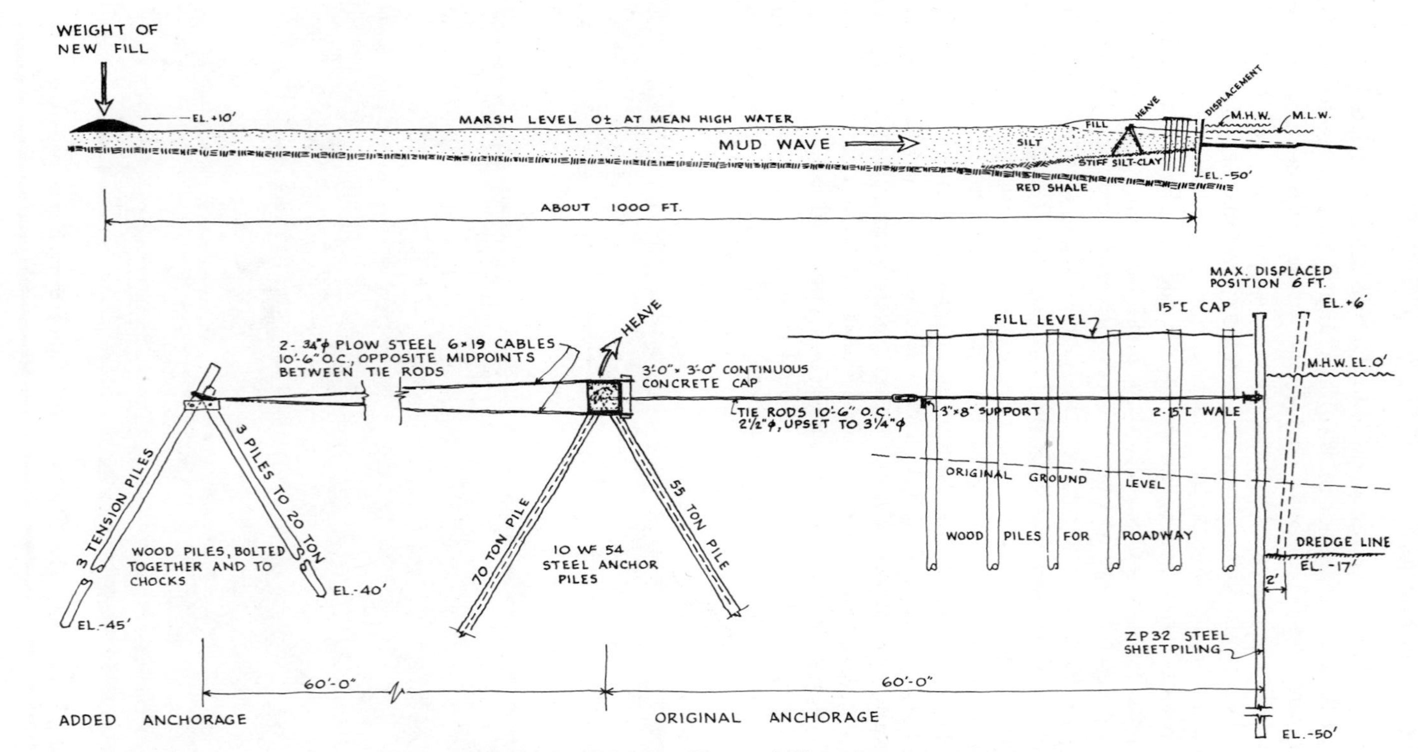

Fig. 3.2 Bulkhead displaced by road fill surcharge in marsh area.

land about 1000 ft away and parallel to the bulkhead had started a mud-wave that floated the anchor piles upward and pushed the bottom of the sheeting outward. There was no toehold in the sloping rock surface, which acted as a lubricated inclined plane and a large mass of the marsh soil was being pushed by the new fill, taking the bulkhead with it and lifting the anchorage (Fig. 3.2).

A timber bulkhead tied to timber anchor piles had been designed for Huntington, N.Y., Harbor in an area of good sand. The design followed standard procedure and had been successfully used for many bulkheads in that part of the island. Between the dates of topographic surveys and construction, a channel had been dug to bring in scows of rip-rap, and the expected toe imbedment of the splined timber sheeting had been reduced from 10 to 4 ft. When the completed bulkhead was backfilled and at the first low tide, the bottom of the sheeting pushed out and the backfill surface dropped several feet and about 60 ft of the bulkhead length was a total failure (Fig. 3.3). A contractor should have the right to assume the existence of conditions delineated by the drawings, but a cursory check in the field would prevent this type of failure.

Sowers and Sowers have reported on a number of sheetpile bulkhead failures, enumerating the causes and lessons to be learned. Bulkheads do

Fig. 3.3 Timber bulkhead failure where channel depth had been dredged.

not strengthen with age, and progressive increase in surcharge loading beyond the design expectation, together with normal degradation of strength from weathering and battering, often lead to collapse. Earth pressures become greater, imbedment resistance is either dug or scoured away, corrosion reduces strength and may even eat away the complete detail connection.

In the Detroit River a pier protected by steel sheetpiling covering old timber bulkheads was used as an assembly plant for ore carriers. These carriers are launched sideways and therefore require deep water at the bulkhead line. Gantry cranes on separate pile supported runway rails serve the yard. After a number of successively larger ships had been built and launched, erection started on the largest carrier built in the Great Lakes. With more than half of the hull assembled, resting on timber chocks on the ground, a longitudinal crack developed in the ground under the ship. To break up the welded hull and provide new piles to support the load would mean at least a six-month delay in the completion of the carrier. A less time-consuming and much less expensive corrective method was devised. All work was stopped and all loose weights removed from the dock area. A line of soldier piles was driven on each side of the dock as close to the sheeting as possible and with deep imbedment below the mud layer. Cables with turnbuckles were threaded in shallow trenches below the hull and tightened after connecting opposite soldier piles. Special support was provided on the launching side so that the load and especially the horizontal push during launching, on the ways would not be transmitted to the sheeting or soldier piles. By round-the-clock operation all corrective measures were installed within a week, and shipbuilding continued. Launching was on schedule and did no damage to the pier. The next ship order was waiting and there was no time to reconstruct the pier as had been planned. After the second mammoth carrier was delivered, the pier was converted to a structural fabrication yard and has so served for 10 years.

Dramatic cofferdam failures are often in the news. Many designs, for bridge piers and for river construction of locks and dams, are based on a normal high water with expected flooding in unusual flood. If not, the unusual flood will always occur. The main pier cofferdam for a bridge in Jacksonville, Fla., collapsed in 1952 after the subgrade excavation had been done and a 20-ft deep tremie seal had been placed a week before the failure. Steel sheeting had been driven 15 ft below subgrade for a 54 × 107 cofferdam, in 20 ft of water. There were two levels of timber bracing and the tremie seal of 4096 yd^3 required a 40-hr continuous pour. The top bracing struts at water level showed signs of distress as the interior was dewatered 20 ft to the next tier of bracing and 15 ft

above the tremie seal. Before repairs could be made, one wall buckled and the entire framework collapsed inward. An enclosure cofferdam was built and tremie sealed around the base after the original sheeting was cut by divers above the lower bracing level, which remained intact. The weakening of the top bracing level was blamed on timber erosion by the clamshell dredging.

In 1958 in the St. Lawrence River, some 10 miles below Montreal, a bridge cofferdam in 30 ft of water collapsed as the ice broke up and developed pressure against one wall. The cofferdam had steel sheetpiles and steel bracing. The ice pressure buckled the steel struts. It is not economical to design for such pressures, but protective measures for avoiding ice buildup are feasible and must be part of a cofferdam design set in waters that may freeze.

Cellular sheetpile cofferdams if properly filled have considerable surplus strength and are not time dependent. However, they are vulnerable to flooding which upsets the strength of the filling material. In 1964 Christmas-week floods took out a cellular cofferdam in the Snake River; repairs had hardly been completed just before an April flood breached the cofferdam and one cell let go, flooding the 15-acre work area with 50 ft of water.

3.3 Piles and Caissons

Piles and caissons must be considered as structural elements used to transmit loads to the soil. The structural value of a pile or a group of piles is often not provided and failure results, either of the pile or of the load being supported, or both.

Timber piles must be expected to deteriorate in a soil layer subject to alternate wet and dry conditions, and they are food for certain marine animals and fungus. Creosote and other deterrents are necessary for long life of the timber used as a pile. Timber always covered by water has unlimited existence, but is rapidly weakened when the water level recedes. The danger of water lowering has been recognized for many years. In 1918 the Chicago River Warehouse suffered cracked walls when river level dropped. The almost-50-year-old building was jacked up and the piles repaired by concrete caps replacing the rotted tops. As areas become heavily populated, ground water recedes because of subsurface utility and transit construction and the heavier demand for domestic and industrial water. Therefore no untreated wood piles should be considered as permanent support.

Wood borers are normally expected along brackish waters, but they

are also active in inland waters as was found in 1933 at St. Paul, Minn. Timber piles under a four-story industrial building were completely eaten away. The original pile installation was in 1886 for a two-story building, but when two stories were added in 1912 the piles had been inspected and found sound.

In the Cooper River near Charleston, S.C., scour had exposed the tops of wood piles under one side of a concrete pier and crustaceans ate away the support and the pier tipped 19 in., taking the bridge deck with it. A new parallel bridge was under construction and the loose pier was backtied with cables and encased in a cofferdam from which the deck was shored to a reasonable grade and alignment. Borers remove very small amounts at a time, and eventually the remainder of the pile cannot withstand the loading, causing sudden failure, such as happened in 1958 when a 100 ft section of a Brooklyn, N.Y., pier collapsed under a load of cargo being unloaded from a ship.

The British Columbia Research Council has developed a sonic unit to be used by divers on exposed lengths of timber piles and determine by the resistance to wave propagation how much destruction exists within the pile.

Structural design of columns normally assumes axial loading. Walls supported on piles in a single line must be positioned to avoid eccentric loading. Many wall bearing buildings are placed on piles located at the center of the foundation wall width. With the exterior face kept plumb, the building loads are considerably eccentric to the single line of piles, and walls rotate outwards with major cracking near the return walls. A long length of such walls in the Bronx, N.Y., required shoring by jacked A-frames to avoid complete collapse, and addition of new piles and pile caps to provide stability (Fig. 3.4).

Steel pipe piles driven to high bearing value should be checked for continuity and integrity before filling with concrete. Unexpected intermediate resistances in the form of boulders or other buried obstructions, such as were found in the fills placed above a swamp crossed by the Long Island Expressway at Alley Pond, Queens, accordioned 24 ft of pipe into 4 ft of crushed steel with practically no change in diameter. In the glacial moraine hillside not far distant from that bridge location, 60 ft long H-beams driven for sheeting of a subway excavation were diverted by the boulders and became structurally useless. One beam actually bent into a horseshoe and the tip came out of the street surface without any warning.

The experience with mandrel driven shell piles for a hangar at the San Francisco airport in 1963 indicates some of the troubles to be guarded against when installing high-capacity piles in soft layered soils inter-

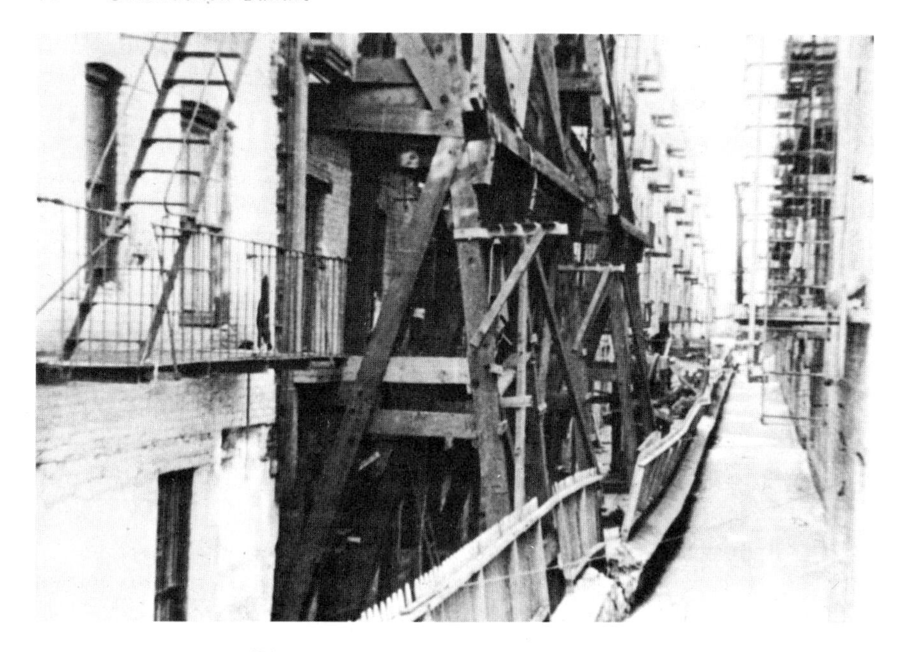

Fig. 3.4 Wall bracing frame to resist tipping while piles are added.

spaced with granular soil lenses. Piles were 90 ft long with several segments of tapered interlocked shells and a bottom closed end section of 10¾ in. pipe. With a hammer of 32,500 ft-lb rated energy and an 80-ft mandrel, extended to the bottom of the pipe, driving was performed to obtain 70-ton bearing values. Driving lengths required to indicate the desired resistance did not agree with computed lengths based on soil test values, but the load tests confirmed the driving resistance criteria. In some of the piles, with apparent satisfactory penetration resistance and electric light examination showing straight and clean piles to the bottom, measurement showed loss in length up to 20 ft. The visible bottom was not the bottom of the pipe but at a squeezed or bent section. Satisfactory piles left overnight before concrete filling were found squeezed almost completely at the bottom of a 30-ft-deep soft soil layer. About 5 percent of the driven piles required replacement because of shell or pipe failure, which was reduced by adding an outer collar at the critical depth over the shell before driving to prevent such squeezing.

South Dakota Highway Department in 1963 used 48 in. prestressed cylindrical piles, shell thickness of 5 in. and up to 174-ft length for support of the bridge crossing the Fort Randall Reservoir. Pipe was cast on the site of 7000 psi concrete and stressed with 42 strands 7/16 in.

diameter. Piles were jetted and driven into position, the shells were then filled with sand to the point of contraflexure and concrete above. During low water in the reservoir, damage in the form of vertical cracks and bulging of the walls was noted at 20 ft from the top of the piles, about 10 percent of the 276 piles were damaged. Bridge completion was delayed by investigations and corrections which had to wait for low-water conditions in September.

Because of improper installation of extrusion piles for the foundation of a five-story concrete office building near Montreal in 1958, local failure in the pile shafts resulted in settlement and tipping of the almost completed structure. The building was 130 × 75 ft in plan, of slab and girder design. Piles were uncased concrete shafts with extruded bulbs to carry 125 tons. When about 90 percent of the dead load was in place, a workman noted a "bang" and the elevator installers reported that the shafts were out of plumb sufficiently to make the cars inoperative. An immediate inspection revealed extensive cracking in the basement floor, a settlement of 8 in. at the north end and a lean of the north wall 4 in. outward. Within 10 days, 45 steel tube piles were jacked to rock refusal and with steel grillage horseheads the building load was picked up. First assumption was that the rock below the piles had failed; core holes were made and 10,000 ft^3 of slurry was injected, but settlement continued at the rate of ½ in. per week. H-beams with brackets were installed along the north wall and 12 more tubes were jacked within the building, all within five weeks of the first settlement. Settlement reached a maximum a few weeks later and the building was 18 in. average out of level, without a single sign of distress in the concrete floors or columns. Borings to check the rock showed no change in thickness or deformation of the sedimentary rock sheets, no grout in the seams. Most of the grout had become a cover layer on top of the rock. Careful check of design and materials used indicated no possible clue. The total weight of the building, some 10,000 K was located with less than 6 in. eccentricity from the center of gravity of the pile locations. Piles of this design had been successfully used for large loads and the dead loads averaged only 167 K per pile with little variation. Rock depths agreed closely with pile lengths and there was no lateral movement to indicate slippage along rock seams. Elastic squeeze of the rock seams could not explain the magnitude or the differential amounts of settlement. There was no unbalanced loading that could shear the piles. Groundwater and soil tests, including concrete made up with groundwater, which unexpectedly tested higher than with city water, showed no possible concrete corrosion failure.

Nine of the piles were then completely exposed for inspection within

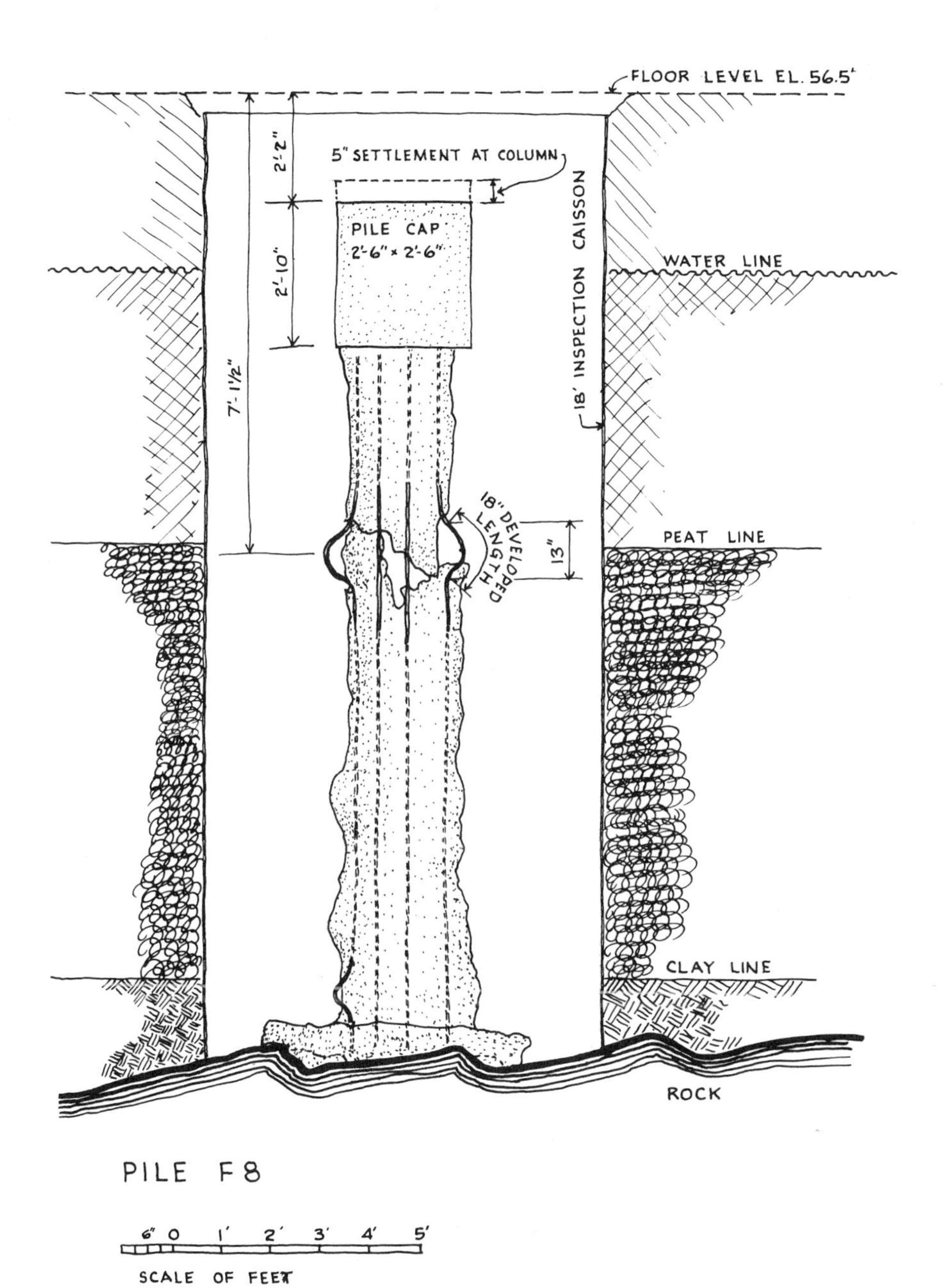

Fig. 3.5 Diagram of extruded caisson pile removed from settling building.

split-pipe hand-dug caissons. Piles had been specified 18 in. diameter, 3000 psi concrete with a cage of five #6 vertical bars within a ¼-in. spiral on 8-in. pitch. The pile casing was driven to a resistance of 10 blows to the last ½ in. of penetration, and concrete placed therein was compacted by a 3-ton hammer as the casing was retracted. Exposed failures in the piles were all similar, at 3–13 ft below the pile cap. The concrete was crushed, with seams of mud enclosed and the vertical bars bent into ears. The amount of compressed length of the bars equaled exactly the settlement at that location for each pile (Fig. 3.5).

The explanation for failure was evident. The casing had been retracted above the concrete being placed and mud had squeezed into the shaft to seriously weaken the strength of the pile. As soon as one pile failed, probably near the north end, adjacent piles became overloaded and failed, at the section of concrete which had been necked most by the squeeze of the mud. The building was completely demolished, new caisson piles with metal liners left in place were installed and the same building reconstructed. A large economic loss resulted from a small lack of proper supervision in the pile installation.

Uncased concrete cast-in-place piles have given trouble when installed in plastic soils. In 1925 piles driven in the soft clay at Plainfield, N.J., were found to be sometimes necked and even completely sheared from the lateral pressure set up by driving adjacent casings. Similar piles in 1942 at Maspeth, L.I., in saturated varved silts behaved the same way and the supporting structure required extensive underpinning. N. B. Hobbs reports similar experience in South Africa, with the explanation that upward flow of artesian water caused a layer of fine sand to absorb the cement slurry of the concrete, when the vibratory motion of the casing extraction caused volume changes in the sand. Dunham also cites an example of pile failure when 14 in. casing tubes produced piles necked down to 8 in. within the soft clay stratum when 173 piles were driven through 5 ft of hard clay overlying 9 ft of soft clay and then 6 ft of silt. Uncased piles must be installed with special control to guarantee the structural integrity of the structural shaft. In soils that squeeze readily a precored hole must be provided within the plastic layer to avoid lateral displacement and vertical heave of neighboring piles (Fig. 3.6).

Piles are structural members inserted in the ground to transfer loads to layers of sufficient bearing or friction capacity. If the pile cannot be inserted to the desired depth, a different type of foundation or even the abandonment of the site, is the proper solution. In the planned development of a full block of land along the east shore of Manhattan for a power plant pile foundation with units of high capacity was necessary.

Fig. 3.6 Lateral squeeze from piles driven in clay.

Extensive full scale research by all available pile drivers and by coring, proved that no economical procedure would penetrate the 75-ft layer of disintegrated mica schist overlying the hard rock. The site was then abandoned as unsuitable for the desired foundation, after an expenditure of more than half a million dollars.

Cored and hand-dug caissons are of little value unless the bottom is at the resistant layer of soil chosen for bearing and the shaft is constructed of solid continuous concrete. A building in Los Angeles was to be partly supported on hard ground below a loose fill. After the holes were dug, they were inspected and approved by a county official. The contractor only checked the depth for determination of concrete requirements and

did not inspect the bottom of the excavation. Subsequently, the building settled, and on exposure, it was found that three caissons were founded on the loose fill, the excavation either had not been carried to proper depth or else the loose fill fell into the completed hole. The courts held against the contractor on the basis that he should use reasonable care to determine that the caissons as built penetrated to the load-bearing ground.

In 1966 three incidents of seriously faulty caissons were discovered in Chicago. At the 8-ft core drilled caissons foundations for a 100-story office building, some four units were found to contain earth intruded into the concrete, in one over a 14 ft length starting 50 ft down. Steel erection was stopped when a caisson showed distress under the weight of the lowest column section, about 12 tons, and all caissons were investigated by coring. In a 25-story concrete building a 4-ft diameter caisson settled 1¾ in. before the frame was topped out, causing the column to drop and several floor slabs to crack. With repair costs many times that of properly built caissons, purely from an insurance risk consideration, one would not expect lack of quality control on such important items of a building assembly.

In each defective caisson, as determined by drilled cores, 12,000 lin ft of 2-in. cores in the former case, all defective concrete was removed and replaced. In the concrete building the column loads required temporary shoring before such concrete replacement.

In a similar incident in 1961, at a building in the Chicago Loop area, an inspector's question as to proper contact between caisson and the rock was settled by coring the full depth, only to find hollows in the concrete near the bottom of the shaft. All caissons were then cored and 20 percent were found unsatisfactory. The corrective measure ordered by the architect was to add two 18-in. caissons along the side of each questioned caisson, which were 48 and 60 in. diameter. Almost five years of litigation resulted in an agreement to settle all claims out of court, especially to avoid adverse publicity. With such examples in the record, test coring of all caissons became an added expense in such foundation work.

The design methods and construction technique of pneumatic caissons have long been established and the necessary continuous control usually is provided and results in a successful foundation. Failures in recent years are rare, yet a single omission of the sequence of control can lead to a mishap, such as the mysterious explosion in the caisson for a bridge foundation in the St. Lawrence River at Three Rivers, Quebec, in 1965. The explosion heaved men and materials 100 ft into the air and tore the caisson in half, floating one 13-ft cell 15 ft above the others. Men were

working at 40 psi in the 52 × 132-ft caisson, built pneumatically to better control position. The bottom had penetrated 35 ft into the clay bottom with 50 ft of water in the channel. The blast came at 4 p.m. as the shifts were changing after a 90-minute work period. Cell N4 was ejected upward, and tiled on 1:4 slope, breaking free at the third joint, some 25 ft below the dished head, and damaged the two adjacent shells which did not fail. The foundation work inside the 24 cells would have been completed in a few days. The force of the explosion threw the entire foundation out of line and at least six of the cells filled with water. The immediate area is known to contain gas pockets in the subsoil and some local residents have even tapped such gas pockets for heating and cooking.

3.4 Sewers and Tunnels

In spite of continuous research on sewer design for over 50 years at Iowa State University and many other research centers, and wide dissemination of experience, sewers still continue as subjects of failure and often as campaign material in election contests. A sewer out of service directly affects the voter, and bond issues earmarked for sewers are usually local assessments. The problem is encountered everywhere, but especially in metropolitan areas.

A complete investigation of the Homewood, Ill., sanitary sewer system in 1932, then about five years old, consisting of 30 miles of concrete pipe from 6 to 24 in. size, showed infiltration of 100,000 gal per mile of sewer each day. The disposal plant could not handle the volume and in addition to owners complaints, land owners below the sewage outlet threatened suits against raw sewage discharged into the outlet creek. Exposure showed open joints, longitudinal cracks, and some shattering of the upper part of the pipes. Trenches had been dug too wide and backfill control had been neglected.

At a town neighboring Homewood, 6000 ft of sewer in 9–20 ft depths in treacherous soil, was partly placed on a concrete cradle, but was generally omitted to reduce costs. Pipe breakage was so serious that flow ceased. Investigation uncovered 51 locations of cracked crushed pipe. Some sewer failures in Los Angeles developments resulted in 1932 on an agreed strengthening of both concrete and vitrified clay pipe so that they would withstand up to 20-ft depth of cover without a concrete cradle. The added pipe cost was a small part of the cradle.

In the Borough of Queens, N.Y., sewer cost and replacement have been a major campaign issue in the past 50 years. In recent years the city

has taken over the control of such construction to avoid the sewer scandals and is enforcing tighter specifications and more realistic designs. The original greater costs are a good investment, among the savings being the cost of grand jury and court actions.

Usually the blame is on too wide a trench and lack of proper bedding. At Middletown, Ohio, these explanations were given for the failure of 1200 ft of 36 in. vitrified clay-pipe sewer laid in 1958 with the aid of a portable sheeting box 7 ft wide and 16 ft long.

Buckling of a 66-in. corrugated pipe sewer outfall line in Salinas, Cal., in 1958, affecting 2000 of the 7500-ft length was blamed on faulty backfill. The pipe of 10- and 12-gauge metal was laid through sandy clay, river sand, and muck, in trench kept dry by wellpoints. Crushed rock was placed ½–2 ft below flow line for stability, with native soil above as backfill, there being no special specification on this important item.

Change of surface grade must take into account the effect on buried structures. An old brick 8-ft circular sewer in Atlanta, Ga., designed for 15-ft cover was asked to sustain 40 ft of new highway embankment. The shape changed to elliptical with large longitudinal fissures in the crown. Some 600 ft had to be shored and rebuilt as a gunited 6-ft pipe with two layers of mesh reinforcement. Smaller sized pipe under such changed condition of loading requires full replacement.

Pile-supported rigid pipe lines crossing roadways in soft soil areas will locally stop the natural soil consolidation under the pavement, as happened at the Long Island Expressway in New York City. Combined with high temperature, a section of 9-in. thick double reinforced roadway slab heaved up 2½ ft and closed up all traffic in June 1962. Compatibility between the rigid pavement on a flexible base and the rigid sewer below was established by a concrete arch cap on timber sills spanning the sewer with open space above the sewer to allow for future settlement of the pavement.

Large-sized corrugated steel culverts designed according to extensive research and development are still subject to distress, whenever the desired uniformity of soil pressure is not obtained. Causes are nonuniform compaction of backfill and very high eccentric load applications, as happened in Crestwood, N.Y., when a wheel load of a concrete ready-mix truck caused failure of a 3 × 4-ft corrugated drain with 3-ft cover and when the pipe was reconstructed under most careful control, identical failure resulted from the first concrete truck crossing (Fig. 3.7).

The imperfect imbedment design for buried pipes was developed to reduce maximum loading from surcharge and cover. It permits the use of lower strength pipe than is required by the assumption of loading by the full weight of soil cover and is effectively used in many installations.

Fig. 3.7 Steel culvert failure. (Redrawn from Robert F. Baker, *Engineering News Record*, August 15, 1963.)

Whether such saving is worthwhile seems questionable in the light of the failure of concrete drains 6, 7, and 8 ft in diameter with deep imbedment in a high roadway fill built as part of the Cannonsville Dam development in New York. The pipes cracked and distorted even before the pavement was constructed, not only under the traveled way of construction equipment, but also beyond the points where such surcharge could have an effect. Maximum diameter steel pipes were assembled within the distorted concrete tubes to provide a restricted maximum flow, because of the high cost of removal and replacement. The expected hanging up of the vertical loading on the sides of the excavated trench, thereby reducing the load to be carried by the pipe,

did not occur. The proof strength of the pipe indicated that full load was being transmitted to the buried sections.

An 18.5-ft diameter corrugated-steel plate culvert was installed under a roadway fill 82.5 above the top in Montana during the winter 1963–1964. A few months after completion of the filling, ruptures were noted in the bolted ring joints and sidewall displacements up to 5 ft. Failures were not only under the maximum fill depth, but also extended under the side slopes where depth of cover was less than two diameters of the pipe. The pipe had been designed by the ring compression method and a suggested use of imperfect bedding was discarded as unnecessary. Fabrication and backfilling were under careful control. The reports on this failure were opened for discussion and all data made available in reports to the Highway Research Board, a commendable action. There was no agreement on the proximate cause of failure, which varied from computed overstresses in the bolts from bending moment combined with shear stress to laboratory proof that the high strength steel bolts had undergone hydrogen embrittlement and stress cracking in the saturated soil. Reconstruction of a similar steel pipe, with larger sized bolts of a softer steel required the complete removal of the fill. The refilling followed the imperfect bedding method using a 30-in. layer of hay a short distance above the pipe and over the full 18.5 ft. The reconstruction was satisfactory and instrumentation proved that less than the total load above the pipe was being carried. It is interesting to note that the original design expected a factor of safety of 2.5 for the completed installation.

Large culverts built in nonporous soils must be designed for possible uplift from hydrostatic pressure. One such concrete box culvert built in San Juan, P.R., in 1964 in a soil of poorly stratified red to white silty clays and fine sands, failed during a heavy rainfall, with 6 ft of water in the 30-ft wide, 9½-ft deep rigid frame structure, by the floor buckling for several hundred feet, while the vertical walls remained substantially fixed in position.

The failure is similar to that of a concrete four-pass aeration tank at the Jamaica Bay sewage treatment plant in New York in 1965. The soil was a hydraulic fill sand on top of meadow mat and the tanks were designed on the assumption that an earlier construction program at that plant had removed the impervious meadow mat. Adjacent operating tanks, some 35 years old, were known to permit some raw sewage leakage into the subsoil. Sufficient accumulation of this waste liquid had sealed up all filter capacity of the soil, and upward pressure developed to the extent that the new tank floors were heaved and displaced and the intermediate walls were crowned up, a maximum of 32 in. in the 200-

ft length (Figs. 3.8 and 3.9). End and sidewalls had been backfilled and were held by the soil. When the floor was broken by the distortion, blows developed, the major geyser being almost 2 ft in height. With the tanks filled to that level, the floors receded to almost the original position and the walls returned to a straight line. A few months later with the tanks pumped out, a similar but much smaller distortion again receded upon flooding. Corrective work consists of pressure relief valves in the floor, an added pile foundation for the support of the tank walls and a new floor since all metal water stops had been sheared.

Tunnels, of all engineering structures, are most often plagued by small troublesome and large catastrophic failures. The operational shock of

Fig. 3.8 Upheaval of aeration tank on clogged sand bed.

removing macroscopic volumes of soil or rock from the internal structure of a stable mass necessarily upsets the state of equilibrium. To aid in obtaining prewarning of incipient rock movements in the tunnel workings, F. J. Crandell devised a microseismic detection apparatus which charted the rate of strain change in the bedrock. By calibration with laboratory determination of the microseismic rate under crushing strain for the kind of rock encountered, some indication of proximity to rock failure can be obtained. Instrumental recording of subaudible sound waves emitted by the strained rock will now supersede the audible sound of popping rock.

Sudden release of internal strains cause rock movements which cannot

be predicted from the geological investigations. Such falls can come near the portal or more often under the highest rock cover. In the diversion tunnel for the Green Peter Dam in Oregon a mass of rock broke through the steel roof supports when the heading had only advanced 60 ft. In the tunnels carried into the Cheyenne Mountain in Colorado the complete geological investigations did not reveal the cross-bedding of seams in the center of the mountain and a large cavern formed which required extensive structural support where cross-drifts met and also some million dollars cost of rock bolting to stabilize incipient spalling.

The successful completion of a tunnel in which things went wrong is a monument to the stamina and determination of the workers in a recog-

Fig. 3.9 Squeeze effect of rising tank on perimeter through grating cover.

nized hazardous occupation. Some tunnels get more than an average share of the possible troubles, whether foreseeable or not. The 8-mile long Shimizu Railway tunnel in Japan was holed through in August 1966, after three years of hardships endured by three contractors who divided up the work. This is a parallel tunnel to the 6-mile long old Shimizu Tunnel, completed in 1931, after nine years of work, at the cost of nearly 50 lives. The contract for the center section of tunnel, which required an inclined access shaft and which had 4000 ft of rock cover as it crossed the mountain ridge, was exposed to the greatest hardships. Water came from the rock seams in one 150 ft stretch at 215 psi and up to 15 cfs requiring six 400-hp and three 500-hp pumps to eject the water

through the access shaft. Sealing the joints with silicate and epoxy resin grouting finally stopped the flows, with water pressure buildup in the rock to 615 psi. Under the highest peak elevations, rock pressures caused squeeze of the walls and falling rock which required steel bents at 5-ft centers with full lagging cover. At 800 ft from the tunnel portal a hot spring at 122°F required boring a ventilation shaft to permit men to work even a reduced shift of 4 hr. Complaints of nearby hotel resorts that their supply of natural hot springs was being depleted then required chemical sealing of the flowing seams. An unusual trouble was electricity leakage from the operating tunnel making it hazardous to set off electrical detonators and requiring electric insulation for all ducts and rails carried into the tunnel.

Lebanon also had an eight-year job to hole through a 9.9-mile 10-ft diameter water tunnel in 1965 to connect the Awali and the Litani Rivers. Construction was stopped in 1959 when the heading advancing in limestone entered loosely cemented sandstone and a flow of 265 gps brought down the roof, filling up almost 2 miles of excavated tunnel. The water flow increased to a maximum of 1500 gps, and then leveled at 50 gps, after some 130,000 cu yd of sand was deposited in the tunnel. Work was completed after treating the problem as a soft ground excavation.

Another seven-year-long tunnel job was the twinbore Wilson Tunnel through the Koolau Mountain near Honolulu, driven through the well-known geological volcanic ash. The tunnels each carry two lanes of traffic and are 26 ft wide. Total length is 2775 ft, originally (1953) shown as 1925 ft in rock and 850 ft in earth, on a 6 percent grade. The work started on one tunnel in rock with a three-platform jumbo on which nine drills were mounted as a full-face operation. By June 1954 the rock section changed to earth, continued as a full face method with 8-in. I-beam arch ribs spaced 2–3 ft, excavating the sides ahead so that the posts could be set first.

On July 10, 1954, several tunnel supports collapsed 60 ft back of the heading, soil fell in and buried the equipment and filled 170 ft of the tunnel. A crater formed in the mountain 120 ft above the tunnel. Water flow increased, a second mud flow started 130 ft back from the heading, followed by a third flow on August 14 when 200 ft of tunnel was filled and five men were killed. Work was stopped on the earth section and continued in the rock section while the city and the contractor searched for consultation advice.

With an election close at hand and Wilson's job as Mayor at stake, concerted efforts were made to clear the city of any responsibility. Some, but not all, the tunnel consultants contacted by the authorities signed a report placing the blame on the contractors' methods, but the electors

defeated Wilson in November. With 1800 ft of the 2775 ft clear, the 18-in. concrete was placed in the rock section and probings were made by a pipe 200 ft into the cave-in dirt. The contractor claimed that a tunnel could not be driven in the earth section and offered to complete the work as a wide open cut or else a concrete tunnel in open excavation to be later filled over to avoid possible slides in the deep open slopes. The open cut would be 200 ft deep maximum, 50 ft wide at the bottom and 650 ft at the top.

The contractor claimed that the extra costs of the city experts' recommendations would pay for the open cut for two tunnels (the city contemplated a parallel tunnel in the future) and the lined second tunnel in the rock section. To avoid any further soil flow, the cave-in section was bulkheaded with a 3-ft concrete wall and some 1500 cu yd concrete was force-pumped behind the bulkhead to fill the space between the roof and the debris of the earth section tunnel. The new mayor felt that the contractor should continue the work in spite of the city consultants' recommendation that he be terminated. The resultant litigations must have been valuable to the legal profession; claims and cross claims between city and contractor, and a $1.5 million damage suit by the contractor against the consultants, for malicious and false statements, made "in bad faith and in reckless disregard of the truth and facts."

In May 1955 the city ordered the contractor to complete the work, but he insisted on an agreement on extra costs and liability for the failures, claiming the design was deficient. Some nine months later an agreement was reached with some interesting clauses:

1. The contractor will assume the cost of reexcavating the 500 ft of cave-in.
2. The city will assume the extra cost for a heavier design of temporary and permanent tunnel lining.
3. No penalties for delay in completion of the contract.
4. Both the city and the contractor will replace their tunnel superintendents.
5. The city will purchase any steel ribs on hand which will not fit the enlarged tunnel section.
6. All cross claims are waived except those arising from damage suits of the families of workers killed in the cave-ins.

One year later an exploratory drift had been carried through the cave-ins and to the end of the tunnel, approach road contracts were let and more than $1 million in damage suits were filed against the city for the five deaths of workers. The earth section was redesigned to incorporate a curved invert, the 18-in. wall thickness was made 48 in. at the

maximum cover and 32 in. elsewhere in the earth section. Excavation was changed to side drifts with very close spacing of steel arch ribs. Progress was very slow and concrete was placed in 3-ft segments immediately behind excavations. Initial progress was only 6 in. per day as the saturated soil was being removed. The first bore was completed in May 1959 at a cost of $1 million above original contract of $4.5 million. Soon after resuming work, the same firm received a contract to build the second parallel tunnel and completed it one month after the first was completed, at a cost of $4 million. The tunnels are open and in heavy use, the normal increase in traffic required the twin bores. Since different facets of the construction industry will draw quite different conclusions as to the cause of the failure, no attempt is made to point to the lesson to be learned from this incident.

Mud flows in the approach ends of the Tenche Tunnel in Colombia under the Central Andes Range, although over a small part of the total length, seriously slowed up completion. The residual soils vary from hard decomposed rock to soft fine-grained silt, weighing from 105 to 80 lb per cu ft. Four major flow slides occurred in the 920 ft of earth section of a 10.7-ft circular tunnel, which slowed progress to 15 ft per week and required forepoling to hold the soil and 4-in. I steel supports on 12-in. spacing to hold the roof. Experience at the Miraflores Dam tunnel, at the same project, was similar. British miners, experienced in London clay, were brought to do the excavation and control the beginning of soil flows, which caused serious mud flows and formed huge caves in the ground above the heading. It was empirically learned that the top of the heading must be kept sealed at all times and all excavated areas must be packed tight and wedged against the arch ribs.

Natural gas in shale rocks and man-made fumes have caused serious accidents in tunnel work. At the El Colegio Hydroelectric project in Colombia pockets of methane gas in the shale were released into the free air pressure in the tunnel and caused five explosions, the most serious one came after the tunnel was lined and grouted. In a Japanese irrigation tunnel at Monomura 25 workmen died of carbon monoxide poisoning after the foreman brought in a gasoline powered generator because electric extension cord was not available. He ordered the men to continue work after they complained of dizziness and was therefore charged with professional negligence leading to the deaths.

A sewer tunnel was rocked by a series of explosions in 1962 in Philadelphia. Although the 6-ft diameter tunnel was being dug in ground known to be impregnated with petroleum wastes and the contractor did periodically check the exhaust air, sufficient seepage gas accumulated to be set off by an electric spark from an open switch or a

defective connection. Sparkproof lighting and additional ventilation was ordered before work was continued.

Tunnels in sand require fairly simple repetitive steps of protection and excavation. But a sewer tunnel in Brooklyn going through coarse dry sand was suddenly confronted with live free flowing sand when they crossed under a series of railroad tracks. The vibration of the passing trains converted the stable sand into a fluid mass, requiring a complete change in excavation methods.

The incidents described are only a few cases to indicate the types of trouble causing failure in tunnels. There are many such cases and they have spurred on the development of tunnel coring and boring machines to guard against sudden change in material and to provide protection during excavation of the heading. These machines, when they work, outproduce any other excavation method. But there are limitations, only to be learned from failure to perform. In 1964 the London Victoria subway line was being extended by a mechanical mole within a shield at a depth of 70 ft in firm clay. When water laden gravel was encountered, the tunnel collapsed and buried the equipment. Progress in the clay had been 3½ ft/hr, but the loss of time in retrieving the equipment and passing the gravel layer seriously reduced the average progress.

Similarly, a 71-ton "mole" was brought to bore a 12-ft diameter water tunnel in schist rock under the Narrows in lower New York Harbor. The machine had produced well in several western locations, but a year's work and repair only produced 300 ft of progress and the machine was withdrawn to make room for men and drills in 1965. Eventually these devices will be developed to cope with nonuniform soil and rock layers, arrayed at any angle to the cutting face.

3.5 *Dams*

The spectacular and devastating collapses of dams, sufficient to unleash the force of large volumes of contained water, are the subject of much publicity and discussion. Of course the number of safe and harmless installations is predominant, and relatively few of these ever reach public knowledge and then usually only at the time of dedication when some political figure takes part. In early medieval days philosophers argued at length as to the size of the fraction which is significant to cause rejection of the whole; in other words, what is the tolerance to be allowed in excellence. The agreed rule was then set that 1/60th of the whole was the largest permissible "minimal," so that the general class of structures known as dams have a satisfactory record if less than one out

of 60 misbehaves. The rule should not be blindly applied, since not all dams are equal in size, value, or in potential harm if they fail. By any form of application the record of dam failures indicates an excess of cases beyond reasonable tolerance. Failures have occurred in masonry dams where the foundation was sufficient, but most of the failures are either foundation failures or from lack of strength in the man-made soil or rock embankment.

In one of his last pronouncements Terzaghi (1962) discussed the basic reasons for the failures of the Malpasset Dam and Wheeler Lock from foundation inadequacies that eluded detection despite extensive site investigations, considerable testing, and the application of the most advanced analysis and theory. He states:

"All foundation failures that have occurred in spite of competent subsurface exploration and strict adherence to the specifications during construction, have one feature in common: The seat of the failure was located in thin weak layers or in weak spots with very limited dimension. . . . If the failure of a structure would involve heavy loss of life or property, the structure should be designed in such a way that it would not fail even under the worst foundation conditions compatible with the available data. If no such consequence would ensue, the cost of the project can be reduced considerably by a design involving a calculated risk. . . . With increasing height and boldness of our structures, we shall find out more about this interesting subject, but always by hindsight as we did in connection with these failures."

These statements are not very encouraging to one looking for a factor of safety in his design.

Twenty years earlier Terzaghi wrote:

"Since aboout 1928 the attention of research men has been concentrated almost exclusively on the mathematics of stability computations and on refining the technique of sampling and testing. The results of these efforts have been very useful and illuminating. At the same time, they created a rather indiscriminate confidence in laboratory test results. In some incidents confidence is warranted, but in many others, it can lead to erroneous conclusions, because the attention of the investigator is diverted from what is really essential."

The reported lack of confidence in the field of large dams must result in a demand for a more realistic and rational margin of safety in foundation design. Following the Bedford Hills Dam failure in December 1963, the U.S. Federal Power Commission (F.P.C.) ordered a strengthened safety program to avoid similar incidents. At five-year intervals, all

F.P.C.-licensed dams are to be inspected by independent consulting engineers for any indication of structural distress. The California Department of Water Resources ordered an inspection of the 971 dams within their jurisdiction. UNESCO is also exploring a program of international dam inspection.

The reported results of these programs (to April 1965) only cover the concrete disintegration and expansion of a 163-ft high-arch dam, but the existence of such programs shows the small reliance generally placed on the design factors of safety. The record is not too good. A 1961 Spanish study listed 1620 dams, of which 308 had serious accidents in the period 1799–1944; foundation failures caused 40 percent and poor construction with unequal settlement caused 22 percent of the accidents.

At the closing session of the Eighth Congress (1964) of the International Commission on Large Dams (ICOLD), President-Elect J. Guthrie Brown said:

"In the past, dam builders have acquired very substantial knowledge about all the factors that contributed to the safety of the dam structure themselves, but they have had very little reliable or accurate information about the strength and nature of the underlying strata on which the dams are supported. . . . All knowledge gained in the studies of dam design is of no avail if the foundations on which a dam is to rest are in doubt. The recent failures have all been concerned with foundation movements, or geological slips. It is from the geologist and the combined science of soil mechanics and rock mechanics that the engineer must obtain the details and scientific information he requires to confirm his own experience and judgment of the foundations on which rest the safety of his dam."

Is the price of too rapid progress worth the risk? Has there been an unnecessary number of spectacular dam failures, and is there a correspondingly similar although unreported record in other phases of soil, rock, and foundation works?

In the United States 1800 dams were built before 1959, and 33 failures have been recorded, mostly from inadequate foundations. Seven of the failures came in one week before 1918, so that it can be argued that the designers did not have the advantage of modern soil mechanics. But even 1 percent failure would involve huge losses in life and property of the 6549 existing dams over 15 m (116 over 100 m), 773 under construction (65 over 100 m), and 962 projected (74 over 100 m) listed in the world-wide survey of 1964.

There were foundation failures before 1925, but the last 40 years have had their share, despite the progress in soil investigation and control,

theoretical analyses of internal stress, and better graphical representation of probable failure pattern and necessary resistances to avoid movement.

Structural Failure of Concrete and Masonry Dams

In 1959 a 112-ft high masonry and concrete dam across the Tera River in Northwest Spain failed when 17 of the 28 masonry buttresses gave way as the crest was topped by heavy rains. The structure was built in 1957. The flood hit the village of Rivadelago located at a lower elevation of 1700 ft, destroyed 125 of the 150 buildings and killed 140 people. Litigation resulted in the conviction of four engineers on the staff of the power company for neglect and insufficient supervision of construction and the owner was required to post a bond for $1.7 million to cover awards in damage suits.

Late in 1966 the concrete Ambursen Buttress dam at Mill Brook near Pittsfield, Mass., built in 1909 to a height of 40 ft, failed by disintegration of the concrete slab between three buttresses. Ice buildup between buttresses had been a constant trouble in erosion of the concrete. Only local road and bridge damage resulted from the 9 million gal of released water.

In 1965 the 163-ft-high concrete arch dam across Maltilija Creek in Ventura County, Calif., was found to suffer from exterior cracking, interior swelling, and general disintegration of concrete in the upper 20 ft of the 16-year-old structure. The defects were discovered by the state inspection of dams following the Baldwin Hills Dam Collapse in 1963. The concrete for a height of 30 ft and over a length of 280 ft was removed to reduce maximum stress conditions as a precaution against collapse.

Foundations

In 1926 a concrete dam near Dolgarrog in North Wales failed through scouring out of a 30-ft section of the clay under the footings. The flood overtopped and caused complete failure of an earth-filled dam with concrete core wall some 2 miles downstream. Large masses of earth and boulders cascaded into the town and did a lot of damage resulting in the loss of 16 lives. The flood submerged the generators and entered the plant of the Aluminum Corp. and caused an explosion of the reduction furnaces. The concrete dam was only 18 ft high and was planned to have a solid concrete foundation 14½ ft wide and 8 ft thick, imbedded into the clay a minimum of 6 ft. Water depth was only 15 ft. The failure

took out the bottom part of the dam, the footing, and a large mass of the glacial clay below. Investigation showed that the footing had been built only 18 in. into the clay, that the concrete contained large pieces of granite rock with little mortar covering. The previous year the reservoir was dry for a considerable period allowing the clay to shrink and crevices to form. Sudden rains filled these gaps in the clay and started the undermining of the shallow footing.

The catastrophic failure of the St. Francis Dam of the Los Angeles Aqueduct in 1928 was the result of a weak foundation; but the left section was also affected by a hillside slide. Although arched in plan, the design was a concrete gravity section, with a maximum 205-ft height and a 176-ft-wide base. The center 100-ft-long section did not move; the west side went downstream much further than the east side. Foundations rested on rock, schist at the left bank and in midstream and a red conglomerate on the other side. Excavation into the rock was 30 ft deep at midchannel and to lesser depths toward the higher ground. The flood moved down the San Francisquito Canyon at 15 mph, scoured out the power house and other structures, took several hundred lives. Fragments of the dam, some with rock adhering to the base, were moved half a mile. The debris came from 600 ft of the 700-ft-long dam. The bedrock had not been grouted and certainly the foundation was better at the highest sections.

An expected foundation failure under the earth dike of the Pendleton Levee, Ark., was carefully instrumented and analized in 1940. Knowing that the foundation was a soft clay with marginal stability, the work was scheduled so as to observe and measure the actual failure, which came when the triangular section had reached a height of 32 ft. The width of the levee base was 320 ft. The failure came on the landside of the fill as a vertical subsidence of 6 to 10 ft, 50 to 70 ft wide, a 10-ft horizontal movement of the rest of the fill toward the landside, and an upheaval, 4 to 5 ft high, just beyond the landside toe. Failure was explained on the theory that all maximum shear resistances did not mobilize simultaneously along the internal slide surface and where the local strength was not sufficient, failure started and progressed successively along such surface. Normal design procedure based on the physical characteristics of the soils involved as determined in the laboratory, indicated a factor of safety of 1.3 at the point of failure, if all maximum resistances are mobilized. In the discussion of the Pendleton Levee report it was pointed out that the failure cause was a lubricated horizontal shear surface in the clay. Terzaghi referred to four similar failures in earth dams which occurred at horizontal clay surfaces: the Lafayette Dam in California in 1928; the Clingford II water supply embankment near

London in 1937; the Marshall Creek Dam in Kansas in 1937; and the flood-protection dike at Hartford, Conn., in 1941. Loading such a clay layer introduces pore water pressures that degrade the physical strength of the virgin soil.

Collapse of the Malpasset concrete dome-shaped dam, near the French Riviera in 1959 was caused by a rock shift under the left bank foundations. The structure was 218 ft total height above rock trench bottom with 186 ft of head. The wall was 4.9 ft thick at the crest, 22.6 ft at the base, and 700 ft long at the top. Collapse was total, the released flow down the narrow Reyran River Valley demolished everything in its path and killed 421 people in the town of Frejus, 4½ miles downstream. The 62,500 cu yd of concrete was placed with the utmost care and control of the most experienced technical talent. Construction started in 1952 and required over four years. Investigation after the failure disclosed a thin clay seam about 1½-in.-thick in the rock below the arch abutment. Since experts reported that the seam could have been predicted from a proper boring program, legal action was taken against the engineer who had accepted the design, on the charge of involuntary homicide through negligence. Three years of litigation ended in his acquittal. The government had paid $24 million to companies and individuals after the disaster, but some $40 millions in civil suits were brought to trial in 1965 and also criminal actions against four engineers variously involved in the design, construction, and instrumentation of the completed structure were dismissed in 1966. The damage claims are not yet adjudicated.

During the widening and deepening of the Wheeler Lock in the Tennessee River, the 27-year-old shoreward wall slid into a cofferdammed work area in 1961. Two 40-ft-wide gravity sections of the 360 ft length moved bodily 30 ft, with lesser damage to the rest of the wall, tying up the river traffic, and stranding some 100 barges upstream of the lock. The local geology had been carefully investigated, consisting of limestone, fine to somewhat coarse-grained crystaline and somewhat argillaceous, with a 6-in.-thick shale layer at a small depth below the base of the wall that failed. Later exploration found a thin bed of slippery clay on the top surface of the shale, probably the result of decay and hydration in the pulverized contact. The excavation for the old wall had included 6-ft-wide shear keys at both heel and toe carried down almost to the shale layer. When the concrete blocks moved, they carried along rock above the clay seam. None of the borings made for the new lock, including three 36-in. holes made with a calyx drill, showed evidence of the seam; it was too thin to indicate a change in rock structure. Foundation pressures varied considerably during the filling and empty-

ing of the lock. Such fluctuation on a shale level could well be the cause of local disintegration of the rock structure. As is described later in this chapter, sedimentary rocks are not always stable under loadings differing from the virgin ground condition.

Piping at Contacts

Shrinkages of man-placed fills, chemical and physical erosion, and loss of soil and rock along the contacts between concrete and natural or artificial materials open up channels for water seepage with continuous increase of the gap. Eventually the flows become large and fast enough to erode an appreciable opening with serious result.

Recent examples of such piping and erosion leading to failure are the 43-ft-high homogeneous clay dam near Mercer, Pa., with a concrete spillway 308 ft long. In 1966 a 60-ft length of the spillway collapsed when water began piping through the embankment next to the left wing wall of the spillway.

In 1967 the 259-ft high concrete gravity dam on the Ebro River in Spain developed leaks at both abutments. This Mequinenza Dam had been called unsafe before. The construction engineer in 1962 during construction took the owner-builder to court on the charge that the specifications were not being followed. The rock abutments are the site of a number of worked out lignite mines. The seepage losses on the two contacts are evaluated at $50,000 energy loss per day and amount to over 1000 cfs.

A similar seepage closed down the Fontanelle Dam in Wyoming, when a hole developed near the right abutment of the 157-ft-high irrigation and water supply earth-fill dam. The flow was about 150 cfs, but it washed out a hole in the downstream face that was 80 ft across, 150 ft high, and 60 ft deep. A hydroplant located downstream to use surplus water could not be put into operation because lowering of the reservoir was essential to the repair program.

The ends of two concrete arch dams failed in 1926 where soft rock was eroded at one abutment, but the main body of the dams remained intact. The Moyie River Dam in Idaho was 53 ft high and following a heavy flood storm, the stratified rock of various hardnesses collapsed at the abutment contact. The Lake Lanier dam at Tryon, N.C., failed similarly two days after the first filling of the reservoir. Some ice had formed and when the abutment section fell over, the exposed rock was blocky with soft seams and decomposed rock pockets.

The purpose of a dam is to impound and hold water. If the reservoir bottom is porous, the purpose of the design is not met. The then largest

pumped-storage plant, 1965, at Taum Sauk in the Missouri Ozarks has an upper pool blasted out of a 900-ft-high mountain and enclosed by a concrete surfaced rock-fill dike. The 39 acres of floor are surfaced with 4-in. asphaltic concrete directly on the rhyolitic porphyry rock. Several times the basin has been dewatered to locate the leaks, repairs made to the floor, and cement grout injected into the seamy rock. Maximum leakage had been double the reduced amount after the repairs, some 20 cfs; to repump the lost water costs $25,000 per year in power.

There is a leaky reservoir in the Ptolemais region of Greece. Built as a seasonal storage for industrial water across a river that is dry in the summer, with a 5-mile pipeline connecting several plants, the reservoir remains dry even at the time of spring runoff. The cost of waterproofing the reservoir floor may be more than the provision of deep wells for the manufacturing plants. The May Dam in Turkey stands at a dry reservoir; the bottom holds no water.

During Construction

Earth-fill dams of large size built before 1930 were normally hydraulic-fill slowly depositing graded materials. The development of better compaction procedures resulted in dry-fill mechanical placing and consolidation as the standard method. Hydraulic fills often broke through the edge dikes; relatively dry fills have had their slides, sometimes of very large volumes. A few of the notable examples are listed as descriptions of what went wrong.

The Calaveras Dam in California made engineering news for a decade, and was completed in 1926 as an earth and rock dam 220 ft high above bedrock with a volume of 3½ million yd^3. The lower half was built by hydraulic filling until in 1918 some 800,000 yd^3 of the upstream embankment and part of the clay core slid into the reservoir. Repairs were made to part height by adding rolled fill, and repairing the cut-off trench fill. Additional clay fill was rolled between broken rock berms to complete the dam.

The Fort Peck Dam on the Missouri River in Montana was scheduled for completion in 1939 and represented a project of over $108 million. It was then the world's largest earth dam requiring more than 100 million yd^3 of earth and gravel in its construction by hydraulic placing methods. After an exhaustive geological investigation, a steel sheetpile cut-off had been driven to the top of firm shale. A major slide occurred suddenly near the right abutment in 1938, after which a reevaluation of design assumptions and methods placed the blame on a buried bentonite seam in the shale beds. The addition of fill load on top of the

normal weight covering the saturated seams resulted in a release of water that could not readily escape through the overlying impervious shale and this lubricated the sudden movement of the soil above. The reconstruction and completion of the dam was on the basis of a modified design with flatter slopes and reduction in shear stresses within the dam. The new core and the topping off of the dam were constructed by rolled-fill methods after the additional upstream berm and the replacement of the slide area were replaced by hydraulic fill.

In 1956 some 2 million yd^3 of fill slid out of the east abutment of the Oahe Dam at Pierre, S. Dak. The slide developed slowly, moving 8 to 12 ft in six months with vertical displacement up to 30 ft in the rolled-fill embankment. The area of the slide was 2500 ft wide by 2000 ft in the direction of movement, but the volume which had to be removed and then replaced, with flatter slopes, was only a small part of the 78 million yd^3 embankment. A combination of existing fault planes and bentonite seams in the underlying shales, triggered perhaps by nearby excavation operations along the face of the abutment and aggravated by subsurface water pressures, was blamed for the slide occurrence.

Rolled-fill earth dams containing clay or materials which can alter into a plastic mass develop high-pore pressures which provide temporary artificial supporting strength. In the West Branch Dam near Warren, Ohio, a 1000-ft stretch of a multipurpose embankment 9900 ft long and 93 ft high sank about 1 ft in 1965 when such a saturated layer of clay some 30 to 90 ft below the stream beam shrank under the new applied loading. To remove the water rapidly, electro-osmosis was resorted to, with anodes introduced into some 1000 drilled holes.

When the volcanic ash hydraulic fill for the Alexander Dam on Kauai, Hawaii, was about 80 percent completed in 1930, a trapezoidal section having reached 95 ft of the 120-ft total height, half of the material sloughed out. The explanation was that the soft grains had crushed under the weight of the fill and reduced the expected porosity, slowed drainage and the side was pushed out by the liquid pressure of the added fill (Fig. 3.10).

The Waco Dam, Tex., was almost topped out to its 140-ft height in 1961, when longitudinal and transverse cracks appeared in a 1500-ft section of the embankment. The crest of the dam sank 19 ft and the downstream slope shifted 26 ft. There was little movement at the upstream slope or toe. The fill was rolled local soils over a bed of cretaceous shales and limestones. A restudy and evaluation of rock conditions after the slide, revealed some weak shale layers with intrusions of clay and the capability of builtup pore pressures under the new loading. A repair program consisted of strengthening the fill with added berms and

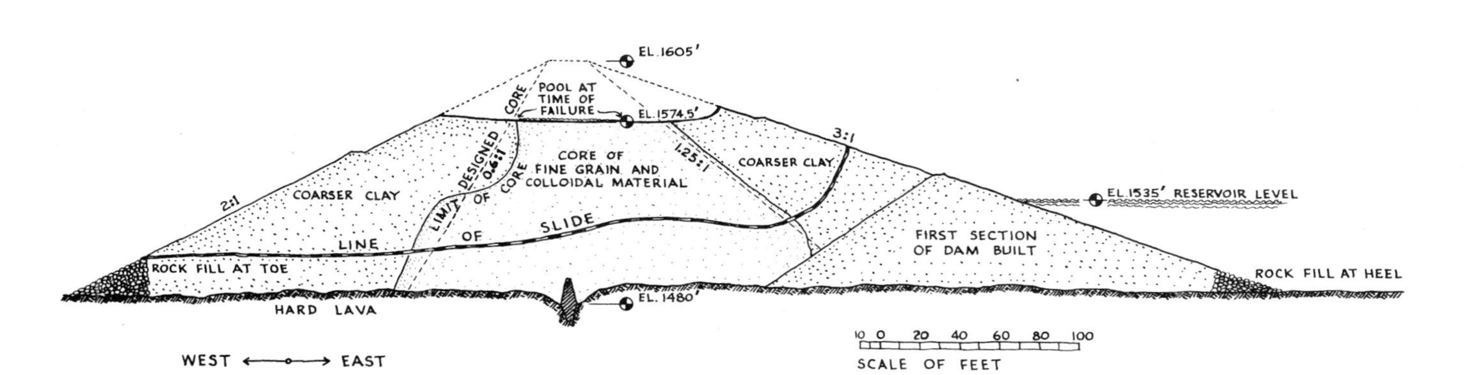

Fig. 3.10 Failure of hydraulic fill dam using volcanic ash. (From *Engineering News Record*, May 22, 1930.)

sealing the cracks with a sand-bentonite-cement grout. It is to be noted how similar are the explanations of causes of such dissimilar incidents of failure.

Completed Dams

Failures of existing dams, some new and some old, are not construction failures when unexpected floods top the dam and erode either the body or the foundation. Rather, they are caused by either an unrealistic decision on spillway capacity or an unwarranted increase in storage volume. Failures have occurred when a properly designed and properly built dam is subjected to a shock loading from slides falling suddenly into the reservoir and starting a flood wave. In 1960 a small dam in Spain near the Bay of Biscay collapsed from such a flood wave and caused the death of 20 people in the village downstream.

The Vaiont concrete arch dam north of Venice, Italy, 858 ft high, suffered only minor damage at its crest when a landslide in 1963 pushed enough water over the dam and into the outlet stream to demolish a town of 4600 and several villages. Almost 3000 lives were lost. The slide contained over 300 million yd^3 of rock, almost filling the reservoir in a period of 15 to 30 sec. The air shock wave preceding the wave of water caused complete destruction of the dam installations. Slide potential of the rocky enclosure of the reservoir had been recognized, but no added resistance against sliding had been provided.

In the report on the Vaiont Dam Dr. Claudio Marcello, 1964 retiring president of ICOLD, stated that "Borings along a foundation may yield but a discrete sampling, and nature is not, unfortunately, so obliging as to comply with any continuity rule which could allow interpolation of results in between." In the report on the Wheeler Lock Terzaghi similarly argued that no method of exploration will always reveal the existence of minor geological details such as weak spots in the rock.

If we now admit that a true determination of subsurface conditions is not possible, who must take the risk of the underground, when either exposure during construction or future structural distress shows incorrect assumptions of material strengths? The contractor may be, and often is, legally forced to assume the risks on the subsoil. If the contractor can prove concealment of the true conditions, however, he is usually free of damage. If he can prove faulty assumptions, he can allege either incompetence or fraud on the part of the designer. The indications are that a reasonable and sufficient margin of safety, even if the budget cost must then be exceeded, is a prime requirement in every design, to protect the owner and the designer. Contractors are aware of the danger

in the underground and, recently, more are insisting on "changed conditions" clauses in the construction agreements which financially protect them in any eventuality. The designer must also look for protection if he cannot rely on subsurface investigations to provide a sufficiently accurate evaluation of soil and rock strengths and if, as a result of using such evaluation, his computed factor of safety is not a true one.

Foundation design is not an easy or simple routine task; one always deals with the unknown. It has no regular rules or patterns, so that a continuous mathematical expression is not possible. Discontinuity is the normal condition. In the New York area heavy construction contractors believe in and guide their operations on the assumed "Finnigan's Law": If there is any chance of a mishap, it will happen; if something can go wrong, it will. That most foundations are successful speaks well of the skill of the engineers and contractors. In addition to a wealth of experience, some gained at high cost as a result of failures, eternal vigilance is necessary if the changes from expected conditions, which can rapidly eat up the margin between safety and failure, are to be noted in time.

A list of notable dam failures from 1883 to 1928 appeared in *Engineering News-Record,* **100,** No. 12, 472–473 (March 22, 1928).

3.6 Rock as an Engineering Material

Failures in work involving natural rock always stem from the assumption that the rock is a homogeneous continuous solid. The high cost of flawless precious stones should be ample warning that continuity and homogeneity are unusual conditions in rock structure. Normally one expects to find the softer, seamier rocks on top of a more massive bedrock. But this is not always true, as is demonstrated in the northwestern corner of Westchester County, N.Y., where the cap rock is an emery of great hardness and toughness, with no seam structure, but the lower rocks are seamy and brittle granites. Cavity formations in sedimentary rocks have given considerable trouble. In the New York subway contract along St. Nicholas Terrace near 155 Street, where borings indicated dense schists and all buildings were required to be underpinned to rock bearing, the excavation of rock in the subway cut uncovered an unusual condition. For a length of over 200 ft there was a mud-filled seam in the rock up to 20 ft in height within the bedrock. All along this exposure, the rock then had to be underpinned to the lower rock as a protection for the buildings and for the subway structure which had been designed on the assumption of a structural section in rock. Borings were 300 ft apart and had missed the discontinuity completely.

Even rock tunnels encounter discontinuities in "solid" mountains. When the West Haven, Conn., twin tunnels for the Wilbur Cross Parkway crossed the contact between the sandstone and schist, trouble was expected, prepared for, and overcome. However, the contact, which is at about 60 deg from the vertical, still leaks through the concrete lining. With one tunnel finished and the twin carried northward to the last full heading, a holing-through party was arranged to celebrate Connecticut's first tunnel, with the switch operated by a high-ranking but mild-mannered political figure. The shot went off, quite successfully, but the vibration opened a seam above the tunnel portal and the entire face of the hill slid down to engulf the tunnel entrance. As he stated, never in his political career had he made so much noise.

One of the most carefully explored mountains prior to tunneling operations was the location of NORAD in Colorado. It was chosen because of rock density and continuity as exhibited by the core tests. At the intersection of the galleries near the middle of the mountain, however, the rock fell apart and a domed cavern, twice the desired height formed. Considerable permanent steel supports and a close pattern of rock anchors were necessary to provide internal stability.

A different difficulty resulting from misinterpreted borings is the case of a shield-driven tunnel under the East River, designed on the basis of river depth soundings to locate rock levels. The profile was chosen so as to get a full earth section except for the shore approaches, and the contract was so let. Construction soon proved that the borings were wrong, the rock protruded above subgrade, irregularly and in varying amounts, making progress difficult and costs almost double of an all-earth section excavation under air pressure. When some of the abandoned boring casings were encountered in the heading, the mystery was solved. The ends were curled into spring spirals where they had hit sloping rock surfaces and the depth to rock was incorrectly taken as the length of imbedded pipe at ultimate resistance.

Borings made for a new third tube are from an anchored Texas Tower frame to permit cores into the submerged rock and determine more accurately the rock profile.

Bedrock may be mineralogically continuous but have open internal seams, often filled with water-deposited soft muds which seriously affect its engineering value. An example of such condition was the exposure of the fine-grained gneiss and schist at the Manhattan approach to the New York Battery Tunnel. Two roadways were built with an intended rock core separation about 30 ft wide. Exposure showed the rock to be dense and hard but completely broken up into blocks with open gaps between and lying on a sloping bed so that considerable steel bracing was

needed to salvage the lower blocks for imbedment in a concreted core wall. The existence of such seams places an indefinite load-bearing value on the rock and practically eliminates the use of uplift or lateral anchors, and even rock bolts are of questionable value.

Natural and man-made cavities in the bedrock can develop into large scale surface settlements or local sinkholes. Drifts from a 130-ft-deep coal mine carried under Spring Creek near Indianapolis caused settlements up to 16 in. in a center pier and abutments of a bridge crossing the creek with serious damage to the deck. To avoid interruption of use, piles were jacked to rock at each support and the concrete spans, each weighing 220 tons, were jacked to a level condition and underpinned. Strangely enough, there was no further settlement.

Continuous ground settlements and even longitudinal stretching of the earth's surface of 4 in./100 ft in the Nottinghamshire coal mining area in England resulted in the development of flexible building frames pin-connected to a floor mat and with special ship lap wall panels. Diagonal bracing between columns with springs attached keep the steel columns parallel and the entire building adjusts itself to the changing ground surface. The design was developed in 1959 by W. D. Lacey, Architect for the County, and has proved quite acceptable compared to rigid buildings on deep foundations.

When some column footings of a large one- and two-story factory located in the southwest corner of Virginia, founded on spread footings on soil and on limestone rock and some on drilled caissons to rock, settled and moved laterally during a cement slurry grouting of cavities, a reevaluation of the program resulted in a complete change. Borings showed that the grout was not consolidating the surface soils or closing the rock seams but disappeared into lower cavities. The solution and pressure effect of the liquid grout was to break up the skeleton framework of the rock and cause loss of support.

Grouting was stopped, certain floor areas where spans were favorable were designed as supported slabs on the column footings. The rest of the area was covered with a harrowed-in skin of limestone dust and fragments which was consolidated before placing the floor. All piping under the floor was encased in concrete to protect against leakage. The perimeter was surrounded by a concrete-surfaced rain trench to avoid rain infiltration into the ground. Columns that had settled were shimmed level and in the areas of movement, column-base anchor bolts were left without nuts to prevent any future footing settlement placing downward loads on the columns and distorting the frame. Use of the building for 10 years shows the remedial work to have been sufficient protection against the ground conditions.

A spectacular sink-hole accident was reported by J. E. Jennings et al. at the 1965 International Conference of Soil Mechanics and Foundation Engineering. A three-story crusher plant at a gold mine some 50 miles west of Johannesburg, South Africa, located in dolomite rock collapsed completely into a sinkhole without warning. Ground water has been lowered 400 ft and surface settlements of 20 ft have been measured. The largest recorded sinkhole in these dolomites is 300 ft in diameter and 120 ft deep. Any man-made change in conditions of the subsurface should be a warning of possible subsidence and in the limestone-dolomite rocks of change of water movements that will dissolve the supporting rock.

Internal seams in the rock volume, often altered by the release of the confining pressures resulting from excavations at upper levels and by later modification through change in water pressure and flow, are the controlling factor in rock strength and stability. Tests on core samples which do not exhibit the extent of the seams or tests which do not take into consideration future change in weathering can mislead the designer into serious error.

Even a minor earth tremor can liquefy a saturated mud seam and cause a major rock slide such as that which wrecked the Schoellkopf power plant in the Niagara River gorge in 1956.

Tree roots have tremendous expansive power. A number of rock cliffs in the northern part of New York City have failed from ice and water pressure after the roots have opened natural seams. One such fall was of a 3000- to 4000-yd^3-volume that crushed a one-story factory building. The fall pulled out part of the rear yards of the apartments on top of the cliff but fortunately did not undermine the foundations.

In another location trees again opened up seams in mica schist. The oriental sumac, "The Tree That Grows in Brooklyn," is a very bad actor in this respect, and a section of rear yard rock fell against the lower building, crushing the rear wall. In the cleanup and repair all tree roots were treated with ammonium chloride injection to kill the growth, seams were cleaned out, crossed with rock bolts, and grouted tight.

Tunnel excavation is also vulnerable to rock movements along internal seams that are cut by the excavation progress. In some rocks there is warning of popping or shear creaking. Recent developments in geophysical instrumentation prewarn the possibility of rock movements. This will supplement the less reliable, visual and audible sensing or even the hunch of experienced tunnel men.

Rock is an aggregate of fused or compressed discrete mineral particles. Elastic properties determined from core samples are only reliable for low stresses. For high-stress conditions the seam structure and the anistropy of the rock must be considered, as well as the plastic creep

from crack and crystal contact adjustment with time. Comparison of test results obtained on core samples and in large scale tests in tunnels and in deep dam excavations by Rocha, Serafirm, and Da Silveira (Portugal) were reported in the Proceedings of the Fifth Congress on Large Dams, Paris, 1955. Such data should be carefully reviewed before accepting strength and elastic modulus values of rocks as determined from core samples in the laboratory.

Even in the best granites, as for instance the quarries on Vinalhaven Island in Maine, plastic flow has been found in the rock at some 150-ft depths. In one case, the solid sheets buckled upwards with serious popouts and the fractures made further excavation wasteful since so little usable stone was recovered.

Similar rock movements, first an elastic rebound and followed by continuing plastic deformation was reported in the shale exposures at the Oahe Dam in South Dakota, when some 12 ton/ft^2 of overburden was removed. A similar large release of vertical load over a limited area in deep excavations will alter the physical characteristics of the bedrock and cause change in shape, opening of seams, and potential zones of infiltration and softening.

The movements, indicated and measured, of the dolomite cuts made about 1890 for the wheel pits to harness Niagara water power required continuous addition of bracing at the bottom of the 180-ft-deep cuts. With the advent of electric power replacing the belt-transmitted water-turned shafts, plants were constructed on the banks of the Niagara gorge. Some rock movements were reported by A. D. Hogg in the construction of tunnels and open trenches for the Ontario Hydro development on the Canadian side in 1958.

After the Schoellkopf power plant was partly wrecked by a slide on the New York side, blamed on a slight earthquake tremor, the Niagara Power Authority built twin concrete conduits 46 ft wide by 66 ft high set into open-cut trenches with 50 ft of rock between. The cuts were up to 165 ft deep in almost perfectly horizontal layered limestone and dolomite. Preslitting the rock by blasting within closely spaced line holes before general excavation was used to contain the vertical sides.

Within the line holes, the drilling pattern was on a 7-ft grid, loaded to break the rock within 6 ft of the walls and the outer 6 ft on each side was slashed by light charges. Lifts were about 12 ft with a final cut of 6 ft to reach subgrade. A rectangular pattern of holes was drilled in the bottom for embedment of rod anchors to resist hydrostatic uplift on the empty conduit.

Considerable difficulty was found at the greater depths from continuous rock movement (Fig. 3.11). Line holes went out of position, anchor

Fig. 3.11 Subgrade upheaval in deep rock cut.

holes distorted, the formwork for the floor squeezed enough to crush 12- × 12 timber sills, rail supports for the sidewalls were continuously displaced, subgrade heaved and cracked and even the 5-ft-thick slab built with a center line construction joint heaved to open the joint a full 2 in. Actual measurement of movements indicated an initial 3 in. at center line, which increased in one month to 8½ in. with a 2½-in. heave along the sidewalls. Local areas went up even more with loosening of the subgrade sheets and raveling of the side walls. Replacement of the load by the construction of the conduits and backfill to the surface reestablished stable conditions but during the construction period, the possibility of movements and compensation for them must be considered as a necessary part of the work program.

Squeeze of rock in deep tunnel construction is a recognized phenomenon, as is the accompanying flaking, popping, and spalling of the rock exposures. In the French half of the Mont-Blanc Tunnel more than 100,000 rock anchors were inserted to prevent rock bursts in the over-

stressed granite. About 20 percent of these were placed in the face of the tunnel as a safety measure before progressing the work. Rock bolting can be successfully used to prevent separation and fall, but only after the exposure has been made, when the rock structure has been already altered. If the rock could be sewn together before exposure and change in position, much of the separation could be avoided.

The excavation for a subway in a built-up area affords an opportunity to examine rock conditions previously tested, and used to support loadings. Certainly the mica schists change radically in a few years. This is probably not the case with igneous rocks unless seams permit internal weathering. In the construction of New York's Sixth Avenue Subway several startling examples of rock change, even when sealed by concrete footings, were uncovered. At Forty-Second Street an office building constructed in 1917 with all foundations on "hard ringing rock" required underpinning to sound rock as exposed in 1936, to as much as 25 ft in height. The design engineer of the building personally examined the condition and brought his original field notebook to show that each of the piers had been examined and tested. The maximum live full load bearing was 20 ton/ft^2. The rock had softened at a rate of about 1-foot depth per year, so that removal was by pick and shovel.

The foundations of a theater built in 1930 on "hard" rock, when exposed in 1937, required some underpinning, about a foot average to remove the soft schist found directly below the concrete.

The Queensboro subway crossing Sixth Avenue was built as a twin tunnel in rock, using timber supports later encased in concrete and grout. Construction reports showed hard and dense schist, but seamy and blocky. Ten years after completion the roof was exposed in the Sixth Avenue excavation, and no hard rock of any kind was found. The grout encased timbers were removed after excavating the dense but soft disintegrated mica schist, using a small power shovel with no blasting or chipping.

Other rocks also misbehave on exposure. In Denver, a shale formation excavated for a concrete reservoir was noted to swell as much as 4 in. following heavy rains. Foundation and wall supports required change in design to tolerate such heaves. Similarly, east of Cleveland, the shale contains as much as 4.5 percent pyrites which expands upon oxidation and heaves footings, floors, and walls. The only cure is to prevent oxidation upon exposure, by immediate covering with concrete or sealing the surface with bitumen.

Blasting effects and damage are covered in Chapter 5.

Natural rock is the best of possible foundation materials, but problems of design and construction must be approached with an open mind and

full awareness of the fact that all rocks may be equally rock, but for construction use, all rocks are not equal. Nor, after operational shock, are they as stable as before construction started. When one considers that soils came from the rock of the mountains, and mountains continuously wear away, the "stability of solid rock" takes on a new complexion.

4

At Surface Construction

Slopes and Slides; Subsidence; Retaining Walls and Abutments; Bins and Silos.

Surface topography is never naturally stable. Gravity aided by wind and rain, moves the hills into the valleys, and the valleys into the oceans. Continuous tectonic and intermittent seismic action form new hills and the cycle of change continues. Man locally interrupts the natural cycle of change by excavating into the ground with slopes in the natural soil, by filling on top of the ground with slopes outlining the embankment, by covering the ground with impervious blankets, by stepping the surface and holding the upper level with retaining walls, and by storing earthy materials within bins. In some of these constructions natural soil strengths are degraded or the necessary strength of the soil in the relocated position is not provided, and a failure results. Such incidents in connection with dam construction have been included in Chapter 3; all other cases are described in this chapter.

With the official pronouncement of the availability of the science of soil mechanics some 30 years ago, one might question why there are so many incidents in recent years in the phases of construction which should be controlled by that science. It is impossible to state that soil structure failures have either increased or decreased numerically or in percentages since the recognition of soil mechanics as a guide to design and to performance of earth structures ("erdbau"). Certainly, designs have become more daring, of larger scope, and size and performance is

at greater rate, with larger dilution of quality control and less direct contact of the designer with the execution. Each of these factors is an incipient invitation to a failure. They are not limited to earth structures, but apply equally and often in greater degree to many superstructure difficulties.

A study of aerial photographs of a site will disclose prior massive earth movements, in the form of displaced contours and areas of light colored newer growth cover, that surface surveys and borings may not detect. The lack of earlier slides does not guaranty that stability will not be impaired by new excavations or loadings. The existence of prior movements requires special precautions in the new construction since the self-healing of slip surfaces seldom replaces the original strength of the soil mass.

4.1 Slopes and Slides

Typical of the extensive damage which can result from the development of areas with prior landslide activity, usually for residential use, is the case of the Palos Verdes Peninsula in the County of Los Angeles, Calif. Roughly about 165 acres in a rectangular area from the ocean up hilly slopes became active in 1956, moving ½ to ¾ in./day. More than 100 homes were affected and damage claims for more than $5 million were made against the county on the allegation that fills placed at the top of the slide area for a road had triggered the slippage. Many suggested corrective measures were tried, including the embedment of reinforced concrete caissons, with half their length below the slip surface, without much success. Within two years maximum horizontal movements were measured at 35 ft with an additional one inch per day. The caisson shear pins in one year had bent with incipient concrete failure at midlength and the tops were displaced 4 to 12 ft in the direction of the slide. Buildings that showed slight jamming of doors and windows were complete wrecks within a day. While litigation dragged on the movement continued, although at a lesser rate down to ¼ in./day in 1963, when for decreased value of the vacant and of the leased land the courts awarded $3.9 million plus 7 percent interest from August 1956, to the homeowners, utility companies, and developers. By the time the appeal was denied by the California Supreme Court in January 1965, the interest totaled $3.5 million. Los Angeles imposed some strict control on future development of similar slide-prone land as a protection against revenue loss from shrinkage of real property tax base resulting from casualty deductions.

In 1955 a landslide destroyed part of the old town of Nicolet, Quebec, situated on the south shore of the St. Lawrence, an area of sensitive marine clays. The slide was 1000 ft long and 800 ft wide, destroying homes and the local cathedral. The rapidity of flow is shown by the formation of a crater 20 to 30 ft deep, 400 ft wide, and 600 ft long in less than 7 min, as a succession of slices moving into the river. The slide area had been a flat-top mound about 25 ft above high water level and 50 ft above the bottom of the channel. On each side of the mound there was evidence of much earlier massive slides and some small slides had occurred at the edges of the mound in 1946 and 1950 (Fig. 4.1). In spite of these warnings subsurface drainage was inadequate and the slide area was fully saturated before the winter set in. Erosion of the unprotected bank of the river reduced the resistance, and saturation of the soil increased the weight and reduced the internal strength of the mass.

G. G. Meyerhof made a comparative study of the Nicolet slide and seven others of similar magnitude during the period 1946–1957 in soft quick

Fig. 4.1 Nicolet slide in sands eroded by river.

clays in Canada, Norway, and Sweden. An analysis of each of these areas would have indicated susceptibility to sliding, with saturation and erosion of the toe by stream or ocean action.

Massive natural slides have recently been reported from overseas. In 1961 at Kiev, U.S.S.R., a landslide moved houses and barracks into a ravine and killed 145 people. In June 1961 some 8000 slides hit terraced suburbia developments around Tokyo, Yokahama, and other cities, after heavy rains, and killed 270 people. The 2500-year old city of Agrigento, Sicily, lost 50 acres of developed homes and apartment houses from a slide of a thin layer of rock-overlying clay. The blame was placed on overloading the hillside with buildings up to 10 stories high. Only a month before the slide, after years of discussion, the city passed an ordinance banning new apartment construction. Complaints of water flooding the city from unknown sources were being studied but no report had been issued as to its source or effect on the hillside.

Slides have wrecked power plants located into steep hillsides. In 1956, probably triggered by a slight earth tremor, 50,000 tons of rock demolished two-thirds of the Schoellkopf station at the bottom of the Niagara gorge. Some little warning was had when water started to leak seriously into the building, but the rock sheared off vertically in the almost 200-ft height. Similarly, an earth slide coming about a month after an earthquake that shook the area near Seattle, Wash., in 1965, seriously damaged the lower Baker Powerhouse, then 40 years old. Melting snow in 1967 saturated a Japanese mountainside and the resulting slide dammed the river downstream and flooded the Odokoro power plant in Niigata. A 600-ft-long barrier of earth and rock, 130 ft high dammed the stream with danger to downstream developments should the impounded water give way suddenly.

At the open pit Beattie Mine in northwestern Quebec in a glacial-varved clay area, the excavation in 1943 was 1000 ft long, from 40 to 200 ft wide, and 300 ft maximum depth. Excavation consisted of 85 ft of clay overlying 15 ft of sand-till before exposing the rock. Some small slides had occurred in 1937 and in 1942. On the night of June 15, 1943, 1 million yd^3 of clay filled the open excavation and flowed into the underground tunnels. The flow was initiated by the collapse of a pillar in the pit, followed by a rock fall. All the clay on the north and east went into the pit. By 1946 some 2 million yd^3 of material had been removed, including several smaller slides. Investigation of soil properties showed the clay to be normally consolidated and easily subject to earthflows.

Landslides in clays have been studied for many years. In 1846 Alexandre Collin wrote a book on his experiences in analyzing and trying to control slips in clays, marls, and sandy clays. Fifteen slides are described

in the constructions for the Canal de Bourgogne, which occurred in the period 1833–1846. Slides were from 10 to 40 ft in height. Also in this period, he describes the investigation of eight other slides in France, one 60 ft high. An English translation of this book by W. R. Schriever of the Division of Building Research, National Research Council, Canada, was issued by the University of Toronto Press in 1956. Quoting from the translations, Collin states (page 9):

"That in clay soils the sliding surface is not pre-existent;

"That within the prism of failure of such soils the surface of rupture is not a straight line;

"That it is a curve;

"That consequently Coulomb's theory is not applicable to such soils, and

"That the line of rupture and of spontaneous sliding appears to be a cycloid or a curve of that family."

He further notes that the surface of a clay embankment after a slide occurs is not permanently stable unless one of three conditions is fulfilled (page 139):

"The cohesion of the central immobile mass is sufficient to ensure stability of the curved slopes; or

"The mobile mass permanently covers the surface of sliding to counterbalance the component of gravity and thus resist new rupture; or

"There is substituted for the mobile soil mass, after its removal, masonry or other work whose arrangement is such as to produce, statically speaking, the same effect, that is to say a material equality between the forces which maintain the system and those which tend to destroy it."

Two classical examples should not be omitted in a discussion of earth slides. The intermittent side slope failures in the Panama Canal, especially at the Culebra Cut has required removal of volumes often estimated as exceeding the original canal excavation. More recently, almost 50 years ago, the major slide in Pittsburgh that covered the main line tracks of the Pennsylvania R.R. was corrected, on the advice of the then leading authority on such happenings, by excavating all that wanted to slide.

The trace of slide effect in clay banks is easily recognized by its curvature and by the formation of vertical cracks in the surface. When such indications appeared in the rear yards of some fine homes located on a bluff of glacial sand overlooking Long Island Sound, and the line of failure crossed one building and cracked the walls and floors, investigations to determine causes and develop corrective measures led to an interesting finding. The sands had been clean and fairly coarse. To permit the desired

Fig. 4.2 Slide in sand modified by irrigation and topsoil intrusion.

landscaping and vegetation, a thick layer of topsoil had been placed with subsurface irrigation. Over a period of years topsoil was added to maintain the lawns and the high porosity of the subsoil required almost continuous watering (Fig. 4.2). The result was to saturate the sand with the finest top soil particles, of clay size fully saturated, and so the 30-ft-high bank acted as if it were a clay mass. The toe was exposed to tidal undercutting and fissures formed at the top with circular interior slip surfaces. The recommended cure, after the affected building was tied together with tensioned cables, was either to eliminate the lawns or to construct a masonry retaining wall.

Rock exposures are also weakened by weathering, often speeded by tree root expansion from the landscaping plantings. Moisture accumulation in the expanded seams becomes further expanded ice that will loosen the rock. In February 1957 some 1000 tons of rock fell out of the slope along the New York State Thruway, closing all three southbound lanes north of Yonkers. During the construction of an interstate highway in western North Carolina, completed in 1966, repeated rock slides in a sidehill cut added 800,000 yd^3 to the 4 million yd^3 of contract excavation. One slide of 250,000 yd^3 took a slash almost 500 ft high into an apparently solid-rock benched slope cut into the Smoky Mountains (Fig. 4.3).

Small slides are quite common in foundation construction, and are often

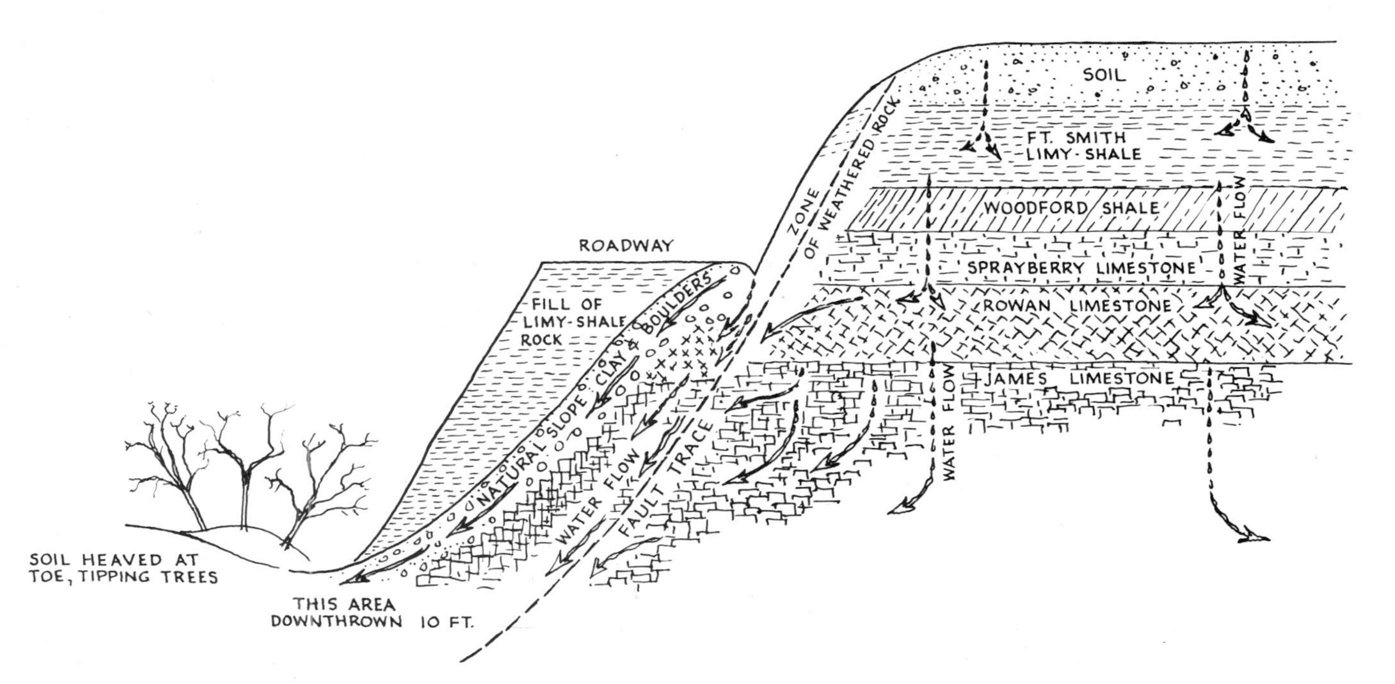

Fig. 4.3 Typical rock slide along highway. (From W. H. Rowan, T. R. Pierce, and R. Webb, "Failure of a Side-Hill Highway Fill," *Civil Engineering*, May 1963.)

the result of carelessness or an attempt to save the cost of temporary shoring. A clay bank collapsed in 1965 at a post office construction in Columbia, S.C. Of eight construction workers building formwork for a 20-ft-high retaining wall, seven were killed. The original soil bank sloped at 60 deg from the horizontal. To clear space for the wall footings, the toe of the slope was cut back 10 ft and protected by 4-in. lagging against some timber piles. The slide snapped the piles and smashed the formwork.

A similar accident in Seattle occurred in 1960 when workmen trimmed the toe of a 36-ft-high sandbank for wall footings. The slope of 3/4:1 had been stable without shoring until the footing pits were cut into the toe.

Massive avalanche slides of man-made mountains of debris are examples of poor control and regulation. In 1965 an avalanche of stored cinders buried a housing development in Kawasaki, Japan. Torrential rains had saturated a 50-ft-high mound of fly-ash, 300 ft long, 230 ft wide, which then flowed down a slope and demolished 15 homes. Storage of cinders and fly-ash in the critical area had caused some alarm and the trucking company had erected a barrier of sand bags and wooden poles, all of which sheared and was carried by the slide.

In 1966 came the tragic coal waste slide that took 200 lives at Aberfan, Wales. A conical pile of coal waste slurry had accumulated from a coal mine some 4 miles away and had become 60 ft high and 450 ft diameter. Heavy rains saturated the mass and some 2 million tons broke out and slipped 500 yd into the town within a few seconds, engulfing the school and 18 houses. In spite of several minor slips the warning of the danger was not heeded. Some drainage channels cut into the storage pile could have averted the disaster, but of course a less hazardous location for the storage would have been safer.

Typical of slides in rough terrain road construction is one described in connection with a Tennessee project in 1963. The section was part-cut and part-fill on a steep hillside. The basic rock was a weathered limestone overlain by a few feet of plastic clay soil. The cut was in this material and used for the fill but additional fill was taken from an adjacent cut in limey-shale. The plastic clay original surface was not removed and the fills remained stable for only a few months, until the buried clay layer became rain saturated and the fill slid down the hill.

The four-lane scenic road running south from Tijuana built along the Pacific Coast in the Baja California Peninsula along the steep slopes was similarly undermined by heavy rains in 1966, just before opening for traffic. Cracks in the pavement extended 1300 ft with the downhill side dropping as much as 6 ft.

Rocks that rapidly decompose on exposure to air will provide nice looking slopes when trimmed but soon show slides and settlements even in flat

slopes. Such rock is found in the Tennessee Appalachians, and the cure was to locate the roads on top of the mountain ridges and avoid cutting into the sidehills.

Failure in considering climatological effects on the subsoil has resulted in many pavement failures. In every instance lack of moisture control is the proximate cause. As shown by a few examples, different soil conditions and rain intensity areas require quite varied protective measures. Copying a successful design from one locality is hardly a guarantee of sufficiency in another area of different soil composition and rainfall intensity.

In 1955 the New York State Thruway 2 miles west of Albany, N.Y., was washed out by the erosion of some small springs at the foot of a moisture-saturated fine sand embankment. Soil loss was progressive and the entire pavement under two lanes was eaten away. The soil is stable under normal rains but cannot withstand steady contact with flowing water.

At the Florida Sunshine State Parkway, the lime rock formation used as the base near Fort Pierce had become oversaturated before the pavement was placed. The asphalt surface prevented drying out and the rock became greasy and the pavement broke up under traffic movement. About 11 miles of roadway required extensive repair. The saturated rock was scraped off and replaced with dry material. Similar subgrade disintegration affected parts of the London to Birmingham expressway in England, where asphalt surface cracking, local settlements up to 6 in., and shoulder failures required closing of the road to traffic in 1959. The damage was noted in areas of the clays and limestones of the Great Oolite soil series. Heavy rains during the construction period brought compaction problems and saturated the base.

Continuous disintegration under traffic of the easterly end of the New York State Thruway Berkshire section after two years use required extensive reconstruction in 1960. The native soil was glacial outwash layered clayey silt and fine sand covered by harder till. In the saturated state excavation was very difficult and actually could not be done until a dry summer period permitted equipment to operate on it. Rock surface just below this layer prevented drainage and excessive pumping action by the traffic ruined the 9-in. reinforced concrete slabs. Some 14,000 yd^2 of pavement was removed and many other areas were strengthened by pumping asphalt emulsion to fill the gaps between concrete and subbase. The pumping of the slabs accumulated water under pressure which bubbled out when grouting holes were drilled through the pavement. The fine grained soils had travelled upward and sealed up the porous base layer.

Failure to provide a base layer that will remain porous proved disastrous in the United States-built highway in Cambodia. About a third of

the 132-mile road, where a mixture of natural local laterite with 5 percent sand and 10 percent crushed rock was substituted for the specified crushed stone base material as an economy measure to save long hauls from stone quarries, had to be repaired. Local heavy rains soon sealed the base and unrestricted traffic loads with minimum maintenance of drainage facilities broke up the asphalt surface.

Roads on expansive clay soils, found in many parts of the world require special attention to moisture control. Serious failures are reported in Mysore State, India, in parts of South Africa, and in North America where excessive differential swelling and shrinkage of the subgrade soil causes bumps at all cross drains and cracking and heaves under rigid slabs. Lime stabilization is a possible remedy with strict control of moisture content before covering the base.

Concentration of water drainage will ruin the best made fills in fine soils of nonuniform porosity. In the arid Western regions, where valley-aluvium loose silts dry out and form interior seepage channels, the first rain melts away the clay and silt particles along the channel faces. Soon thereafter interior cavities form and the grain structure collapses, opening up gullies in the slopes and sinkages under pavements.

This same type of internal structural change resulted in large scale erosion and collapse of the slopes and shoulders of many miles of railroad trackage built in 1949 at an ammunition storage depot located in the marls of north central New Jersey. The marls and the local silty sands have large percentage content of very fine flat particles. The trackage was an extension of existing facilities, built in 1942 with similar soils and identical geometric design. Erosion of the 1942 construction was minor in nature. Further study however revealed a significant difference in construction method. The 1942 fills were placed by small truck loads and bulldozer, positioned in horizontal layers with fairly uniform compaction. The 1949 fills were placed by 25-yd^3 scraper pans on rubber tires. The loaded tires compacted grooves in the fill which became buried drainage ditches for rain water runoff. Where a weak spot existed in the side or bottom of these grooves the water flow would pipe out through the fill and the material above fell down to form open gullies.

Since it was impractical to remove the drainage troughs in the fills, a full-scale research program was carried out to determine the best material for covering the slopes as a protective measure against the erosion. After adjacent sections were covered with various density and grain-sized materials, from straw to crushed rock, the test track was covered with artificial rain at the rate of 1 in./hr. Physical and photographic observations soon eliminated most of the test materials. A lightweight graded blast-furnace

Fig. 4.4 Piping failure of slope of railroad embankment.

slag was found to be the best cover since it held moisture and had some tensile restraint to blow out forces, and was not so heavy that it sank through the saturated silts (Figs. 4.4 and 4.5).

Another case of a fine soil base failing under concentrated water flow is a somewhat different use of the material. A large tank farm of volatile fuels consisted of steel tanks with hemispherical bottoms. Each tank was surrounded by a concrete ring wall, within which a hemisphere was carefully shaped and lined with a fine uniform grained sand as a base for the steel tank. The installation in Central New York State was completed in 1952 with some 40 separate tanks connected by valved piping. It was soon noted that the piping fittings were straining under outside forces and that all the tanks were gradually and continually leaning to the east. The prevailing winds being westerly, the tanks were rolling on their bases as a pneumatic bearing, after the rain runoff from the tanks had saturated the fine sand and formed a liquid lubricant. The highly volatile contents of the tanks required a solution with minimum fire risk, and the cause of the difficulty eliminated any liquid form of grouting. Any addition of water would only aggravate the condition and cement or chemical solidification not uniformly covering the entire base surface might cause local plate

Fig. 4.5 Tests of slope protection for embankment.

buckling and tank failure. The accepted and successful cure was to stabilize the soil cushion by intrusion, forcing dry limestone dust into the wet mass by wooden rods activated by hand power and after some consolidation was accomplished, by the aid of an air hammer at the end of a long broom stick. The blend of dry limestone dust and saturated fine sand after three or four passes around each tank perimeter, provided a stable base, the exposure at the perimeter was then sealed and waterproofed against further infiltration. All tanks are slightly out of plumb, but in the same direction so that the piping required little adjustment.

4.2 Subsidence

Even without earthquake impulse, land masses are always rising or falling, usually at a very slow rate. Land subsidence is sometimes generated by man-made change in natural conditions. Heave also results from change in water content of the soil or exposure to deep frost. When these changes take place in built-up areas, damage and even collapse of the surface structures can entail serious losses. The long-time subsidence of Mexico City correlated with the pumping down of the water table over many

years is still unsolved, although pumping is now carefully controlled and water supply sources beyond the basin of the City are replacing the local wells.

The subsidence of the Terminal Island in the Long Beach, Calif., harbor had reached 4 ft in 1945 and was continuing at a rate of 1 ft/yr. The earliest estimate of 8 ft maximum subsidence was soon exceeded. Causes were generally listed as changes in the land surface from the filling and dredging to provide the industrial complex, pumping of water from the subsurface for industrial and domestic use, lowering of the ground water table during dewatering for construction of the dry docks, and pumping out oil from the sand layers bounded on top and bottom with impervious beds of weak shales. From 1900 on a sand spit had been built up by adding soil and structure loadings, but no appreciable subsidence was noted until 1937. Two general layers of water bearing sand, up to 900 ft below tide water, have been used as sources of potable water. Increased pumping followed increased development and must have caused some shrinkage of the soil layers. The water pressure levels dropped 30 ft in the upper layer and 60 to 70 ft in the lower level, all causing increase in weight on the subsoil and some shrinkage from the dewatering. Dewatering for drydock construction temporarily lowered the water table 75 ft, which was felt 2000 ft away and caused settlements of 6 in. of the surface. Upon completion of the dry-dock, the water table came back to normal, but the settlement remained.

Oil wells pumped from sands at 2500-ft depths started in 1937, with production for the Long Beach area of 100,000 bbl/day for about 10 years. Increase in rate of subsidence correlates well with the oil pumping. The subsidence required major alterations to the waterfront structures, subsurface utilities, and foundations. By 1959 a bowl-shaped depression had developed over 20 miles in diameter and 26 ft deep at the center, and some $90 million had been spent by the City, the Navy, and oil companies to repair damage (Fig. 4.6).

To stop further settlement and as a by-product to recover oil distributed in the sands and not readily pumped out, a major recharging operation with a projected cost of $30 million was started in 1958. Well water is cleared of sand and chemically treated to prevent bacteria growth and inhibit corrosion and then injected into the subsurface layers at pressures from 850 to 1250 psi. Starting with 14 mgd, the capacity of pumps and injection wells was increased to three times that volume. Pumping program is to continue until all oil is squeezed out and sufficient pressure is induced to hold the surface stable.

Another dramatic subsidence was the sudden split and settlement of a 2200-ft length of the fill placed across the Great Salt Lake in Utah to carry

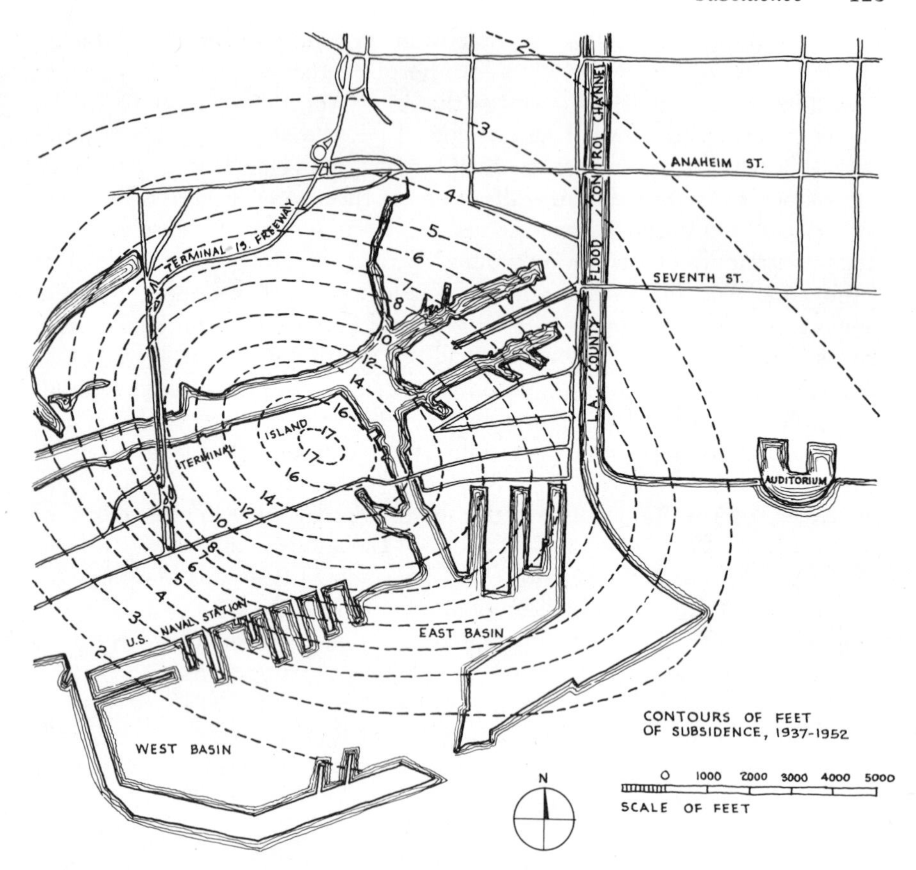

Fig. 4.6 Large-area subsidence from deep pumping. (Redrawn from *Engineering News Record*, February 5, 1953.)

the Southern Pacific Railroad. Design of this fill was based on thorough soil investigations. Yet the load of some 2 million yd³ of material, partly placed in water, caused a 10-ft layer of salt to give way and failure came six days after completing the fill. Reconstruction in 1958 was on the basis of a modified design with wider base and flatter slopes. The rest of the 13-mile fill had a few minor slides, but stability was soon reached when the underlying clays compacted and gained strength under load.

Typical of the troubles encountered in heavy clay soils on which light structures are placed is the report of the National Research Council of Canada, Division of Building Research in 1963, from which the following is quoted:

"In Winnipeg wetter than average weather conditions prevailed during 1962, following two years of below normal precipitation. Two private dwellings constructed at the end of the dry period in the fall of 1961 have since experienced upward movements. In less than a year a maximum heave of 0.89 in. and a maximum differential movement of 0.77 in. have developed in the foundation walls of one of these. As yet there is little visible damage to the superstructure. A basement floor slab has undergone a maximum heave of 1.50 in. As would be expected, the greatest heaving is taking place in the subsoils that have the highest clay fraction and the lowest initial water content. The greatest differential movement is being measured in the house having the widest variation in subsoil types, ranging from silts of low swelling potential to clays of high swelling potential.

"In Regina the walls of a conventional house basement constructed in the fall of 1961 have undergone a maximum settlement of 0.44 in. and a maximum differential of 0.37 in. The floor slab has experienced a maximum heave of 0.86 in. Again, in Regina, the continued upward movements of several basementless houses built during the summer of 1960 were measured during 1962. A slab-on-ground house showed a maximum upward movement of 1.06 in. with a maximum differential of 0.43 in. A house founded on 6 ft under-reamed (belled) piles and 2 ft reinforced concrete grade beams continued to heave, reaching a maximum displacement of 1.53 in. and a maximum differential of 0.66 in. A third crawl-space house, founded on shallow spread footings and a 4 ft concrete foundation wall, moved a maximum of 1.42 in. and developed a differential distortion of 0.75 in. During the same period other basementless houses, founded on 16 ft cast-in-place piles and 2 ft grade beams, heaved a maximum of 0.51 in. with a maximum differential movement of 0.16 in. where an intentional void had been provided under the beams. Where no such void was provided the maximum measured heave was 0.67 in. and the maximum differential movement 0.36 in. Reduced evapotranspiration under these foundations, together with the shading effect of the house and irrigation of lawns and gardens, have resulted in appreciable increases in soil moisture around the houses.

"During the summer of 1962 an interesting case of severe heaving of a grade floor slab in a single-story industrial warehouse in Regina was encountered. Vertical ground movement gauges, building movement points, a deep bench mark, and a neutron moisture meter access tube had been installed during construction of the building in the late summer of 1961, and readings had been taken periodically since installation. By the middle of August 1962, a maximum heave of 3.5 in. had developed in the central portion of the slab and above the centreline of a construction trench in which subfloor plumbing had been placed. This heave resulted in a maxi-

mum angular distortion of 1/30, or 1 in. in 2.5 ft and created serious damage to interior partitions and problems of materials handling equipment. It was feared that damage to the superstructure could result if interior load-bearing partitions or the reinforced concrete grade beams were lifted off the cast-in-place concrete pile foundation.

"The high temperature of the floor slab at the point of maximum heave indicated that a hot water leak had developed in the plumbing system, and excavation proved this to be true. The trench backfill, which consisted of a pit-run sandy gravel, facilitated migration of water along its length; to a lesser extent the pit-run gravel subgrade aided in further movement of water under the slab. A check on water consumption for the building showed that an increase of some 7500 gal or 1200 cu ft over normal usage had occurred during the period in which heaving had been observed. This represented approximately 7 in. of water over the entire area affected. Before plumbing repairs were completed the gravel subgrade had become fully saturated; the neutron moisture meter indicated an average increase in soil moisture above construction conditions of 10 percent in the top 2 ft of natural soil underlying the gravel fill, 5 percent in the next foot, and a 2 percent average increase over the next 5 ft. The rebound gauges, installed at various depths below the slab, indicated a vertical dimension change in the soil of 5 percent in the top 2 ft, 3.33 percent in the next foot and 1.25 percent over the next 5 ft. The observed heaving correlates well with movements expected from the interpretation of laboratory swelling tests on undisturbed samples. Observations are being continued to study final equilibrium conditions of volume and moisture distribution.

"This case has focused attention on the need for special precautions in installing plumbing in or below grade slabs on potentially swelling soil. It is preferable to place such plumbing lines in concrete-lined gutters with removable covers or in other conduits having drainage facilities. In many cases it may be possible to install plumbing lines entirely above the floor slab, thus assuring easy and immediate attention should leaks occur."

4.3 Retaining Walls and Abutments

The historical background of soil mechanics is in the ancient attempt to build stable retaining walls; up to 50 years ago the greatest part of the literature in that field concerned itself with the determination of earth pressures and the analysis of retaining wall failures. Recent theoretical and empirical development have resulted in safer designs and relatively fewer failures, but some still continue to plague the succession of false prophets who announce the complete solution of the earth pressure problem. In the

upper hilly part of New York City, where many retaining walls separate the stepped levels of adjacent yards, the annual spring thaws were accompanied by retaining-wall failures. Many of these were not structurally designed and some did not even have a draftsman's pictorial representation as a guide to the builder. Controls have been tightened somewhat and time has eliminated the most vulnerable cases. Fatalities, narrow escapes, and masonry cascading into adjacent occupied property were the source of much litigation before a retaining wall was recognized as a structure that required investigation of future safety, even though it was not in itself occupied by people.

A few of the more spectacular wall failures are described as examples of the errors in such work. In 1925 Los Angeles built a multiple-arch retaining wall, 450 ft long and 60 ft high to hold a fill for the relocated and widened Cahuenga Avenue leading from Hollywood to Lankershim. The design was chosen over that of a bridge to span Indian Gulch, since it provided space for disposal of fill from an adjacent excavation and the arch design would impose smaller loads on the soil level chosen for bearing, about 15 ft below the ground surface. Buttresses were spaced 30 ft apart and the arches were 18 in. thick at the top and 24 in. at the spread footings. Arches sloped 4½ horizontal to 10 vertical and were capped with a concrete beam and railing. The Santa Barbara earthquake on June 29, 1925, came at the time the fill was half-height and one of the end buttresses, founded on rock moved about 2 in. with only slight changes in the buttresses on soil. The movement thereafter, unexpectedly continued and was 8 in. after 18 months. Six months later it was 15 in. downward and 17 in. outward. Analysis indicated that slippage was along a steeply sloping buried rock surface, 40 ft below the ground level. Adjacent property was purchased and the arched wall was completely filled in to the very top. Even with the downhill side backfilled, some further movements continued.

A stone masonry gravity wall built in 1890 to retain a steep hillside and Geohring Street in Pittsburgh, Pa., extending up to 35 ft above Itin Street, by 1940 was leaning 30 in. at its greatest height. At this time reconstruction was essential and a scheme was devised which permitted maintaining traffic on the streets. Sheathed pits on 23-ft centers were cut into the high street and concrete counterforts built therein to subgrade levels of the wall. The wall was then cut away in pieces and replaced by a stone-faced reinforced concrete wall connected to the counterforts. All new work was carried to shale rock bearing.

In 1957 the Union Pacific Railroad built a 70-ft-high concrete retaining wall along the Albina Yards at Portland, Ore., to make room for more tracks. The wall was 7 ft thick at the base and tapered to 1.5 ft at the top. On

April 4, 1958, a 310-ft section of the wall fell over, releasing the backfill over the ground completely prepared for the new tracks. The wall cost $450,000 but the cost of resulting damages was even higher. The project was abandoned, additional 6.2 acres of private land was purchased to permit a 1½:1 slope cut into the hillside, some 1000 ft of a four-lane highway was moved back from its damaged position and rebuilt. During the reconstruction a steel sheetpile protection was installed to control the hillside before some 165,000 yd^3 was excavated.

An expensive experience resulting from a weak design, which the city engineers had considered inadequate, but a permit had been issued. The original design was for a 63-ft height on a 35-ft. base. Failure started by the stem shearing off the base and then falling over. The required reinforcement of # 11 bars on 4½-in.-centers was found in the wrecked wall, but located with 1½-in. instead of 2-in. cover as detailed. The required shear key had been installed. The design had been based on assumed soil backfill of 110-lb/ft^3 density and an angle of internal friction of 40 deg. The material actually placed in the fill was found to have a density of 150 lb/ft^3 and an angle of internal friction of 11 deg. The theoretical earth pressure value was thereby increased to 250 percent of the design value. The actual earth pressure required to cause failure was considerably less than this value. Control of backfill to insure compliance with the design requirements is as important as for the concrete construction.

In 1962 a retaining wall consisting of cast-in-place drilled-in piles carried to the top of a rock cliff, the concrete face making up 19 ft of the 75-ft-high enclosure of the One Main Place project in Dallas, Texas, fell into the Plaza after a fissure cave-in 20 ft wide and 15 ft deep developed in the upper street. The sequence of events was first a report of leaking gas, a small drop in the pavement, the street cave-in two hours later, and the fall of 75 ft of wall at 18 hours later. The wall consisted of 30-in. drilled cores spaced 5 ft on centers, carried about 5 ft into the rock. A continuous concrete base beam at the bottom of the piles is anchored to the limestone rock by prestressed diagonal tiebacks 15 ft long and spaced 5 ft apart between the piles. The space between piles is covered by a sprayed 3½-in.-mesh reinforced mortar. To permit street and utility restoration and to protect the existing buildings, a similar concrete wall was built in the opposite side of the street, but with cores carried 15 ft into the rock and the tiebacks eliminated and the street backfilled. Geological investigation disclosed some faults in the rock which were then crossed by drilled anchors, and a new wall was then built to enclose the five cellar levels.

The failure of a 12-ft-high abandoned concrete wing wall protecting the west side of the White River in Indianapolis, Ind., in 1927, which killed a man, was at first considered quite a mystery, since the fill had been mostly

removed while an adjacent 50-ft length still backfilled had remained intact. The structure was an L-shaped reinforced concrete wall which had been replaced by a new flood protection wall and apron. Disclosure of actual conditions showed that the failed wall had streaks of disintegrated concrete near the base on both faces, with the reinforcement exposed. There were indications that considerable lime had been used in the concrete mix (Fig. 4.7).

A most interesting wall failure from lateral pressure occurred in 1942 at Chambersburg, Pa. A cold storage building for apples in service for many years had 8-in. concrete walls covered with brick facing and tied to the floors with the slab steel extended. Normal storage was in crates, which became expensive and hard to get because of war conditions. In 1944 apples were stored on the floors uncrated and the lateral "apple pressure" forced a 50 × 100-ft section outward with a grand spillage of apples.

Bridge abutments have their share of failures. The common fallacy of not isolating the wing walls retaining earth pressure alone, from the abutment wall with the bridge load aiding in stability is often seen by the diagonal cracks which nature provides to separate the incompatible units of the structure. Maintaining bridge profiles when an abutment settles (or as happened in one recently constructed overpass, rises) often requires

Fig. 4.7 Several repairs of old bridge abutment.

some ingenious solutions. In 1957 one end abutment of a three-span deck girder bridge carrying the Connecticut Turnpike over railroad tracks settled progressively during construction and reached 17 in. Settlement of the piles due to an unforeseen compressible substratum was noted before the bridge abutment was completed and rather than drive new piles about 100 ft to rock, a jacking bracket was added to the abutment face and the bridge deck was kept level by successive wedging. A battery of nine hydraulic jacks lifted the deck, and steel plates were added under the sole plate. Total settlement was almost 2 ft before the conditions stabilized and the piles reached a gravel layer.

In Chicago one abutment of an old bridge failed in 1959 and the approach street dropped 6 in. Temporary piles and shoring were inserted under the steel bridge and a pile bulkhead placed next to the abutment to hold the street. The abutment had developed a crack 3 in. wide and the settled part was rebuilt on new pile supports.

There have been some reports of bridge abutments apparently carefully designed with long pile supports running through soft clay or silt soils to a proper bearing, that moved more laterally than vertically. The effect on the bridge structure is serious, involving expensive repair and alteration. The reason seems to be the error in the design assumption that loads in the piles (the toe piles are usually battered) are purely axial. The true condition is some bending in the piles, possibly from the lateral pressure of the earth on the high side, causing deformations, and with weak soils there is not enough passive resistance to maintain the structure in a fixed position. Investigation of the stability of the abutment as a retaining wall is of course necessary, but the foundation piles cannot be treated as an independent element; the combination of wall and piles must be checked for stress and deformation under the variation in loadings that exist in a bridge abutment, first without bridge load, then with dead load only, and then with a reasonable live-load allowance. The usual design for full specified live load is safer than any of the actual prior conditions that must be sustained.

4.4 Bins and Silos

Storage structures designed to retain granular materials such as wheat, grains, coal, ashes, and many industrial raw material and by-products have suffered structural distress from unknown or unforeseen pressure distribution, in addition to foundation troubles since they are often located on marginal land. Bins and silos built of timber, steel, and concrete have failed from unbalanced loadings when some units are empty, from un-

expected dynamic loadings during emptying, and from concentration of vertical reactions of the walls when the contents arch in the bin and release the load from the floor support. In very high bins this condition results in practically a thin cylindrical shell crippled by an excessive vertical load.

An early example of bin volume extension, at Fort William, Ontario, was a tile and concrete bin 20 ft in diameter that had a steel extension with a hopper bottom in the top third of the height. It failed in 1910 when the bottom fell under a load of wheat and blew the side of the bin outward. Unbalanced lateral load caused the failure of grain storage bins at Duluth, Minn., in 1903, 33 ft diameter by 104 ft high, when the interstice bins were emptied. In 1954 one of a battery of 21-ft-diameter bins at Buffalo, N.Y., then in service for 36 years, was damaged by the sudden fall of wet corn when some had been drawn out at the bottom. The mass fell and the air pressure forced a section of the 8-in.-thick wall outward. The hole was 11 ft wide and 30 ft high. Foundation failures have caused a number of bin storage collapses, some of which are described in Chapter 3, and also some expensive foundation repairs such as the case of the 56 bins, 115 ft high, at Port Arthur, Ontario, which contained 2.5 million bushels of grain in 1959, and the smaller facility of 20 bins, 19 ft diameter and 120 ft high, 800,000 bu capacity, in Fargo, N.D., in 1955.

Shock loading from dust explosion, which is the result of imperfect control of bin operations, is too expensive a condition to be considered in an economical design. An explosion in Philadelphia, Pa., grain bins in 1956 caused the steel and timber frames to collapse, but the concrete bins survived. A similar explosion wrecked one cell 100 ft high at Wichita, Kan., also in 1956.

Silos for storage of anhydrite, three 30-ft-diameter by 120-ft-high concrete bins with 8-in. walls, completed in 1966 for the Solway Works at Whitehaven, Cumberland, England, had a freely supported steel cone 30 ft up the silo as base support for the rock. After a few months use one fully loaded silo shattered when the steel base sheared through the concrete corbel supports and dropped the 3000 tons of rock to the foundation mat. The fall damaged the electrical-mechanical systems of the facility and the other two bins could not be emptied, even though they were considered in a critical state, until emergency repairs permitted use of unloading equipment.

Troughs and hoppers for storage and loading of coal, ore, and aggregates must not be used for materials seriously different in physical character from the design assumptions. A number of these have failed when the contents became water saturated. Support frames must allow for the maximum loadings as well as possible damage by trucks or equipment. A coal pocket, 30 by 125 ft and 35 ft high built in 1929 along the east shore of

the Harlem River in New York City and supported on a minimum steel frame with a shortage of bracing to resist unbalanced horizontal load or truck impact, collapsed. The criminal investigation, since loss of life was involved, centered on the sufficiency of the design, with the result that a jail sentence was imposed on the designer.

Surface construction appears to be just as prevalent a location for construction failures as the apparently more hazardous underground work. The explanation lies probably in the average greater experience of the organization undertaking the design and construction of subsurface operations. As a corollary it then follows that surface construction requires more careful control and a realization of the inherent dangers.

5

Manmade and Natural Disaster

Demolition Collapse; Deterioration Collapse; Overload Collapse; Alteration Collapse; Fire Damage; Explosion and Vibration Damage; Collision Damage; Wind Damage; Towers and Masts; Storm at Sea; Storm on Land; Lightning Damage; Rain Ponding Effect; Snow Load Failure; Earthquake Failure.

The natural cycle of evolution requires continuous aging and disintegration of all assemblies into their primary elements which then combine under climatological and seismic change into new assemblies. Man, in his wisdom, speeds up both the generation and degeneration cycles. Nature often interrupts normal change by ruinous wind and storm, lightning, floods, and earthquakes. Engineering design and construction must provide resistance for the high static and dynamic loadings accompanying these "Acts of God," but, often through ignorance or by knowledgeable acceptances of the risk, restrict the range or spectrum of these natural forces. Acceptance of risk beyond a conventionally acceptable force level may be good economics, but the failures that result often turn into disasters where a better choice of tolerable level would have been warranted. There is nothing like a natural phenomenon, a typhoon or earthquake, for bringing to view the errors, omissions, and shortcuts in construction performance. Every discontinuity, every weakness, every incompatibility of deformation becomes the focus of a structural failure, with immediate loss in total strength and reduction in available resistance. Failures are therefore common when structures are exposed to recurrent storm or seismic

action, and collapse is mysteriously noted during a phenomenon of lesser intensity than an earlier one of similar but greater magnitude, which the structure safely withstood.

Lack of proper precautions for expected and normal conditions, usually from absence of proper control, will cause catastrophic failure as happens during demolition of obsolete or partly damaged structures, when crucial supports are removed out of sequence, from change or overload during alterations, from permitting uncontrolled fires and explosion or from collision between a moving object and a static structure. These failures are often avoidable and cause complicated litigation and sometimes criminal legal action on the basis of negligence. These are man-made disasters.

5.1 Demolition Collapse

A structure being assembled is seldom as stable as when completed; likewise a structure being demolished undergoes critical stages, especially where deterioration and the usual series of minor alterations gradually weaken the assembly, and when the margin of safety is critically small. Demolition of a building must be planned in the light of that warning. Too often bracing members and bearing partitions are indiscriminately removed when the building is stripped of equipment. The remaining shell is prone to collapse as bearing walls are eccentrically loaded. A series of such accidents in 1966–1967, many with loss of life, was instrumental in the recommendation by the Corporation Counsel of the City of New York that no demolition permit be issued unless the wrecker has studied the construction records and examined the building to determine the nature of the existing structural supports, and a plan of demolition sequence be available which indicates sufficient safe support at each step. Statements to that effect will have to be certified by a professional engineer or architect.

In 1960, while wrecking a six-story factory building in New York, N.Y., a two-tier tank assembly above the roof collapsed and plunged through the building to the cellar level, carrying a large part of the wood floor framing with it. There were two tanks in a vertical array, a steel tank sitting on steel beams which rested on masonry walls and an upper wood house tank on a steel frame with legs supported by the lower tank beams. The top tank had been removed and the only possible cause was the damage or weakening done to the lower tank supports when the upper framework was burned away, since all components salvaged from the wreckage seemed to be sound and well maintained.

In 1966 the collapse of an old building in the theater section of New

York City caused much adverse publicity for the construction industry's apparent lack of control of the work. A belated investigation of the records, after the accident, revealed many warnings of an incipient hazard. At the site bedrock originally existed above the present street level. In 1870 a one-story blacksmith shop was set up at grade with an earth floor, brick walls, and wood roof. Two years earlier the adjacent land had been improved with a commercial building and a cellar excavated in the rock. The excavation was not cut neat at the common lot line and the brick cellar wall was backfilled with local excavation material. One wall of the blacksmith shop rested partly on this fill, the other walls were on bedrock.

Successively, the one-story blacksmith shop became a factory with a concrete floor, increasing from 60 to 100 ft, then a second story was added for commercial use, then a courtyard was cut into the east wall above the second floor level with the interior walls carried on steel beams and a third story added with four apartments on both the second and third floors. Later a penthouse apartment occupying about 60 percent of the area was added. Fire-retarding enclosures were added around all kitchens and ceramic tile floors placed in the kitchens and bathrooms. All of these changes and addition of load were made without modifying the original roof beams, which became the second-floor support for the apartment partitions, and without strengthening the original 12-in. brick walls or the foundations. The assembly stood and was used continuously, the apartments, for 45 years. But when the site was acquired for a new office building, the older building was demolished first and its cellar filled in for temporary parking use. Then demolition began on the one-time blacksmith shop, with its three stories of additions. When the bathroom and kitchen fixtures were broken out and removed, some partitions were disturbed. The entire building suddenly collapsed, trapping five men. The two stories of residence use with matching partitions had been acting as a space frame. The removal of some vertical bracing caused all the load to transfer to the second floor beams which deflected enough to pull the walls inward. The second floor fell flat on the concrete first floor but drifted eastward some 16 in. because of the greater mass along the east wall courtyard. The old lime mortar masonry just crumbled and fell in rubble sized units in all directions. After cleaning up the rubble, part of the foundation wall was shown tipped toward the cellar of the adjacent building where some of the filling had been removed. The first conjecture that such excavation had caused the building to collapse was not substantiated from the shape and direction of fall of the floors and masonry. The focus of failure was within the building and slightly away from the direction of the excavation.

In 1967, during demolition of a six-story movie house in Syracuse, N.Y.,

some of the interior floors were cut away, removing the bracing beams of the street wall. The 67-ft-high wall buckled and fell into the street. Debris crashed onto passing automobiles, killing one motorist and injuring several others.

Buckling failure of unbraced walls is a commonly overlooked hazard in demolition work. One of the large cases was the north wall of the Hippodrome on Sixth Avenue, New York, which was demolished in 1936. This massive well-built brick wall some 60 ft high and about 3 ft thick at the base, with no window openings, was scheduled for large-scale removal after the interior of the building was gutted. Cables were tied to several bulldozers at about mid-height, and some holes were drilled at street level and loaded with small charges of dynamite. The plan was to pull the wall over and let it break up in the 200-ft-wide cleared lot. Fortunately all traffic was kept clear of the street. The wall jackknifed and almost the entire mass fell opposite to the desired direction, onto the street pavement. The shock did considerable damage to the subsurface utilities but the fall did break the masonry into sizes that a large shovel could handle to load the trucks.

5.2 *Deterioration Collapse*

Not very many structures collapse from normal deterioration; usually they become obsolete and are removed before the weakening becomes critical. The few exceptions are usually timber structures where the vulnerable components are covered or hidden from inspection. Such a case was the collapse of the wood truss roof over the Cincinnati, Ohio, Opera House in 1897, then 27 years old. The ends of the 77-ft spans were bedded in the masonry walls and had rotted. Similarly at Portland, Ore., 76-ft timber trusses covering a two-story building were covered from view with a metal ceiling. The timber shrank in the confined space and six of the nine splices in the bottom chords failed, collapsing the entire roof. The necessity for periodic inspection of structures, even where special access must be provided, is too often forgotten.

5.3 *Overload Collapse*

The necessity for posting permissive loading on all floors, as is now common practice in most cities, becomes clear when one examines the record of structural failure from overload beyond the ultimate support value. All of the recorded incidents occurred 30 or more years ago, but the various

types of cases should remain as a warning of what precautions are necessary before introducing heavy loadings into existing buildings or on existing bridges.

When overloaded, wood-floor buildings with beams imbedded into lime mortar brick walls are vulnerable to the twisting action of the deflected beams which can buckle the wall and cause collapse. This happened in 1921 when a five-story warehouse in Newark, N.J., was loaded with wastepaper and the rear wall buckled and fell. In 1928 a two-story warehouse in New Orleans, about 100 years old, was loaded with leather and the 8-in. brick party wall separating the warehouse from an empty building was pushed outward. Both buildings collapsed. Earlier, in 1912, a Philadelphia, Pa., three-story dye plant, loaded with 30,000 lb of yarn on the second floor, buckled the wall. Both a 10,000-gal roof tank and 80-ft smoke stack along that wall fell, wrecking the entire building.

The residence floor loading usually required by code is seldom reached but unusual use may cause failure, as happened in Erie, Pa., in 1911 when 50 people congregated in a kitchen during a funeral and in 1914 in Aurora, Ill., when a floor collapsed during a wedding dance.

Shock loading can disturb timber beams to the point of failure. In 1914 a Dallas, Tex., warehouse for metal rods set into racks lost a section of the wall when a rack failed and the rods fell to the floor causing large deflections. In a New York printing plant a 600-lb bale of paper dropped on the fifth floor and the building collapsed. It was then 20 years old, having been completed in 1871. The amount of safety factor was inadvertently tested in a fruit warehouse in Minneapolis, Minn., when a 40-year-old building designed for 100-psf loading, the floor consisting of 2-in. planks on 7 × 12 beams spaced 42 in. and spanning 20 ft, collapsed under a load of 240 psf.

The normally built-in factor of safety will permit a reasonable overload, but sometimes such overload reveals noncompliance with the construction plans when the actual provision may just be enough to prevent collapse under the design loading. This happened in Minneapolis, Minn., in 1902 when a new six-story warehouse, with floors of 5½-in. plank and on 12 × 20 beams spaced 7 ft centers, was loaded with 220 psf on the top floor. This was only 10 percent above design load and after the load fell through all floors into the cellar, it was determined that the beam hangers were critically overstressed. Similarly, when the upper deck of a Long Beach, Calif., pier leading to the second floor and auditorium collapsed under a slight overload, the supporting girders were found somewhat decayed but, more important, the designed 12 × 14 girder was actually only a double 2 × 14, less than one-third of the design value.

5.4 *Alteration Collapse*

Changing times often require alteration of physical conditions and change in occupancy of buildings and sometimes even bridges. Many building codes carefully spell out the necessity for permits based on application to remove or modify any structural element. It is remarkable, however, how little awareness exists of the importance and care needed in changing an apparently safe structural assembly. In 1870 an entire block of five-story buildings, interconnected by arched openings in the brick bearing walls was built by "Boss" Tweed for the A. T. Stewart Department Stores in lower Manhattan. To provide natural light into the interior spaces, an open well was framed out by timber headers, covered over by a skylight at roof level. Timber beams of excellent long-leaf yellow pine spanned about 25 ft, and the openings were 6 ft^2.

As the mercantile center moved uptown, the block was broken into separate buildings by closing up the arched doorways, and was sold to various owners, who closed up the light shafts because electrical lighting was now common and the full floor area could be used. One building was occupied for manufacturing envelopes, using bales of paper from many sources, often of a quality rejected for printing paper. In 1944 paper was scarce and a carload had become available and was taken for stock storage. As much as possible was placed in the basement, stacked to about 3 ft below the ceiling. The rest of the load was placed on the street-floor level, filling in the usually vacant space around a paper cutting machine. This machine had operated without incident for a number of years and was located directly above the covered light well. Later exposure showed that all the light wells had been filled in by a pair of 2 × 4s and a single layer of flooring planks.

Because the cutting machine, the nerve center of the entire operation, was blocked by the storage of paper, the manager sent the 35 workers home at 5 p.m. and cancelled the customary overtime work for that day, leaving only the watchman in the building. The watchman took the elevator to the top floor, and stopped at each level to check that all lights were out, leaving the elevator at the cellar level, two tiers below the street, and left the building by way of the cellar emergency exit in the sidewalk. He reported later that nothing unusual had been noticed and he shut the iron trapdoor and went around the corner to get supper. When he returned, the entire inside of the seven-story building had collapsed, with almost the entire front wall and ornamental cast-iron assembly hanging up in position by their connection with the adjacent buildings.

Because the property was owned by an estate in trust to a minor, the legal determination of the proximate cause of failure was necessary to provide an explanation of the capital loss. The wreckage was demolished floor by floor to examine each level and evaluate the loadings thereon. The cause was the added loading at the altered floor area, collapsing the floor down to the basement stock of paper and releasing a line of interior partition and successive floor failure. Because there was fortunately no injury to workers and the loss was only a completely depreciated building, the trustee absorbed the cost of demolition and agreed to a termination of the lease without any claim. The tenant, strangely enough, suffered no loss because he salvaged most of the stock, and in the six-month period of removal paper costs rose so sharply that he was profiting by the incident. The realization that the collapse could have involved a tragedy with serious loss of life, sobered all concerned, and no one made any claim or even considered legal action.

Bearing-wall failures from added story loads have occurred, usually in lime mortar masonry, which loses strength with age. In 1909 a 50-year-old three-story Philadelphia building was increased in height by two stories. Half of the building collapsed under the added load. In 1920 a nine-story wall-bearing apartment building in New York was modernized by altering the lower two floors for stores. The removal of floor bracing against the outside walls started a gradual collapse which soon became total. In 1927 a two-story building in Kimberly, Wis., where the interior walls were pierced by fireplace flues, was being increased in height by steel framing. Four columns placed on top of the masonry wall broke through the masonry shells around the flues and collapsed the structure.

The Los Angeles, Calif. Elks Club was being altered in 1929 and a door opening cut through a partition, which was actually a reinforced concrete girder spanning 60 ft and 15 ft deep. The area collapsed. Two steel trusses were added to relace the lost support over the room below. In 1912 an alteration of a two-story building required replacement of the wall and the wood beams were placed on temporary shores. Construction material placed on the floor caused the shores to tip and the floor collapsed. A bowling alley in Buffalo, N.Y., showed some vibration during use, and the floor was altered by reversing the beams but the headers were disregarded and the repaired floor then collapsed.

Brownstone residences in New York, and similar buildings in other cities have often been altered into separate apartments. One such project in 1963 involved combining two adjacent houses into four modernized apartments on each of the four floors. During the construction, without advising the architect, it was decided to lower the first floor, originally at stoop level, to match the sidewalk and reduce the cellar height to a mini-

mum. The 8-in. brick party wall at each floor had a 2 × 4 flat timber sill under each level of floor beams. In effect there was a 2-in. timber separating the masonry at each story. The outside walls had been chased out to receive the beams at the lower level. A four-beam section of floor was cut loose and the operation of prying out the beams, which did not have fire cuts in the party wall, acted like a 20-ft long crowbar, twisting the wall. The entire building collapsed, providing a much-needed parking lot in the neighborhood.

5.5 *Fire Damage*

Fires do not give any warning. Protection against fire damage is a lesson learned from the sad experience of many devastating losses of life and property. The recommendations of insurance carriers are spread widely, easily available, and incorporated in all building codes and regulations. Yet fire damage is still an extensive drain on the economy, usually caused either by a disregard for necessary precautions or from the optimistic faith that a fire-resistant enclosure reduces the vulnerability of inflammable contents. And the high temperatures developed by the burning of furniture, fixtures, and flammable goods will do serious damage to fire-retarded and even fireproof construction.

In 1910 the Chamber of Commerce Building, in Cincinnati, Ohio, 100 × 150 ft in area and 200 ft high, with granite walls and tile-arch floors between concrete-covered steel beams, burned with an intense fire. The steel roof trusses melted and fell to the cellar, collapsing all floors. There are too many similar examples from before the times when fire-protection was a well recognized technique; only a few more recent ones are described here. In 1946 the 22-story La Salle Hotel in Chicago lost all the wood trim and fixtures, although the fireproof steel frame came through without damage. The same year the Winecoff Hotel in Atlanta, Ga., a 15-story steel fireproofed building with one open unprotected stairway, was similarly gutted with all trim and fixtures lost. In 1953 the General Motors plant in Livonia, Mich., 1.5 million ft^2 of open space, 866 ft wide, with a metal deck and steel-framed roof, was a complete loss from fire.

The McChord Air Force Base steel-trussed hangar, Tacoma, Wash., was badly damaged by fire in 1957. Built in 1939, the twin hangars provided an open floor area of 194,000 ft^2 and were each covered by 11 three-hinged steel arches 275-ft span and 90-ft rise. Walls were of brick with steel-framed window panels above. Fire was confined to burnable contents and a wood-framed office and shop building located in the center aisle between hangars. Steel members were badly distorted and one truss fell

when the rivets at the main leg sheared. Fortunately the main door held the truss from falling to the ground, which could have triggered a chain reaction collapse of the entire roof. Repairs were made by local shoring and flame straightening of members bent beyond tolerable dimension with only 46 members requiring replacement. Of the 7327 members in the roof structure, 1486 required straightening or replacement. After heating the bent member, concentrated on the convex side, sometimes with jacking, wedging, or sledge blows, the cooling resulted in a straightening to approximately the original shape. The structural steel repairs alone were performed at a low cost of about 1 dollar per ft^2 of floor area, which amounted to a third of the total repair cost.

In 1964 the easterly end of the Forty-Second Street Shuttle subway at Vanderbilt Ave. in New York City went afire and developed enough heat to buckle the steel columns, requiring extensive shoring, emergency repair and reconstruction to protect the underground utilities and replace this heavily-traveled rapid transit link. Combustible items included some subsurface stores and a wood-plank station platform. Two trains, a total of seven cars, were reduced to scrap. With only 5 ft of roof cover, the street pavement settled 1 ft when the columns buckled, but fortunately no subsurface utility was severed. Twenty-eight columns, not fireproofed in the original 1904 construction, required replacement, as well as considerable roof steel that was replaced by new girders at a lower level because the surplus headroom for train operation was available. The record of repair is noteworthy. The fire started at 5 a.m. and was controlled in 2 hr. Within 48 hours thereafter, all members had been shored using over 50 tons of steel, all debris was removed, the platform rebuilt, guards and partitions to enclose work area added, and trains operated on two of the three tracks. Reconstruction was with noncombustible materials.

In 1965 the top floor of the Chicago Civic Center was gutted by fire. Two years later, just before the opening of the National Houseware Show, the McCormick Place Chicago Exhibition Hall, had a fire that made ashes out of $100 million value of displays, collapsed 200 ft of precast concrete outer wall, and made junk of almost the entire structural steel roof framing covering 8 acres of floor area. The fire ruined the structure, insured for almost $30 million, within an hour. Roof steel located 20 ft or more above a fireproof floor is normally exempt from fireproofing protection.

This tremendous loss, together with the awareness of what catastrophe could have happened if the fire had come when thousands of people were at the exhibit, requires a reevaluation of the fireproofing requirements and certainly on the volume of combustible contents within any one enclosure. The floor construction was a concrete waffle slab with 30 × 30 × 14-in. domes and a 4½-in. slab. Even in the area where the fire had melted glass

exhibits, there was no damage to the concrete. Since the contents in the exhibit do not differ much from those in any large mercantile warehouse or sales space, and the large capability of the Chicago fire fighting equipment could not cope with the fire, a serious question arises as to what is really fireproof construction.

5.6 Explosion and Vibration Damage

Shock loading or vibration resulting from accidental explosion or from operational techniques for construction or production operations can cause sudden damage to structures, often with loss of life. Many of these incidents can be predicted and proper precautions built into the operation or necessary resistances included in the construction. A gas explosion at a meter connection to a 40-psi main located within a garage wall, demolished four panels of roof support (Fig. 5.1).

Various kinds of gas explosions have caused major damage and of course there are a great number of homes affected by improper gas heating elements, leaky pipes, and containers. In 1932 the State Office Building at Columbus, Ohio, a 12-story steel-framed building, lost areas of some 10,000

Fig. 5.1 Plaza failure from gas explosion.

ft² of the first and basement floors when gas accumulation under the lowest floor exploded and lifted the slabs which fell as rubble. During the reconstruction in 1965 of a Titan II missile silo near Searcy, Ark., with the warhead and fuel removed but the 490-ton cover door closed, an explosion, apparently of a tank of diesel fuel serving an engine for auxiliary power, started a fire that knocked out the electrical power and killed several people, although damage was limited to a small area. The silo is 146 ft deep and 53 ft diameter. Loss of electricity prevented operating the closure door and very high pressures developed from the blast.

In 1966 a trailer truck loaded with 5000 gal of asphalt was crossing a two-lane steel bridge in Atlanta, Ga., when it struck the steel center divider, and broke through the floor. The span collapsed and fell to the railroad tracks below. The fall snapped a suspended 6-in. gas line and the explosion started a fire visible 20 miles away. The steel span and the truck were completely demolished, but the driver escaped with some burns and minor injury. Also that year a 120-ton cylindrical high-pressure autoclave, used for curing concrete blocks, at a Hamilton, Ontario plant, suddenly exploded. The 106-ft-long and 8.5-ft-diameter vessel slid 150 ft, cutting through a truck and stopping just short of the highway. The 4-ton door tore through several buildings, including the laboratory and flew 200 ft.

Natural gaspipe line explosions have caused some local damage and even more psychological fear, with the result that Federal safety regulations are being imposed on the design and operation of interstate lines. The failure of the steel pipe forms a mysterious scalloped tear of the metal, as if exposed to a shortwave spectrum of intense pressure, with little attenuation but with a corkscrew travel along the pipe (Fig. 5.2).

There are about 1 million miles of pipelines for long-distance transport of gas and liquid fuels in the United States. Some of these lines operate at pressures up to 1000 psi, and only a small part runs through populated areas. The Federal Power Commission issued a report on pipeline failures and for the 15½-year period ending June 1965 2294 incidents were reported. The worst was the gas explosion on March 4, 1965, near Natchitoches, La., which killed 17 people and formed a crater 27 ft long, 20 ft wide, and 10 ft deep. Another major failure was on Thanksgiving Day, 1950, when about half a mile of 30-in.-diameter ⅜-in. wall natural-gas pipeline exploded and ripped out of the ground. Publicity on such incidents has caused Congress, after the President signed a law in 1965 providing for safety regulations for all pipelines other than natural gas and water, to study the necessity for including gas transportation in the regulations. Most states do have such regulations and the American Gas Association is arguing that the USAS-B31.8 Code could be used as a Federal enforcement in the states that have not assumed jurisdiction. That Code does vary

design standards on the basis of population density, on corrosive soil and contents conditions and on range of thermal exposure. Bending, welding, and soil bedding controls in the field as well as the care taken in fabrication and protection of the pipe are of better quality than exists in average heavy construction. The pressure test under water at 10 percent above operating pipe pressures should reveal any deficiency in design or construction. Yet there are some dynamic forces that are not clearly defined and the steel embrittlement phenomenon is partly the explanation of sudden loss of strength of the completed installation.

The effect of sonic boom on buildings is becoming an important study with the imminent introduction of the supersonic airliners. A sonic boom from a low-altitude military plane did damage the Uplands Air Terminal Building in Ottawa, Canada, in 1959. A folded-plate prestressed concrete roof at Ft. Sills, Okla., undergoing a water load test in 1963, seemed satisfactory until exposed to a shock wave during ballistic firing tests, an overpressure that was enough to cause failure. Investigation of the debris indicated omission of some of the column reinforcement, but the structure had been proof-tested under static conditions. Concrete being placed at a Montreal airport in 1965 was found to set poorly and show low strength when exposed to a shock wave from the afterburner at jet plane take-off, before the concrete had hardened.

Fig. 5.2 Gas transmission pipe failure.

In 1960 the 20-year old asphalt pavement on the runway at Denver's Stapleton Field was ripped up by the jet blast of a DC-8 plane during take-off. Similar phenomenon has been experienced at the New York Kennedy Airport where small patches of asphalt cover have been lifted off the concrete runway.

In 1958 some 86,000 ft^2 of concrete-covered metal decking and light steel roof framing collapsed at a paper-box factory in Mexico City. The crash came suddenly, injuring 190 workers, with two fatalities. The vibration effect of the heavy machinery was the explained cause.

Damage to structures from blasting vibrations has been the basis of much litigation and has been so extensive that the General Adjustment Bureau of the Insurance Companies Association has included the subject in the curriculum of its adjuster training school.

Blasting effects are studied from two opposite points of view: (a) the most efficient concentration and location for breaking up the rock in the desired configuration and fragments, and (b) the restrictions necessary to avoid damage to existing structures. If all rock were of uniform density and isotropy, and all adjacent structures with same foundation support, some reliable answers to item (b) could be developed. The availability of new blasting agents and improved drills and equipment requires a continuous study of the item (a) as applied to each problem.

Damage possibly caused by blasting, however, may be the deciding factor in the most economical program for rock excavation. Much is available on each of the two problems in the technical literature, the first as studied by the producers of explosives and of equipment, and the second by insurance carriers, and more recently the Highway Research Board reports with the Bureau of Mines, and a similar joint effort by the National Research Council of Canada with the Hydro-Electric Power Commission of Ontario. A correlation of the two problems to show the economic as well as safety considerations by which each affects the proper solution of the blasting problem, would be a worthwhile contribution in this field.

Buildings damaged by pile-driver vibration while constructing the Philadelphia end of the Walt Whitman highway bridge across the Delaware River resulted in long litigation, finally resolved against the contractor. Before the work began the 30-year-old houses were inspected by the construction company's engineer and were photographed to show existing defects. Some damage resulted when piles were driven to a depth of 36 ft at points as close as 2 ft from one property and from 17 to 29 ft from the others. Trials in the Court of Common Pleas resulted in verdicts in favor of the homeowners, and the construction company appealed to the Supreme Court of Pennsylvania.

That Court affirmed the judgments holding that the construction com-

pany was guilty of negligence, on the premise that it should have known that the driving of the piles with such heavy equipment so close to the houses would cause vibrations that would damage them. The task of writing the opinion was assigned to Justice Michael Angelo Musmanno, not only a judge but the writer of a number of popular books, whose style of writing is livelier and more sparkling than that found in most legal opinions. In this case he said in part:

"A pile driver, when properly harnessed and kept within the channel of its legitimate functions, is, like a steam shovel or a bulldozer, a mastondonic mechanical creature assisting man in laudable conquest over obstacles holding back worthy enterprises in civilization's progress, but a bulldozer with an unsteady hand at the wheel can leave its prescribed route of travel and reduce adjoining structures to tin and kindling wood. The vibrations of a pile driver can be as damaging as the gigantic blade of the bulldozer if the operator ignores the existence of structures within the sweep of its invisible fulminations.

"A giant striding by a kindergarten should tread softly. A pile driver operating in the vicinity of frail or unsubstantial structures should restrain or mask its thunderous blows to the extent necessary to avoid inflicting avoidable damage.

"The fact that the pile driver itself did not come into physical contact with plaintiffs' houses does not exonerate it from responsibility. Its invisible tentacles of terrestrial violence struck at the houses as certainly as a cannon shot hits its target. A pile driver whose operator ignores the presence of dwellings within the periphery of its vibrations is as responsible for the resulting damage as the bulldozer which leaves the road and knocks down adjoining buildings." (*Dussell* v. *Kaufmann Construction Co.*, 157 A 2d 740.)

5.7 Collision Damage

The contact between a moving barge or ship and a structure usually ends as damage to the structure. The Pontchartrain Causeway in Louisiana had been struck by barges, seven times between 1956 and 1966. Replacement items of the precast elements are kept available to permit rapid replacement. In 1958 a 150-ft concrete span at Topeka, Kan., was put out of commission by a barge. The next year the Hood Canal Bridge in Washington was damaged and some of its precast boxes sunk by a similar collision, with consequent serious delay in completion of the project. The fender around the pier of the Mackinac Bridge was dented by a ship's prow and as a

result it was discovered that the mortar intrusion of preplaced concrete aggregate had not provided a solid concrete. In 1964 the Maracaibo Bridge in Venezuela was damaged by a seagoing vessel that missed the main channel and hit a pier.

In 1966 the Raritan River railroad bridge was hit by a barge with serious effect on New Jersey commuter traffic, and the Poplar Street Bridge over the Mississippi River, under construction, had some of its falsework carried away by a string of barges. Engineering construction talent is used to maximum value in the repair and reconstruction of such bridge accidents. A recent example well illustrates this know-how. The Murphy Pacific Corp. built a 2500-ft multispan bridge in 1958 crossing the Sacramento River at Rio Vista, Calif. In 1967 a 10,000-ton freighter knocked down the 108-ft suspended span and an adjacent 36-ft cantilever section, cracked the concrete supporting piers, and damaged an adjoining truss. The bridge was back in temporary service almost immediately with wooden trestles and steel decking. New trusses were fabricated, assembled in full spans with reinforced-concrete deck, loaded on a 560-ton derrick barge, and moved 45 miles upstream from the assembly yard to the bridge site. Meanwhile the damaged piers and truss were repaired. The temporary trestles were removed and the new 36-ft span lifted into position and bolted securely. The sunken lift span was removed from the river bed, loaded on the barge and the replacement span placed into position. Fill-in strips of concrete pavement were placed while the steel was permanently connected. The total job required only 84 hr of continuous work by Murphy Pacific after that company was called in by the California Division of Highways, and for most of that time traffic was using the temporary trestle.

Equipment sometimes gets involved in these physical contacts with damage to both structure and moving item. In 1959 a 525-ft grain ship bumped into the 200-ft lift bridge over the Buffalo River, wrecking the span. Falling water after the crash lowered the stern of the boat, which became enmeshed with the bridge steel. Fuel and cargo were removed to lighten the boat so that it could be pulled free.

Trucks crossing the Muscantine Bridge over the Mississippi River connecting to U.S. Highway 92 on the Illinois side in 1956 smashed into the steel trusses and one span fell into the river, with one truck hanging by its rear wheels over the edge of the adjacent upright span. A tractor being hauled on a flatcar in a 44-car Illinois Central freight train came loose during the crossing of the single track bridge at Fort Dodge River, Ia. The 196-ft span bridge fell into the river 25 ft below, carrying six railroad cars with it. In the same year, 1964, a 28-yd^3 scraper working on an embankment in Seattle went out of control and ran backwards down three blocks,

demolishing two automobiles and ending in the corner of a hotel. Four stories of masonry collapsed into the street exposing the full height of the hotel rooms to street view. The wood framed floors deflected somewhat but stayed in place, even with the loss of part of the return wall where seven beams had been supported.

5.8 *Wind Damage*

Structural damage resulting from storms of hurricane, tornado, or typhoon intensity is to be expected although even these forces can be resisted by proper and well-known design procedures, except possibly for a direct hit. But normal wind pressures should not be the cause of so many incidents of repetitive nature. Buildings, usually those of small weight, and their components, as well as bridges and towers, have failed under winds that are normal in their own localities. Resistance to such forces require appreciation of the fact that wind forces are not uniform horizontal pressures. Uplift, suction, and torsional effects have caused most of the damage.

Parts of roof and wall assemblies have been sucked out by the uplift drag of a wind. In 1958 the three-year-old circular coliseum auditorium in Charlotte, N.C., lost half the aluminum roofing cover when a wind with possible 75-mph gusts passed over the area. The 20-gauge metal was nailed to expansion sockets set in the fiberboard insulation of the concrete dome roof, but there was insufficient uplift resistance. In 1959 the Jefferson Parish courthouse in Gretna, La., lost a 32-ft width of glass and block wall from the two top stories of the nine-story building during a normal windstorm and only two days after the receipt of an engineering report that the fasteners for the exterior block walls inside the glass facing were insufficient and hazardous.

In the new skyscrapers in lower Manhattan, where large glass areas with metal skin walls replace the traditional masonry enclosures, glass panes and in one instance a complete wall panel assembly have been sucked out on the lee side of the towers. There were a few cases in 1962 of glass panes being blown into the floors. The ¼-in. plate glass has been replaced by much stronger 5/16-in. thickness installed in all upper floors. In one office building 15 windows blew out in one year, and then all panes with any minute flaws were replaced. After a year of no failures, two more of the ¼-in. plate glass areas failed in a similar way; a center area about 2-ft^2 popped out of the 7- × -5-ft panes.

Metal panels of skin walls have come off in Philadelphia, Hartford, and New York, all during winds below the intensity assumed in the design. Usual failure is in the connecting clips. In one such incident, where full-

story aluminum panels sailed off the building to the street, an analysis of the details indicated that the manufacture had provided too perfect a seal for the air volume between skin and masonry backup. Therefore all panels, the replacements as well as those that remained, were perforated in the bottom projection to allow pressure equalization during the passage of low barometric storms. This building has been exposed to many similar northeasterly storms in the past 10 years with no further failures of the skin.

Concrete block-bearing walls during construction, with the roof bracing not yet installed and with no temporary support, are large areas of sails ready to take off under wind velocities far below the design assumption. A simple computation will surprise most people at the small amounts of wind load that can cause failure of a concrete block wall 20 ft high, many of which have blown over.

Structures of roof without walls are not common, but like a wing of an aeroplane will be affected by a wind. The roof of a covered open-sided parking shelter in Daytona Beach, Fla., consisting of steel trusses on three column supports, 100 ft wide and 111 ft long, was lifted in part at least by a wind in the summer of 1936, and collapsed as it dropped. Anchor bolts of one column were pulled apart, the center column under one truss was rotated 100 deg clockwise and the center columns of the other three trusses were rotated 170 deg counter-clockwise. Since the structure had no walls, no wind pressure was considered in the design.

The collapse of the Tay Bridge in 1879, a single track crossing for the main line between Dundee and Edinburgh cost 75 lives. As a result of a long and detailed investigation by a Royal Commission, who published several volumes of testimony, the standard for wind pressure design was then set at 50 psf on all exposed surfaces. It required 40 years of research and experience to relax that figure to a more realistic value of wind pressures. The bridge was probably designed to resist a horizontal force of 12 psf on all exposed faces, with no allowance for the areas of the railroad coaches. This figure came from recommendations by John Smeaton in 1759 as the pressure of a storm or tempest. The explanation was that the gale wind caused one or more carriages to swing and hit a critical member of the wrought iron truss. The divers report proved that the train and bridge had fallen together. Later analysis showed that a 40-psf wind was needed to overturn the bridge, but many connection details at the bracing ties would fail at half that loading. Use of such cast iron details became less common after the Tay Bridge collapsed. With a solid wood deck and large sized iron truss members, a gusty wind could overturn the bridge, which had little lateral strength. The exposure of the train-covering part of the span may have given the critical added lateral load.

The most dramatic bridge failure from wind action was the Tacoma Narrows suspension span, in 1940 the third longest span in the world, and in its four months of use, known as the "Galloping Gertie." It was not the first bridge to become a victim of aerodynamic instability; eyebar chain-suspended spans had similarly failed in normal winds more than a century earlier in north Wales. Several cable-suspension bridges without the diagonal stays used by Roebling in the Brooklyn Bridge required stiffening and correction, such as the Mt. Hope Bridge in Rhode Island, the Whitestone and the Thousand Island Bridges in New York.

However the photographic record of the torsional oscillations of the Tacoma Narrows Bridge made by Professor F. B. Farquharson did more to prove the necessity for aerodynamic investigation of structures than all the theoretical reports. The unsupported span was 2800 ft with two 1100-ft side spans. The two-lane bridge was narrow for such a span, only 39 ft between cable centers. Designed at maximum efficiency of materials, to meet a tight budget, but also based on wide experience in suspension bridges, the stiffening trusses were only 8 ft deep. The ratio of span-to-truss depth, a measure of stiffness, was 350, almost double the 168 ratio in the Golden Gate Bridge, which also showed some oscillation under cross winds.

The Tacoma Bridge was a daring design, since most long span designs restricted the ratio to 90 or less. That design, checked by wind tests on a scaled three-dimensional model, expected vibratory oscillations and even lateral deflections of 20 ft. None of the tests predicted torsional oscillation and during its life under severe wind, the largest lateral deflection measured was 4 ft. Before opening the bridge to traffic, diagonal cables were added to connect the stiffening girders to the cables at midspan and dash-pot buffers separated the towers from the floor system, all to absorb longitudinal vibration. A month before the failure back cables were added between the anchorages and the side spans. None of these protections avoided the main span going into a corkscrew motion which caused disintegration. The maximum estimated wind velocity at the time of collapse was 42 mph.

The reconstructed Tacoma Bridge, incorporating the original piers and anchorages, is four lanes wide with 60-ft cable spacing, 33-ft-deep trusses in place of the 8-ft-deep plate girders, adding 50 percent dead weight, and with slotted roadways to reduce wind action. The new bridge was opened 10 years after the failure.

Steel frames for narrow buildings are vulnerable to low wind loads, far less than are used in the design of the completed structure, when erected before the installation of permanent floors and walls to provide mass and lateral stability. In 1958 an 11-story welded steel frame in Toronto,

Canada, 215 ft long and 65 ft wide, was made up of columns spaced on 20-ft centers and girders spanning the full width. Connections had been welded through the ninth floor and all of the 1850-ton frame was erected with temporary cable bracing. Wind bracing across the building was provided with girder-to-column moment connections. In the longitudinal direction deep concrete spandrels were to be provided with the floor construction, which had not been started. With no one at the site, some wind action caused the top two stories at one corner to fall and the entire frame then fell like a pack of cards in the longitudinal direction. Not a single piece of steel in the wreckage could be salvaged; only one of the several hundred welds had failed. Without the concrete spandrels, the very light longitudinal tie beams at each floor level in the face of the walls did not have enough rigidity to brace the 11-story columns.

In 1965 the 270 × 100-ft Federal Building in Jacksonville, Fla., had most of the 11 stories of steel erected and temporarily bolted with some cable guys. Welding of the joints had not started and no floors had been constructed. A wind of possibly 75 mph threw over 300 tons of the steel separating it from some lower portions that stood up. A similar collapse in Hamilton, Ontario, of a three-story 600 × 225-ft framework in 1966 could have been tragic except that some 100 workers were off the job during the afternoon coffee break. The steel was being bolted with no floor or wall construction in place and a 50 mph wind did the damage. In the same month, under the same wind intensity, a section of four-story steel erected for the Duquesne University Science Center collapsed while it was being bolted up. The extensive experience in steel erection should prevent such incidents, but precautions to prevent lateral movement of columns, which immediately bring a condition of gravitational instability, are a necessary requirement in these jobs.

At the Utica, N.Y., airport in 1961, a standard design steel hangar for small planes had been completely erected and the metal roofing was being hoisted in bundles for assembly when a moderate south wind toppled the entire mass in a perfect alignment to the north. The wind instruments at the control tower, 100 yd away, recorded a south wind, building up to 12 knots with gusts of 22 knots at the time of the failure. Roof trusses fell to the ground undamaged; not a single bolt in the splices or connections was disturbed. The columns either sheared or pulled out the anchor bolts from the concrete wall footing. An analysis of the steel framing for wind load stability showed that tipping would occur under a horizontal load of 1.6 psf of all steel exposure. The usual conversion of load from velocity indicates that a 22-mph wind will produce 1.6 psf. The cables sent to the job for use as temporary guys until the metal walls were in place, were found in bundled coils, unused, under the wreckage. The horizontal stay bracing in

the plane of the bottom chord of the roof truss had not been erected, so there was no resistance to lateral load.

5.9 *Towers and Masts*

Towers and masts on land or in the sea are expected to perform their intended function under all storm conditions and are designed to withstand forces estimated to reproduce the effect of the maximum storm. Radio towers erected before 1916 were designed on the principle that high winds occur seldom; the risk was economically sound to design for normal conditions only. As reported by A. W. Buell to the Institute of Radio Engineers in 1928, the failure of towers in tropical storms when their antenna services were urgently needed, caused the United Fruit Company to order designs that would guarantee continuous performance. Considerable research, most of it checked by full-scale exposure to hurricanes in the Caribbean area, showed that three-legged flexible designs were most economical both for guyed and self-supporting towers.

Some spectacular tall radio towers have failed in recent years, mostly under unusal wind conditions. In 1944 an 85-ft FM Radio Antenna tower on top of a 55-story building in New York City's main center twisted about its vertical axis during an 85-mph wind. The corkscrewed steel frame did not collapse and the anchor bolts held securely. The possible damage to life and property if the tower had fallen to the street was tremendous. No one was hurt and the failure was not even noted by the press. The tower was removed and even without this exposure, the building itself takes on some disturbing oscillations during northeast storms, enough to make it impossible to write at a desk located in the top few floors, so that employees are regularly excused during such storm periods (Fig. 5.3).

With the growth of the radio and television industry towers became higher and failures more spectacular. In 1949 the then tallest self-supporting antenna tower 828-ft high at Spokane, Wash., broke in half during a storm. A replacement tower, made up of solid round steel rods in triangular array, reached 608 ft of its 826-ft total, when the top bent over and total collapse resulted. Tower heights soon exceeded the 1000-ft guyed towers completed in 1930 for the Federal Telegraph Co. in Shanghai, China. Four square angle framed towers, uniformly 6 ft^2 for the full height, each with eight tiers of cable guys, and located 2600 ft apart on a square array, carry a cable mat antenna which is counterweighted to maintain constant shape. This structure, now operated by the Communist Chinese, has survived a number of typhoons and is still in service.

In 1956 gusty winds in Cedar Rapids, Ia., toppled what would have

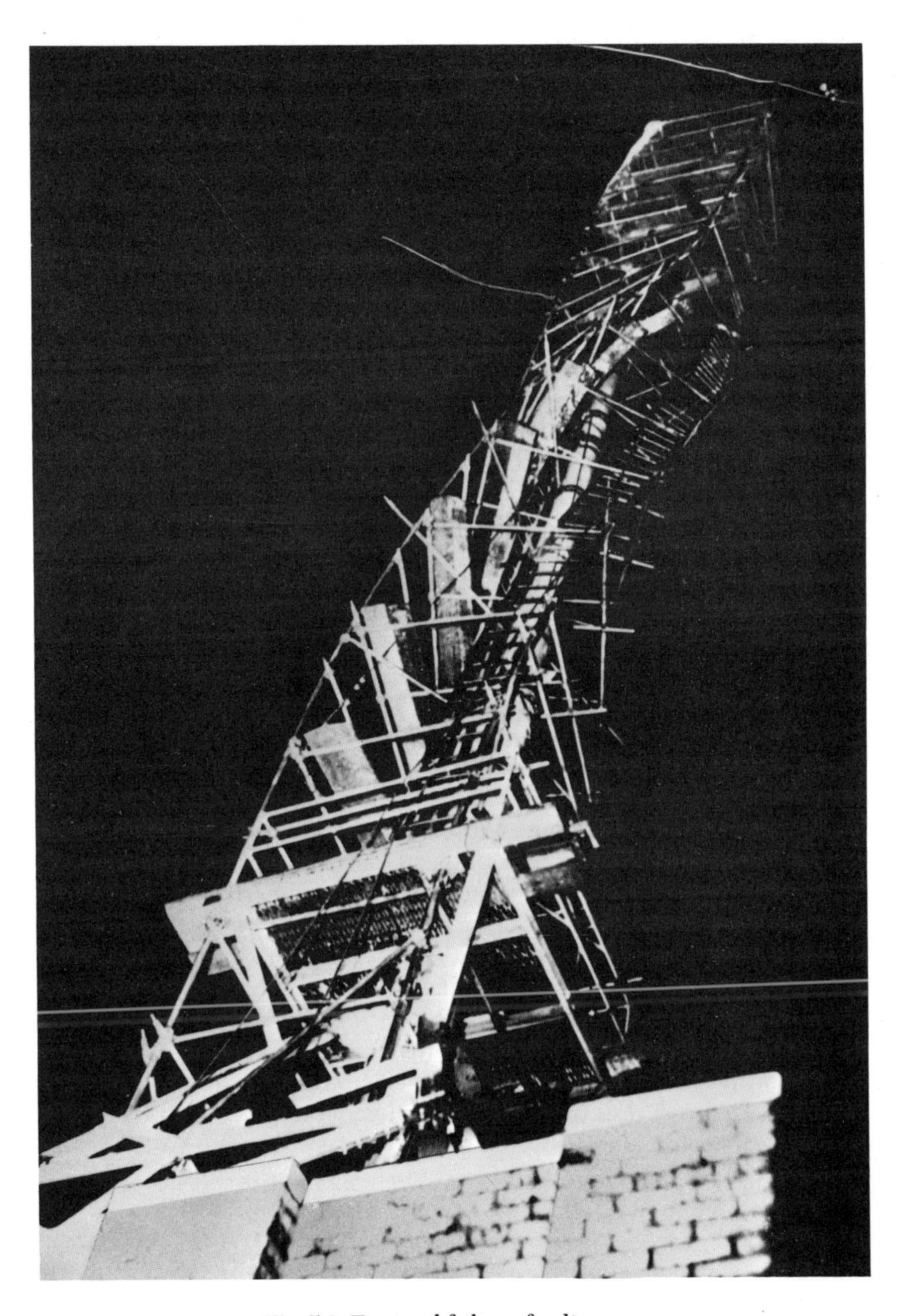

Fig. 5.3 Torsional failure of radio tower.

been the world's fourth highest tower. The TV tower had reached 1250 ft of the 1358-ft design when the top unguyed section fell, hitting guy cables lower down and caused complete collapse. In 1957 at Nashville, Tenn., a 1262-ft tower was completed and fully guyed when a guy snapped under dead calm conditions and the tower failed. The guy was of low carbon 90-ksi yield and 105-ksi ultimate strength steel. The redesign doubled the number of guys and reduced the allowable stress in the main legs.

Still higher towers were being built as the 1839-ft television transmitter at Roswell, N.M., failed in 1966 after only six weeks in use. Although designed to withstand a 100-mph wind, a reported velocity of 45–50-mph gusts snapped off 1430 ft of the tower, with no explanation available as to the cause of failure. Gale winds, gusting as high as 120 mph swept Britain in November 1966, taking as one victim a 950-ft television mast with some finishing work still to be done on the guys. It was located in an isolated site in Derbyshire, but the fall damaged the transmission building.

The failure of three 375-ft-high cooling towers at the Ferrybridge coal-fired power plant in Yorkshire, England, during a 75-mph wind on November 1965, caused a detailed analysis of these hyperboloid natural draft thin concrete shell structures. A battery of six units had been substantially completed, three collapsed in identical form and shape and the others cracked but remained standing. Four of the towers, including those that collapsed, were fully completed. The failed ones and the fourth tower with extensive cracking at the top, were on the leeward side of the battery of eight. The Committee of Inquiry preliminary findings and recommendations for reconstruction were issued in March 1966, after considerable public demand for a complete investigation of the design and construction. The *London Guardian* in an editorial of November 4, 1965, stated in part:

"British engineering is dependable and safe, and its practitioners set themselves high standards. But they cannot deny that the standards at Ferrybridge were not high enough. A gale of wind is not an unfair test. It is a regular event" (Fig. 5.4).

The towers were built under a competitive "design and construct" contract based on a wind pressure spectrum set by the authority. There was considerable precedent for this type of structure and several very similar units, of comparable dimensions, were built within 10 miles of this site. Those units, however, had been designed by consultants with competitive bidding for construction only. Although the Ferrybridge Towers were among the first of the 375-ft size, the specifications permitted a lower pres-

Fig. 5.4 Concrete cooling tower collapse.

sure spectrum than had been used in the design of the similar units. At 300-ft height, the wind pressure was set at 17 psf for Ferrybridge, while 24 was used for three other designs, 29 for two, and 31 and 35 for one each, all built since 1959.

Ferrybridge installation was in two rows of four and two in a north-south direction, with the east row stepped to the north so that its towers are located opposite the gaps of the west row. Each tower is a 5-in. concrete shell, 375-ft high, sitting on a base 289 ft in diameter. Spacing varies between 140 and 190 ft clear between the 10-ft-deep water tank bases. A concrete Warren truss 20 ft high sits on the tank and carries the 20-ft-high lower ring beam, 18 in. thick at the bottom and tapered to 5 in. where it meets the shell. The 335-ft high shell is reinforced with #3 bars at 10-in. spacing set in the middle of the wall thickness. The three tanks that failed left almost identical portions in place, jagged equilateral triangles 90 ft high on the downwind side of the ring beam. Almost all the debris fell

within the towers. The theory that funnelling effect increased wind pressures would be reasonable except that one failed tower was beyond the possible effect, and one that was within the effect, remained standing.

The Committee of Inquiry report was a finding of facts with no attempt to place the blame for the failure. The prime mode of failure was a vertical tensile failure within the lower part of the shell. Shear, buckling, and foundation failure were eliminated as prime causes but forced vibrations could have aggravated the inherent weakness. The wind loads set for design were too low by at least 25 percent. The use of a 1 min average wind velocity in place of the higher 10 sec average was not warranted. The unknown wind pressure effects of tower grouping must be considered, but the failure was instigated by the uplift draft within the towers. There was no factor of safety in the design for the unknowns of wind action. It is inadvisable to concentrate design and construction in one contract. Further study of dynamic effects of the shells, which were reported to flutter before failure, will be of great use in future designs.

The loss was not only the towers themselves, but also the delay in placing the completed power plant into operation by the Central Electricity Generating Board. The recommendations for reconstruction included advice to increase shell thickness with substantially increased reinforcement, on the existing bases. Future towers, however, should be spaced not closer than 1½ diameters between bases. The added reinforced concrete skin on the remaining towers required new piles and these were set into drilled holes to avoid the vibration from normal pile insertion.

Two material handling towers were blown away in 1956. A violent windstorm started the cableway tail tower at the Table Rock Dam in Missouri down the rails and it smashed over an embankment. The tower was 165 ft high and consisted of 675 tons of steel and concrete. The cable carried an 8-yd^3 bucket that was dragged across some formwork in the 450 ft of runaway travel. A giant iron-ore handling bridge near Cleveland, Ohio, traveling along a track with its 450-ft Pratt truss cantilevered 136 ft at one end broke its anchorage to the track during an 80-mph wind and sailed off the track to complete failure.

5.10 Storm at Sea

Collapse of the Texas Tower No. 4 off lower New York Harbor during a southeasterly storm on January 15, 1961, with the loss of 28 lives, resulted in extensive investigations to determine the cause, including a detailed hearing and report by the Preparedness Investigating Subcommittee of the Senate Committee on Armed Services. TT4 was one of three radar

platforms serving as an extension of the early warning system and located some 80 miles at sea. The Committee report dealt with the design, construction, and repairs as well as the safety of the other two similar towers, which have since been abandoned and demolished. TT4 was the last of the three built, and was set in 185 ft of water. The tower had withstood, not without some damage, two hurricanes. The report covers answers to the questions of possible design deficiency, construction deficiency, and repair deficiency in the attempt to restore the tower to its intended structural integrity after the hurricane incidents.

The design had been based on criteria carefully developed as to storm wave action. It had been determined that the maximum storm wave would be 40 ft high and after a feasibility study which was satisfactory, a design was developed of a triangular platform supported by three pipe legs. The record of two oil-drilling platforms in the Gulf of Mexico that failed when waves struck the platform determined the elevation of the bottom of the platform at 67 ft above M.S.L. A study of the storms of the past 20 years indicated a maximum wind velocity of 128 mph (which was measured during the 1938 hurricane) and the highest waves to be 60 ft during an easterly storm in November 1945. Theoretically, one wave in every thousand could be 50 percent higher, but the crest would be only 60 percent above M.S.L., which should have given sufficient leeway for all waves to clear under the deck. Yet the waves during the Hurricane Donna of 1960 carried away the undercarriage inspection structures at elevation 67.

Although all Gulf of Mexico platforms had legs in multiples of two, to prevent collapse should one leg be hit by a boat, it was decided to use only three legs for TT4 since additional supporting members increase wave forces as rapidly as the value of the additional support. A three-legged structure has no factor of safety since the loss of one leg by any cause results in complete and immediate loss of the platform. On that basis it had been recommended in the engineering feasibility study that some fail-safe emergency supports be provided, even with over-strong three main legs. Since it was felt, however, that the tower could be evacuated before collapse, if a collision seemed possible, the recommendation was not adopted in the final design. The depth of water made it impossible to dispense with bracing between the legs because of lateral deformation and drift of the deck, which had to be kept to close tolerance of position at all times.

The final design of the deck indicated a dead weight of 4300 tons and the fittings and equipment added 1200 tons. The three pipe legs are designated A at the south corner, B at the north corner, and C facing the ocean to the east. Legs A and B were used to store fuel oil and C was an intake of sea water supplying the evaporators. Each leg was a 12½-ft-diameter steel

pipe, $^{13}/_{16}$-in. thick, with a concrete footing 25-ft in diameter and 18-ft high located below ocean bottom. Within each leg from the bottom of the platform to 50 ft below ocean level, there was an 8-ft-diameter interior shell, with concrete filling the annular space between the shells. Horizontal pin-connected pipe bracing was located at three levels, 25, 75, and 125 ft below sea level, with tiers of K-bracing in each plane of the triangular base, connecting the mid-points of the horizontal braces with the main posts (Figs. 5.5, 5.6 and 5.7).

The design called for a novel erection method. The underframework was to be assembled horizontally and floated into location where, by means of controlled flotation of the empty mainposts, the frame would be set vertically into the sand bottom. The deck would then be floated into position, making contact with the three posts, jacked to proper elevation, and connected. This horizontal tow and vertical tip-up method was patented; the owner, an employee of the Architect-Engineers, waived his rights for this project. The completely fabricated framework was assembled, its three legs and all bracing in the final vertical and horizontal planes, except that the highest pairs of diagonal struts were lashed down so as to provide clearance for the deck structure to float into position. It was then towed horizontally to the site. A moderate storm delayed the tip-up process, and the two diagonals of the upper panel of bracing on the A-B side, which were temporarily lashed to the posts, broke loose, and were lost. The frame was set vertical and the platform was then towed into position between the three legs which were then only the $^{13}/_{16}$-in. thick steel shells; the concrete filling had not yet been placed.

Normal sea swells, about 3 ft in height, suddenly made the contact between deck and the legs, denting all three legs about 10 to 12 in. deep over exposed areas 10 ft high and 6 to 8 ft wide. Before correction of these dents, the deck was jacked up to the maximum extent permissible by the details. Since tide range was found to be 3½ ft in place of the assumed 1 ft, the legs were sunk into the sand 18 ft in place of the expected 20 ft, so as to maintain desired clearance at high tide. Repairs were then made for connecting new diagonals to legs A and B, working in 65 ft of water, strengthening the dented legs and then placing the concrete stiffening therein.

The facility was taken over for instrument installation in November 1957 and within a few months there were reports of considerable movement under 30-knot winds and 15-ft high waves, with frequencies of 15 to 18 cpm. The estimated design frequency of horizontal oscillations was 37–46 cpm, so that the integrity of the bracing system was questioned and it was decided that the top tier of bracing, partly replaced and all connected in place at the site, was not acting rigidly with the framework. Al-

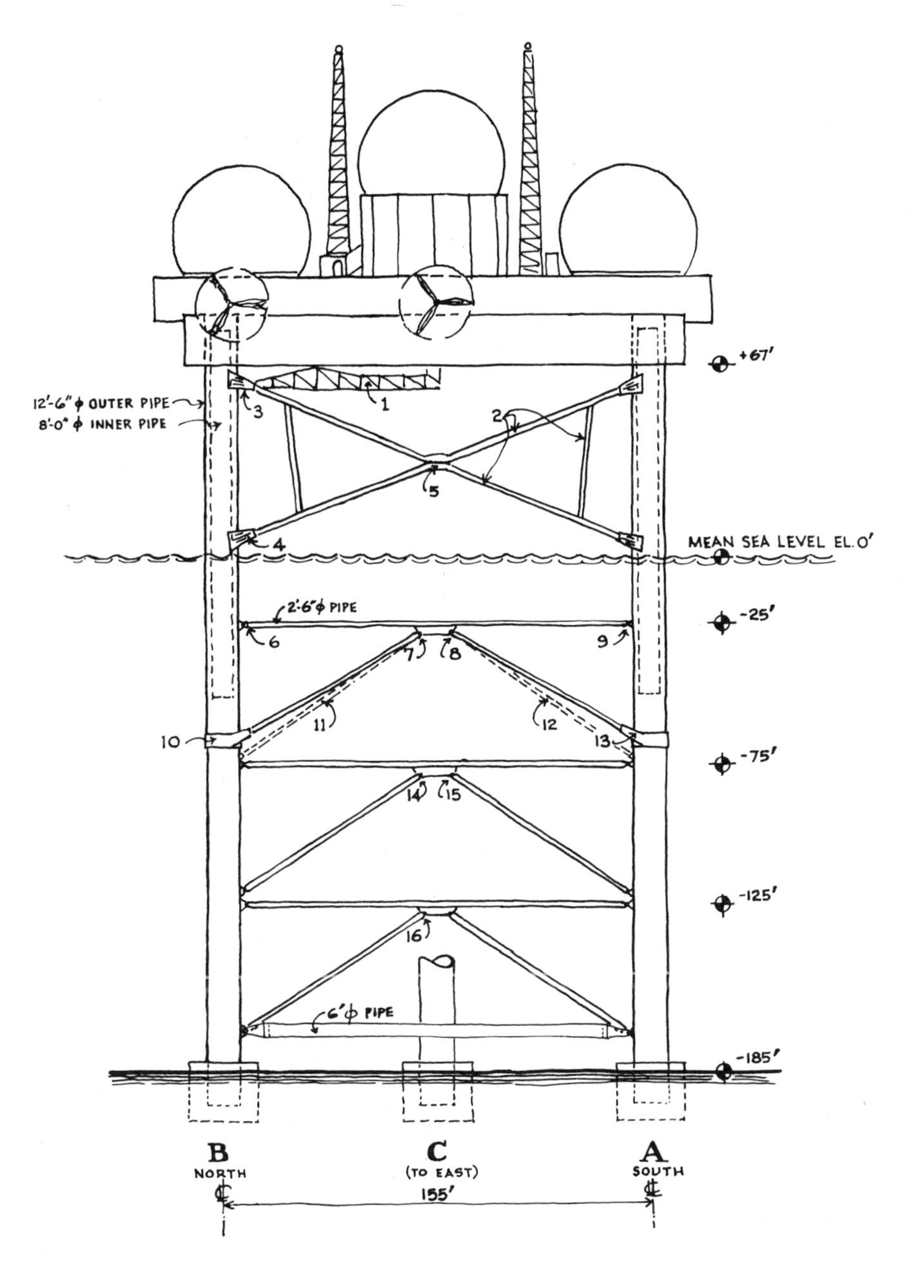

Fig. 5.5 Completed TT4 tower. See Fig. 5.6 for damages and repairs, keyed to this drawing. (Redrawn from diagrams in Senate Investigation Report.)

Date	Damage	Date	Repair
July 1957	Diagonals of upper K-truss tier, items 11 and 12, lost during towing to site	September to October 1957	Collars installed, items 10 and 13, new diagonals fitted (by Navy)
October 1, 1958	Failure of vertical shear bolts in collars at 10 and 13 reported by divers	November 1958 to May 1959	Vertical shear bolts in collars replaced (by Navy) using T bolts
February 1960	Slack in pinned connections, locations 6, 7, 8, 9, 14, 15, reported by divers. Slack greater than observed Oct. '58	May 1960 to August 1960	X bracing system (item 2) installed on all three faces, AB, BC, and AC, above waterline, to reinforce entire structure (by Air Force)
September 12, 1960 Hurricane Donna	"Flying Bridge" item 1, smashed against Caisson B	September 1960 to November 1960	Flying bridge repaired
November 11, 12, and 13, 1960 Divers' Inspection	Stress crack found in primary and secondary members of X bracing system at locations 3, 4, and 5	November 1960 November 14, 1960 to January 6, 1961	All stress cracks rewelded as soon as discovered Collar connection at 5 repaired
	Diagonal member fractured at location 13, collar gusset plates	November 1960	Diagonal removed
	Pin plates at locations 14 and 15, between horizontal and diagonal members, torn loose and moving freely in slots in horizontal members		Diagonal wire rope system designed to reinforce and replace upper two tiers of K-truss. Fabrication started December 15, 1960. Never installed
January 7, 1961	Diver reported diagonal brace at location 16 fractured at upper pin plate connection (Storm of December 12, 1960)		No repair
January 15, 1961	Total collapse in gale (55 knot winds), waves 35 to 40 ft high		

Fig. 5.6 History of damages and repairs to TT4. (See Fig. 5.5 for item and location numbers.)

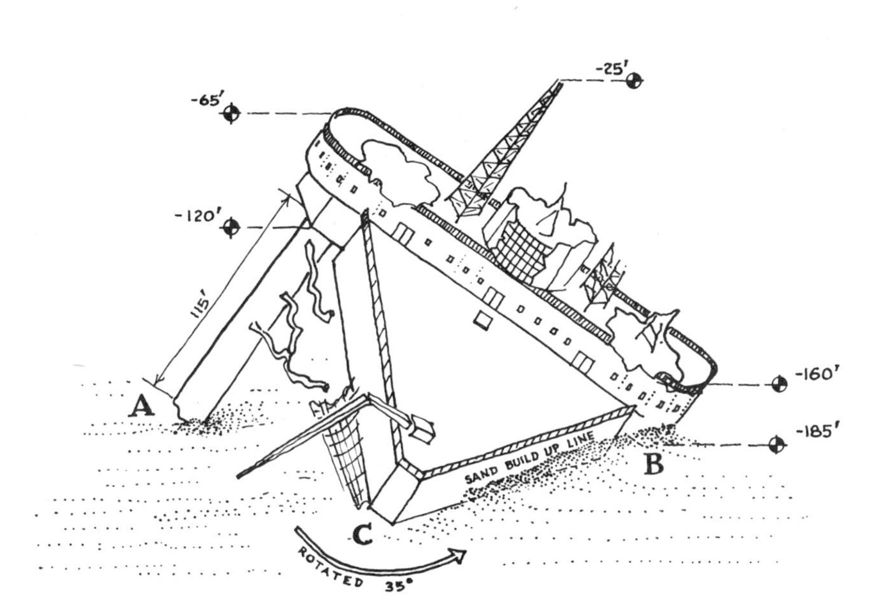

Fig. 5.7 Shape of collapse of TT4 tower. (Redrawn from diagrams in Senate Investigation Report.)

though on this assumption, computations indicated sufficient strength to withstand 125-mph winds with 36-ft-high waves, the tower was evacuated in advance of Hurricane Daisy in August 1958.

Instrumental studies made thereafter during several winter storms with 65-knot winds and 30-ft-high waves indicated that the natural tower frequencies were about 20-cpm translational and 23-cpm rotational, or about a third of computed values, so that the tower was much more limber than desired. The undersea bracing did not seem to operate until lateral movement exceeded 3 in. and rotations were greater than one-tenth of a degree. Hydrophone investigation proved that there was relative motion between legs and bracing, an indication that joints were loose. The greatest tower movements were caused by 10-ft waves, a loading far more critical than any aerodynamic forces.

Divers were then sent down and confirmed that the upper tier of bracing in A-B plane was not functioning properly, with eight bolts loose and some sheared on both legs. Since some increase in hole size had been permitted from the specified tolerances to permit assembly, the resulting rat-

tle space was blamed. New bolts in lieu of the unsatisfactory Dardelet bolts were ordered and all were in place by May 1959, the work being done between several storms. Tower motion was then studied in connection with radome operation and in September 1959 it was reported that the tower movement was less in magnitude than at any time since its construction.

By January 1960 operation personnel again complained of excessive platform motion accompanied by disturbing noises at Post A. Divers investigation showed that pins had loosened to as much as one inch from the ⅛-in. tolerance, and it was recommended that a system of bracings be added above water level to stabilize the position of the legs. This corrective measure had been discussed previously with the notation that above-water bracing would impose greater wave forces which might exceed the strength addition of the bracing. The X-bracing was installed as an emergency measure by August 1960, but no correction work was done at the pins in the below-water bracing.

On September 12, 1960, Hurricane Donna crossed the area and the tower was subjected to winds and waves exceeding the design criteria. Some light steel members on the side of the platform were dented and the inspection scaffold under the deck was torn loose. Bracing on the A-B side as deep as 75 ft below water level was fractured and torn free of the posts. During negotiations for repairs and for an evaluation of stability of the installation, the tower was partly evacuated leaving only a standby complement of personnel. A storm in December 1960 did further damage to the bracing.

On January 15, 1961, before the designed replacement bracing was installed and while a grouting of the legs to add stiffness was in progress, a winter storm with 55-knot winds and waves not exceeding 40 ft collapsed the tower and the 28 men aboard were lost. The aircraft carrier "Wasp" which had been waiting some 30 miles to the southeast, for a subsidence of wind to send helicopters to TT4 for evacuation of personnel, reported a train of three very heavy waves that must have hit the tower. The tower platform was located on the ocean floor 200 yards southwest of its original location, rotated in a counterclockwise direction through 35 deg. The A-leg still had 115 ft attached to the platform, the other two broke at the base of the platform. The footings did not move and leg B bent over, but the other two fractured at the top of the footing.

From the engineering point of view it seems that recognized criteria for the depth at which wave forces act must be modified. Also the height at which wave damage can be experienced must be increased. Theoretically synchronous dynamic action on the legs from waves 120 ft long, which are about 10 ft high, would make the design of TT4 most vulnerable. The max-

imum tower movements were measured during exposure to such waves. The distance from one leg to the plane of the other two was the same as the length of the wave.

Twice in the three years after the collapse, parts of the bracing pipes floated up and were salvaged. The welded diaphragms made the pipes airtight and a close inspection of the ends showed perfect welding but with tremendous distortion of the gusset plates during failure. Clean and shiny fractures on parts of the connecting plates proved that the bracing had been parted by wave action after the collapse. It is surprising that these items, all at least 100 ft below water level, could be so affected by wave action. The interior of the pipes still had the painted shop detail marks and positive identification of the components was possible.

There has been some, but remarkably little, damage to Texas towers in the oil drilling program in the Gulf of Mexico and elsewhere. An East Coast storm in March 1962 did catch a similar tower with its deck only a few feet above normal water level in the Chesapeake Bay. The tower had 100-ft legs and a platform 150 × 70 ft, and was used for driving large concrete piles for the trestle section of the Bridge-Tunnel crossing. The tower went over on its wide side when hit by 24-ft. waves.

5.11 Storm on Land

Design and construction to completely withstand the forces of hurricane, typhoon, or tornado may be possible, but except for some critical installations, it is probably not economically feasible. The extent of damage can be reduced, however, by an empirical analysis of damage sustained from such natural forces. The investigations of the National Bureau of Casualty Underwriters of damage caused during such catastrophes could well be the store of information for such studies. The Subcommittee on Disasters acts promptly to evaluate all damage after each incident but chiefly on a monetary approach of replacemement cost. A parallel study of avoidable damage could point to errors in design and construction that might considerably reduce the amount of damage and insurance risk, but would be a worthwhile guide to the avoidance of failures. The amount of damage is enormous; so also must be the available but not public data. In 1967, before the normal season of wind and hailstorms that usually comes toward the end of March, there were three major storms by February 15. A tornado and hailstorm struck St. Louis County on January 24, with insurance payments in excess of $15 million; a hard hailstorm followed by freezing rain hit Tulsa, Okla., on January 25, with 10,000 structures affected, to a total of $3 million; and a severe windstorm struck a large area from Illinois

to New York on February 15, with an estimated loss of $12 million, half in Ohio.

Winds to 100 mph with barometric pressure down to 29.1 hit Union City, Tenn., in 1952, collapsed a wood truss roof, and blew out the entire west wall of a 280 × 180-ft factory building. The wall disintegrated with the bricks torn apart. Timber truss connections were pulled through the timbers. Suction and uplift forces had not been considered in the design. The 1959 tornado that crossed St. Louis near its center did most of its damage to small houses, but it also toppled radio and TV towers and ripped the end quarter of the roof off the St. Louis Arena.

Grand Rapids, Mich., had 80-mph winds in September 1962. Concrete block walls for a 273 × 67-ft factory building, 12 in. thick and reinforced in alternate courses with welded steel ties, with some temporary bracing holding the 20-ft height before the roof was built, fell as a flat mass of loose blocks. During the same storm period a 75-ft steel-rib barrel-arch roof over a bowling alley in Abilene, Kan., took off and flew 60 ft, to crash into one of the bearing walls. Reports are regularly found in the newspapers of similar instances of wind damage when suction and uplift resistances are not provided.

Storms bring heavy rains and rivers flood, causing damage to bridges and shore construction. After each flood studies are made of the economic waste in not having provided channel protection and flood control. Planned but unbuilt projects totaled less in cost than the flood damage in the northeastern states in August 1955. The damage in Rhode Island alone was estimated at $170 million. The Delaware River lost four bridges connecting New Jersey and Pennsylvania. Some of the reservoirs for flood storage have since been built and saved similar damage, but the necessary protection of this heavily populated area is not complete. Damage to bridges in Connecticut was $30 million and resulted in careful study of protection needed against scour undermining of piers and against debris closure of the clearances under bridge decks with consequent toppling of the bridge.

Construction of twin bridges across the Big Sioux River north of Sioux City, Ia., in 1959 modified channel flow clearances. A flood in 1962 scoured out the soil around the pier in mid-channel even moving the supporting timber piles, and three spans of welded steel plate girders in the upstream bridge collapsed. The downstream bridge was not affected.

A week of monsoon type rainfall in the northwestern states at the end of 1964 resulted in floods that did $1 billion damage, but existing protective walls and reservoirs saved an equal value of possible loss. During the height of the floods every major highway in Oregon was cut, and 24 bridges were lost. Completed only a year, the $2.5 million bridge over the

John Day River near Rufus, Ore., lost its two 200-ft main spans when the center pier, founded on cemented gravel, fell over. The flood waters eroded the gravel base under the pier. Three older bridges in the same area remained standing, even though some of their piers were set on the same gravel layer. There was a report, however, of some preexcavation for the new bridge footing which left loose material around it susceptible to erosion. The choice of gravel rather than a more expensive rock foundation for the new bridge was dictated by the expected construction of the John Day Reservoir which would eliminate high velocity in the channel at the bridge location.

In mid-1965 a series of tornadoes and cloudbursts 60 miles south of Denver raised the South Platte River 15–20 ft above normal and started a sequence of bridge and railroad collapses. At 45 miles south of the city the flood carried away twin 200-ft concrete girder bridges at Larkspur. At Castle Rock, 15 miles downstream, six grade separation structures, four bridges, and two ramps, plus half a mile of a four-lane dual highway were washed out. At Sedalia, 20 miles from Denver, the town was devastated and two bridges lost. At Littleton State Route 88 crossing and a triple link of Route 70 were washed away as well as several smaller bridges. At the center of Denver a mile-wide stretch of warehouse and railroad occupancy was flooded. The piers of a 180-ft span were damaged, although the bridge was not affected. North of Denver one railroad and three highway bridges were lost near Fort Morgan, 80 miles away. A number of towns were evacuated and many railroad lines washed out. A total of 26 bridges and 700 miles of highway were taken out. According to the Corps of Engineers, the flood damage, estimated at $200 million, could have been prevented by the construction of two dams, one of which had been authorized in 1949 at a cost of $26.3 million but was not built because of opposition by real estate developers.

Typhoons coming across the open Pacific hit land installations with winds as high as 200 mph. Such a storm hit Guam in 1962 and inflicted $210 million damage, requiring upgrading in the resistance criteria provided for both private and military structures. Yet many buildings designed for 40-psf pressure on exposed surfaces, combined with 30-psf suction on leeward faces, and 30-psf uplift on roofs survived the storm without damage.

Hurricane damage to improperly designed structures in exposure areas has been well publicized. The high insurance losses have been instrumental in more rigid design requirements. Among the seriously damaged buildings in the 1926 Florida hurricane, were the 15-story Meyer-Kaiser and the Realty Board office structures. The former suffered considerable torsional distress from the unbalance between resistance and wind expo-

sure. The buttresses added to provide such balance on the face of the existing building, which has been through many storms since it was repaired, are still mute evidence that the analysis of the cause of failure was correct.

5.12 Lightning Damage

Protection of tall buildings against lightning is a recognized requirement, but incidents of damage from strikes of lightning on concrete pavements indicates that the reinforcement which is usually bonded by welding should also be grounded at expansion joints. In 1957 lightning destroyed parts of 700 ft of newly laid pavement in Elk County, Pa. An exposed steel bar was struck and all bars in that section were oxidized. Eleven holes about 2 ft^2 were torn in the surface. The 9-in. slab was believed to be so weakened as to require replacement.

In 1965 a bolt of lightning gouged a path in a Minnesota concrete roadway to a depth of 8 in., reducing it to rubble along the sides of the path. Recently lightning struck a pavement in Connecticut, dislodged a piece of concrete that flew in the air and went through an automobile windshield, killing the operator. This fantastic accident was observed and fully documented.

5.13 Rain Ponding Effect

With so called dead-level roofs getting lighter and spans longer, several cases of roof failure from water ponding should warn the designer to take proper precautions. In all these designs the roof leaders are normally located at the columns, where there is no deflection of the framework. Dead weight of conventional sheet-metal decking, roofing, steel joists, and beam framing is seldom over 15 psf. Live loads used in the design vary between 20 and 40 psf. Normal deadweight deflections introduce ponds in each bay which do not drain out. Sudden heavy rains add water that increases the sags so rapidly that failure of some minimal detail occurs and parts of the roof collapse. A number of technical articles have appeared to show that a minimum critical stiffness of the framing members is needed to avoid such collapse.

In 1960 one such building with a 14 psf dead weight, 422 × 505 ft in area, located in Philadelphia was exposed to a short but violent rainstorm when almost completed. A nearby rain gauge recorded 1.1 in. in 10 min and 1.32 in. in 15 min. Steel columns spaced 60 ft apart carried 17-ft long pieces of 21 WF from the ends of which were suspended castellated steel

beams 36 in. deep. Bents of these columns are 30 ft apart and carry high-strength steel open-web joists covered with a lightweight premoulded roof plank with subpurlin spanning 6 ft.

Under 30 psf live load, equivalent to 6 in. of water, all steel members were stressed to about 5 percent above the allowable 20,000 psi. Computed deflection for total load in inches were 0.2 for the subpurlins, 1.6 for the joists, and 1.1 for the girders. At expansion joints an 8-in. high cap acted as a barrier between sections.

Some 12,000 ft^2 of roof collapsed, mostly in one large area adjacent to an expansion joint, pulling the front wall down, and two smaller areas. All panels of the roof had ponds of 2–3-in. depth after the failures. The joists deflected enough to pull the sliding end off the beam at the expansion joint. In addition to rain loading, a suction from partial vacuum inside the building as the storm blew by and a water wave on the roof at the expansion dam probably combined to cause failure (Fig. 5.8).

A similar incident in Elmsford, N.Y., also of a just-completed building, proved the quality of the workmanship. When a section of the roof failed, the shock wave blew the concrete block side wall outward as a single unit onto the saturated clay soil. Although the masonry did crack along several

Fig. 5.8 Long span lightweight roof failure from ponding.

horizontal joints, strangely, the glass in the continuous row of steel sash did not break.

In July 1967 parts of lightweight roof constructions failed in Edison, N.J., and also outside of Boston, when steel joists gave way during heavy rains. High-strength steel joists have higher deflections than the normal types and are more susceptible to such failure.

A peculiar partial roof failure in a large storage warehouse in New Jersey ruined the finish of the concrete floor, which was placed after the roof was tight to obtain better control on the concrete work. The roof deck was an 8-in.-deep fluted pressed metal with welded seams spanning continuously over beam supports. No one noticed that the flutes had filled with water from the successive rains and snow. Some snow accumulation had been melted by flames that dried out the top surface and the roofing was then applied. A wet snow covered the roof and was being saturated by spring rains until the loading was close to 20 psf. The steel decking deflected under the combined water and snow load, the welded seams broke, and the trapped water cascaded down on the newly finished cement floor.

Rain ponding also collapsed the roof over the Playhouse Theater in Wilmington, Del., in 1925 but only because all the roof leaders were stuffed with debris and water was trapped behind high parapets.

5.14 Snow Load Failure

Familiarity with snow and its weight assures that adequate provision for snow load appears in all building codes, but usually only as a hypothetical uniform static loading. The large number of roof failures in the "Snow Country," apparently when the code requirements for snow load was not exceeded, indicates that assumed snow loadings are not level, or may not even be static. The 1953 National Building Code of Canada specified a uniform load equal to the weight of the maximum ground depth of snow recorded for the past 10 years, at 12 lb per cu ft, plus the weight of the greatest 24-hour rainfall. After extensive measurement of actual snow accumulation on roofs, the 1960 Code modified the specified loading to 80 percent of the 30-year ground snow load, plus the maximum 24-hour rainfall, with increased loading by 50 percent at projections on a width of three times the height of the projection but not exceeding 15 ft. The requirements also include study of the effect of unbalanced loads for curved and pitched roofs.

Actual snow load measurements made by Arthur E. Rowe at Cleveland, Ohio, show how much normal requirements in design may be wrong. With

local code specifying 25 psf for roof loading, on a generally level roof, and 18-in. snowfall with 10- to 20-mph winds made drifts 43 in. deep over an area of 25 × 300 ft. At the deepest point the snow weighed 53 psf. New snow left the depths the same but increased the weight to 65.5 psf. The unit weight of the snow increased to 18.7 lb per cu ft.

A tragic snow load failure occurred with the sudden fall of the roof over the arena at Listowel, Ontario, in 1959, while a peewee hockey game was in progress. The building was 240 × 110 ft, with 8-in. concrete-block walls about 20 ft high, buttressed to carry bow string glue-laminated timber trusses of 110-ft span. Roofing was wood on timber joists spanning 20 ft. The building collapsed completely, and so suddenly that the air shock developed by the fall of the roof caused a window frame of the front wall to fly 70 ft outward. The roof had considerable snow cover and at least twice that winter snow overhanging the walls had been removed, but none from the main roof area. The structure was six years old and had been constructed from sketchy plans and as a community project, without any building permit. There was no record of design loads and although the plan called for trusses at 16 ft spacing, the prefabricated trusses used were placed on 20 ft spacing, were 12 ft 9 in. deep instead of the 14 ft shown on the drawing and the top chord had only 10 laminations of planks instead of 11. On the assumption of a uniform 40-psf snow load, the maximum excess overworking stresses in the sizes actually used was only 6 percent, which would not explain the failure. Under a 60-psf load on the entire roof or on one side only, analysis indicated serious overstress of the timbers and especially of the connections in the trusses. Immediately after the collapse snow measurements on the roof showed depths of 24 to 28 in. with some 2-in. ice at the bottom, weighing a total of 55 to 75 psf., with maximum weight concentration on the north side (Fig. 5.9).

In 1966 the roof over a curling rink near Stockholm, Sweden, collapsed, but the splitting of a 2 × 8 noticed by one of 30 players, gave a signal for evacuation, and no one was hurt. The structure was a 50 × 150 wood truss supported roof with no walls. Moist warm air accumulated under the roof and froze on the ceiling, which with the snow imposed a load of 60 psf.

An almost identical failure was the complete collapse of a warehouse at Eau Claire, Wisc., in 1962, observed by a truck driver as he was leaving the building. The first warning was glass breakage, followed by cracking of timbers, then the roof and wood trusses fell down and the walls fell outward. Snow cover on the ground was 32 in. The trusses spanned 60 ft and were spaced 20 ft apart. Bottom chords had been previously reinforced with ¾-in. plywood gussets when the joints opened up some ½–1½ in.

During heavy snows in February 1967 several roof failures were reported in the Midwest and Middle Atlantic areas. In Baltimore 100 per-

Fig. 5.9 Snow collapse of timber arch roof. Copyright by Star, Toronto.

sons at an early morning mass escaped with minor injury when the roof of a Catholic church collapsed. Timber trusses spanned 53 ft and were supported by brick-faced concrete-block bearing walls. Roof sheathing was 2-in. plank spanning 7½ ft between trusses. Because of the pitch in the roof, local rules requiring 30-psf loading were reduced to 20 psf for the roof design. The roof was completely snow covered, but was reported to have survived deeper snows in its 15 years of life, but not necessarily heavier.

Following a two-day 24-in. snowfall, two roof collapses were reported in Lansing, Mich. Over a bowling alley bowstring roof trusses of 170-ft span, carrying some 5000 ft^2 of roof, crashed down during a windy night. The roofing was plywood on timber purlins spanning 20 ft. Design load was 30 psf, but the winds piled the snow in the end bay which was a flat roof supported by the bottom chord of the first truss, which failed under the combined vertical and lateral load. Also at St. Paul, Minn., about 33,000 ft^2 of the education building at the Minnesota State Fairgrounds collapsed under a 5-in. layer of ice. Design load of the new steel girder and precast concrete plank roof was 40 psf.

Accumulation of ice formation on the exposed rock faces of quarries and

deep excavations is a hazard often disregarded. In the deep cuts for the Niagara conduits of the New York State Power Authority ice removal every morning during the winter was a major operation to protect the workers from icefall. In March 1963 at a granite quarry in Barre, Vt., accumulation of such ice on the walls fell as an avalanche and demolished the warming shack in which the workers were taking a coffee break, killing several of them.

These examples indicate the necessity for a more realistic evaluation of the weights of accumulated and drifting snow when saturated by normal melting and freezing as well as by rain.

5.15 Earthquake Failure

Serious earthquakes occur somewhere on earth on the average of more than one a month. Modern communication and transportation facilities now permit prompt on-site inspection of damage and other effects of earthquakes from which valuable lessons can be learned. But classification of intensity by the extent and severity of damage to structures is not altogether valid. When construction is based on new designs that incorporate the latest theoretical and empirical knowledge gained from the effects of prior incidents, an earthquake of a given intensity will cause less damage than heretofore. In effect vulnerable areas are being inoculated against damage, although seismology still has a long way to go before the location and intensity of earthquakes can be accurately anticipated. Very little progress has been made in predicting the time or place or intensity of future earthquakes.

Because damage is used as a criterion of recorded scale or intensity, many shocks centered on uninhabited areas are not included among those evaluated. Some of the damaging and well-investigated earthquakes in recent years were recorded in Seattle, Wash.; Rat Island, Alaska; Niigata, Japan; Anchorage, Alaska; Skopje, Yugoslavia; Bagdan, India; Barce, Libya; Northland, New Zealand; Ghazvin, Iran; Acapulco, Mexico; Westport, New Zealand; Irpinia-Sannio, Italy; and Apulia, Greece; Concepcion, Chile; Agadir, Morocco; Lar, Iran; Mexico City and Acapulco, Mexico; Mazandaran and Hamadan, Iran.

Some earlier well-recorded incidents were Arvin-Tehachapi (1951), El Centro (1940), Long Beach (1933), Santa Barbara (1925), and San Francisco (1906), all in California; Napier, New Zealand (1931); and Tokyo-Yokohama (1923).

Of 83 countries listed with earthquake records, 62 are of major seismicity. Through a great amount of research in these countries on struc-

tural design for earthquake protection, numerous codes, some with considerable authority, have been established. The approach is usually from a theoretical development of minimum forces to be resisted, considering not only quake intensity but also the mass, shape, and type of structure and local geology. Empirical analysis of damage actually suffered seems to indicate that too much reliance is based on the axioms or assumptions (usually greatly simplified to permit mathematical derivation and to save time in preparing the designs) from which the code provisions are logically developed.

The scientific approach to earthquake resistant design is valuable in its broad application if the assumptions are consistent with the local conditions.

A complete theoretical solution of a structural frame is possible as a dynamic analysis of energy absorption by a definite assembly of structural elements, with joint rigidities known, and for a definite pattern or spectrum of earthquake shock. The Japanese have set up programs to solve the problem by computers for a number of shock spectra taken from vibration records of several earthquakes. A more usual design procedure is to select a percentage of the mass as the value of a horizontally applied force, in each of the major axes of the structure, and apply static structural design to evaluate necessary member sizes and connection details.

No one knows what the spectrum of the next earthquake will be. Statistically a 45-sec duration of the intense shock is a sufficient assumption; yet the Anchorage quake in 1964 continued with fairly uniform intensity for 330 sec. Much of the damage at Anchorage would not have occurred if the quake had followed the statistical rule. A spectrum analysis of serious earthquake records shows ground movements equivalent to 0.50 g (50 percent of the mass weight), yet no design code requires more than 0.13 g for areas of maximum earthquake intensity (Zone 3). Actual designs are usually for somewhat less, and many have survived serious earthquakes without damage, although the Caracas, Venezuela area in 1967 suffered serious damage in carefully designed structures.

The engineering approach to earthquake-resistant design must recognize limitations of ground conditions, available materials and labor, historical building customs, proposed occupancy, and the economic value of surplus strengths and factors of safety. The best earthquake-resistant designs cannot escape damage when the soil below the foundations is liquefied by shocks as at Anchorage and at Niigata. Some similar soil movements have been noted in earlier quakes of long duration. The effect of even weak vibrations, such as those caused by traffic or construction equipment, on loose saturated soils is well-known.

Available local material and labor must be used. To insist on special

steel framing in some localities will only result in the continuation of local building practices even though they are unsafe. Wood frame houses may be safer than the rubble stone walls laid with mud joints that collapsed in Libya, but only a masonry house is acceptable as a home in the Mediterranean basin. The houses with mud and adobe walls and heavy earth-mounded roofs collapsed and killed 12,000 people in Iran. Concrete houses would be safer but they are not economically feasible there. Similar earth adobe and brick walls with some timber framing behaved well, however, in the Bagdan shocks. Engineers must follow local procedures in design and upgrade them with available local materials.

The Turkish building code enforced in Palestine before World War I, where buildings were of stone walls with stone or ceramic domed roofs, required that the roof be built first on four or more masonry columns, and then the walls be filled in as nonload-bearing enclosures. When the flat-domed masonry roofs are supported on the walls they tend to push them outward with a complete collapse of the building when subjected to shock. The flexibility of corner columns may be the solution in providing the necessary resistance.

Flexibility in design to take up shock energy is a recognized necessity. The amount of flexibility permitted depends upon the use to be made of a given structure. More is tolerated for a factory or commercial use than for living purposes. Also important is the choice of nonstructural elements that can tolerate the deflections and deformations of the supporting framework. If the partitions, wall coverings, glass panels, and other parts of a building will crack and fall while the framework safely flexes, the structure may not collapse but there could be serious damage to life, limb, and property. Structural materials with some flexibility are available, and sliding connections can be made between partitions and structure so that the partitions will not absorb energy, and in effect, will float during a shock.

The assumption made in most codes that only horizontal forces are to be resisted in earthquake protection is contradicted by the damage that has resulted from seismic shocks. The 1964 French code recognizes the necessity for considering vertical accelerations in the recommended dynamic response analysis. The flexibility of the typical Japanese house, in both frame and partitions, is an example of consistent design. In a structure combining elements of different stiffnesses the most rigid will absorb the shock energy first, and if not strong enough to resist all the forces with no help from the other members will collapse and start a chain reaction of failure if the shock continues. If the entire structure moves as a unit, there is much less danger of collapse.

Local earthquake damage, if not repaired to reconstitute original

strength or if reinforced to give too much rigidity, only leaves a weak link for serious damage in future shocks. Such cumulative damage has been found in Italy and in the 1957–1962 Acapulco shocks.

The economical decision of how much can be invested for shock protection requires further study. The survival of the Latino-Americana Tower in Mexico City, when it went through a serious earthquake unharmed, paid many times for the cost of the original resistance provided. The secondary damages that would be caused by the failure of a dam or a main-line bridge justify greater expenditures for seismic protection for them than for an unimportant structure. Small expenditures for proper construction and uniform stiffness (rather than the normal criteria of uniform strength) should always be economically warranted.

A great deal can be learned from careful analysis of observed damage. From personal inspections, a number of observations and analyses are reported by the writer. At Anchorage proximate cause of damage in the 1964 earthquake can be classified as to soil movement, built-in weaknesses, and shock effects. It is difficult in many cases to distinguish the amount of damage resulting from each cause, and in many structures the plan of what was intended to be built is not available. Enough reliable data have been collected, however, to allow some general rules to be drawn up as guides for the designer.

Built-in weaknesses were fairly prevalent in the structures that showed damage. There is nothing like an earthquake to bring to light any deficiencies in design, any poor construction, and any deformation incompatibility between adjacent elements. Perfection in any construction job is seldom obtained, and in places where labor is scarce and the construction season is short, it certainly should not be expected. Design and details must then be so developed that continuity of the framework, jointing, and shear resistances will not be seriously degraded by normal local construction practices (Fig. 5.10).

Many errors, evident after a failure, were found. Designs that may be economical and architecturally "modern" may depend on perfection in welding and on an unattainable stress pattern. If rigid shear walls are to absorb horizontal forces, they must be so placed that the center of resistance coincides with the center of mass of the building. The provision of shear walls on three faces of the Penney Building, a five-story, reinforced-concrete frame covered on two walls with rigid precast facing, proved to be insufficient protection. The wall fell off the fourth face and torsional forces sheared several structural elements. This building had been designed for Zone-3 earthquake resistance, but did not survive.

Rattle spaces between sections of long governmental buildings were provided so that the vibration of one unit would not hammer against an-

Fig. 5.10 Flat plate lift slab failure at Anchorage.

other. The vertically extended Anchorage Westward Hotel was directly connected to the older six-story unit and considerable damage occurred at the junction with the shear walls of the taller section. The steel columns of the structural frame were embedded in the ends of the shear walls and the incompatibility of the rigid and flexible components caused much distress.

Weaknesses at concrete pour lines were found everywhere. Perfect bond and continuity cannot be expected, and shear keys and dowels must be provided to tie successive pours. This condition was found in both private and military work.

The Air Force Hospital was damaged, and part of the facility was put out of service. The earthquake damage showed up the causes. The concrete frame was designed for Zone-3 resistance, with the center core providing lateral resistance. Foundations were on stabilized gravel fill; the core on a mat and the rest of the columns on spread footings. It is questioned whether the flat-faced smooth-surfaced gravel, locally available, could be consolidated to the desired density. There is no doubt that the center core had a differential settlement of 1½ in. long before the earthquake and that the marble main corridor facing had been displaced. The floors showed the repairs to compensate for the deflection. Such a settlement in a rigid frame threw a load onto the next line of columns, which, with their designed factor of safety, were still sufficient. The connecting beams were very deep to allow ducts to pass through them. In at least two of these beams, where serious distress was noted after the quake, the concrete below the duct can only be described as discrete. Rubbish that had accumulated in the forms had not been cleaned out before placing the concrete. Two seriously damaged columns were found to be composed of a 2-in. concrete shell surrounding an almost solid wall of rods and loose dry gravel within the core. The rods were carrying too much of the load and failed in compression. The diagonal cracking of the concrete-block wall panels filling in between the beams and columns merely proved that concrete block will fail under deformations that a reinforced concrete frame can tolerate.

Vertical acceleration had been completely disregarded in the designs. Earthquakes cause compressional body waves moving radially (called the P waves). Coming at a slower rate are the shear waves with similar propagation but with action normal to the direction of motion and with up-and-down components. At Anchorage there were vertical ground motions, and the change in vertical loading caused failures in columns. There is some difference of opinion as to why a corner steel column at the Cordova Building failed by vertical collapse. This steel I-column was 14 × 6⅝ in. with $5/16$-in. flange thickness. Being in a stair hall, this narrow section was

used so as not to interfere with stair clearances. It is not the column normally used in the bottom floor of a multistory apartment house.

Construction plans were available for every building in the Elmendorf Air Base. Among them were three almost identical three-story-and-cellar steel-framed concrete buildings, which were designed as four separated units. In the lowest floors of one of these buildings considerable distress was noted. The explanation was evident when it was found that an extension for one office area was connected by a rigid concrete corridor which crossed and completely locked an expansion joint in the original structure.

In a steel-framed hangar structure, with distress localized along one wall, the bottom of the diagonal bracing was found to be loose. The bolts were on the floor, and paint covered the holes in the angles. The bracing had never been connected.

Wherever serious concrete cracking was found, the concrete appeared to be of inferior quality, with loose uncoated gravel and insufficient concrete covering on the reinforcement. The low masonry buildings in Anchorage showed very poor mortar jointing and complete dependence on rigid block walls to resist movements far in excess of what such masonry can withstand. Welded connections are certainly not flexible; this was proved in the collapse of a lift-slab apartment building with floors welded to steel columns. Similar failure occurred in two flexible precast roof decks with token welding at the supports. Many more such instances were found.

Not all the damage can be attributed to poor construction or noncompliance with intended design. Structures geometrically close to a cube suffered the least damage, the Penney Building being an exception. Long narrow buildings, even with expansion joints, showed the greatest damage. Such structures had common footings for columns along a joint, and the roof detail seemed to be lacking in freedom. In several of these buildings, with wings separated and running normal to the length, no damage was found in the wings but usually both ends of the front were damaged.

Some interesting and peculiar movements were found in various structures. In a warehouse with circular pipe columns the celotex ceiling, 10 ft above floor level over an office area, had circular holes cut in it for passing the columns. After the earthquake the circles had been squeezed into ellipses with the east-west axis enlarged to 16.5 in. symmetrically at both sides of the column; the north-south axis remained at the original 12 in. Lateral movement at this height was therefore over 2 in. each way and probably twice that much at roof level.

Metal framed stock shelves anchored to the floor drifted 2¼ to 4 in. southward shearing the bolts, when fairly well filled with stock, but did

not move when empty. The horizontal force acting on the lighter empty shelves was not sufficient to shear the bolts.

One freak observation reported was that in a two-story wood frame officers' quarters a heavy grand piano, set against the wall with the rear legs on the floor and the front legs on carpet, "walked" in a series of neat hops to the center of the room without tipping two candlesticks and a tall vase which were on the piano.

The most mysterious movement was the rotational fall of several glass-doored bookcases, full of books, flat on a concrete floor without breaking a single glass panel. One general observation of the writer is that the structural element that suffered the least damage at Anchorage was glass.

At the International Airport two men remained at their posts in the control tower and continued operations while two planes landed during the first few minutes of the quake. One man gave up the precarious job and ran down the stairs, only to be crushed when the tower fell. The other rode down with the falling tower and escaped injury. The tower was a total wreck, although its earthquake-resistant reinforced concrete frame had served well for several minutes (Fig. 5.11).

The concrete communications building and the steel transmission

Fig. 5.11 International Airport control tower at Anchorage.

tower suffered very little damage and would have been able to remain in service continuously after the electric power failed if the racks of wet-cell batteries on open steel-framed stands had not toppled over.

An analysis of the plans of the long storage warehouses, 200 ft wide and comprising four and five 200-ft units, showed the expected action. All units were separated by expansion joints and fire walls. Most had wood roof decking on steel beams supported by pipe columns in a 40 × 33-ft array. Diagonal sway rods are in all exterior panels of the roof in each unit. Fire walls are concrete with brackets for beam seats. Beams at each fire wall are through-bolted. Generally this structure, especially with metal exterior walls, behaved well.

In a modified design several warehouses have 2½ × 6-in. tongue-and-groove decking laid at 30 deg, with no sway bracing. The fire walls are concrete block and the beams rest on the wall. The exterior face of the concrete block walls at each girder support has a block buttress made of hollow blocks with some vertical rods and filled with concrete. In these structures the roof girders moved laterally and kicked the top of the wall out. Sections of the wall at the top fell out in a regular pattern as if a horizontal wave 200 ft long had twisted the roof. In one warehouse the second unit of five fell 12 ft north and the fourth unit fell 12 ft south. The other three units showed no substantial damage, and the railroad track along one side of the warehouse remained straight and level (Fig. 5.12).

In Anchorage, about a mile apart, there were two practically identical reinforced concrete apartment buildings, 14 stories high, that were built more than 10 years before. They had received some minor damage in an earlier quake and had been repaired, but there is no record of what correction was provided. Both buildings suffered identical damage—serious shear cracking of exterior columns and of all wall panels. It is possible that the earlier cracking of the earthquake-resistant designs had seriously degraded their strength.

The few examples given here show that the true character of the forces to be resisted in an earthquake are quite complex; therefore the theory of earthquake resistance must, for the present at least, be based on assumptions to which neither the earthquake nor the structure will conform. The other approach is an empirical analysis attempting to incorporate deformation compatibility in a structure designed for stress sufficiency. Forces during an earthquake are not static, they may come from any direction, may even neutralize gravity, and are irregularly transient. A structure, to be completely damage proof, must either be designed to resist every possible force without overstress to yield point and must not deform beyond the elastic limit, or, as an opposite extreme, must be as flexible as a damp

Fig. 5.12 Warehouse roof lateral drift from earthquake.

cloth with perfect articulation, but somehow maintaining its shape at the same time.

Because damage is the result of excessive differential distortion, its control is a problem in strain rather than stress. Recent earthquake damage investigations seem to have had an effect on codes and design standards. The Navy Bureau of Yards and Docks in 1965 proposed modifications in resistant structural design, giving increased attention to vertical forces, stress path continuity in compatible materials, effective structural separations, interconnection between rigid shear walls and flexible structural frame, and better control of construction performance. An earthquake-resistant design code has also been proposed by UNESCO. The new serious approach to the problem should provide better protection against this ancient and unpredictable natural force for those parts of the world where such damage has cost thousands of lives and uncalculated physical loss.

6

Masonry

Wall Failures; Construction Errors; Aging; Joints and Cracks; Water Tightness; Masonry Cladding; Partitions; Ornamental Screens; Plaster.

One of the oldest construction systems is an assembly of stone or brick set dry or with mortar to enclose space or support bridge decks. Many such structures have lived for thousands of years. This long experience with masonry has not eliminated failures during construction or after relatively short life.

6.1 Wall Failure

The ability of masonry to sustain load is well known but often the error is made that aging will not reduce the factor of safety. Continued use is no guarantee of safety, as was shown by the sudden fall of two arches when a stone pier in the 113-year-old Aricicia Bridge in the Alban Hills of Italy broke apart in 1967. The center part of this highway trestle consisted of three tiers of masonry arches between massive piers. The pier collapsed under the top tier and two spans disintegrated in the top two tiers without warning at night; at least two cars fell into the gap of the roadway. The companile in Venice, Italy, was constructed between the years 888 and 1517—not a rush job—and was under continuous repair. During the work in 1902 it was found that the successive additions of mortar in the grout within the walls was becoming the structural support. Before the condition could be corrected the tower collapsed.

One of the very few masonry crushing failures was reported in 1911 in Boston, Mass., where an exterior 7-in. cast-iron column with a 16-in.-square base plate sank 4 ft into the stone masonry pier, leaving an exterior shell of masonry that looked quite good.

Old masonry walls, weakened by the weathering of lime mortar joints, fail regularly, sometimes with serious damage. In 1922 the Majestic Theater in Pittsburgh, Pa., originally built as a church, lost part of its roof when the end wall, 13-in. thick brick and 50 ft high, bowed and pulled over a 60-ft span timber truss. In 1947 a brick pier of the Empire Apartments in Washington, D.C., collapsed after 54 years of satisfactory support. The structure was seven stories high with walls varying from 13 to 27 in. thickness. The only change in the original wall was a new opening cut 15 years before the failure.

Explanation of steel column settlement became apparent on exposure of masonry pier support (Fig. 6.1). A row of Civil War vintage houses in Parkersburg, W. Va., had been condemned as unsafe for occupancy but demolition was delayed in 1964, by legal appeals against the condemnation. The masonry could not hold out any longer, however, and the buildings collapsed. Fortunately large cracks had appeared at the base of the wall to warn occupants and passers-by to keep clear. A similar three-story building of the same age collapsed in St. Louis, Mo., in 1966, resulting in a local decision to demolish some 150 of these potentially unsafe buildings.

Old walls lose part of their facing or totally collapse from the accumulated effect of traffic vibration on a rough surfaced street. A 70-story warehouse collapsed in New York in 1930 when the wall moved outward and allowed the floor beams to drop. Parts of walls, especially areas adjacent to chimneys, have fallen in a number of old buildings facing railroad trackage. In 1959 a fatal accident at a school in Houston, Tex., was caused by the toppling of a section of brick wall 14 ft long and 8 ft high adjoining a gate. Although the wall was reinforced and tied to the footing, the rather small relative dynamic reaction of the gate action seemed to be enough to cause overturning.

6.2 Construction Error

Masonry block walls such as used for educational and industrial buildings are seldom really stable until tied by the roof and held by the floor. Unbraced wall lengths and heights exceeding normal practice are encouraged by economy, a false economy when the cost of failure cleanup and of litigation is added to the estimate. One common error that is noted is the hanging of work platforms on one side of the wall being built, with through ties and no exterior bracing to resist the overturning moment. A similar collapsing effect comes from edge-supported flexible steel joists being loaded by wet concrete. The slope of the loaded joists at the imbedded

Fig. 6.1 Masonry pier caused column settlement.

wall support, usually not over 4 in., cripples the block masonry wall. The lack of pilasters is then given as the cause of failure. In 1956 the 8-in. concrete block walls of a school building in Orlando, Fla., buckled when the 56-ft span double-T precast concrete roof members deflected from rain loading, and the wall was not as flexible as the imbedded roof members. A

similar failure at Waltham, Mass., in 1959 involved roof members of hollow slab section.

Masonry decorative panels on the walls of shopping centers, schools, and industrial buildings are often eccentrically supported on steel spandrel beams which do not have any counteracting load to provide torsional

Fig. 6.2 Wall collapse from tank-column push.

restraint. The steel beams, usually connected to the columns by not too tightly bolted web clip angles, rotate enough to tip the masonry off the lintel seat, or at best to laminate the wall and cause unsightly and dangerous bulging. Fortunately this usually appears before the building is occupied and corrective revisions are made.

In 1930 in Brooklyn and again in 1936 in the Bronx, N.Y., apartment buildings suffered partial collapse at the roof line from the same construction error. The six-story wall-bearing structures had roof tanks supported by steel beams set into the front wall. Where one such girder came at a window opening, a grillage beam was provided as a support for the girder, spanning the opening. To avoid encroachment on the window height, the grillage was placed on top of the girder instead of underneath, and connected by a pair of field bolts through improvised holes in the flanges. The tank load pulled the assembly and parts of several stories of brickwork into a rubble mass (Fig. 6.2).

6.3 Aging

Temperature fluctuation, rain exposure, moisture absorption, chemical alteration of the masonry units and of the mortar, elastic and plastic strains, all acting in many combinations, alter the appearance, weather-tightness, and strength of masonry. Design details and construction procedures must expect and compensate for the changes occurring during the reasonable life of the structure. Some changes, but very few, can be beneficial. The ancient stone structures, where the lower contact face was carved out so that only a narrow rim of stone was in bearing, with age became tighter as the plastic flow of the stone closed up the irregularities of the joint. Masonry units set in adobe mortar soon took on the appearance of a tight wall, as rain exposure eroded the earthy filler. Weak lime mortar jointing, although increasing in strength with age for several years, absorbed the expansion of the soft clay brick when it became saturated and the wall exposure became tighter. The recently adopted cement or masonry mortars when stronger than the brick do not permit such compensation and when the brick expands, the mortar cracks or crumbles, and if the mortar is too strong, the brick spalls and cracks.

All masonry units, from sponge-like porous cinder concrete block to almost impervious glazed brick, will absorb water in varying degrees. But the less porous the masonry unit, the larger the percentage of the moisture on the wall face that goes into the mortar joints. Saturation brings possible frost damage and where conditions permit, sulfate reaction. Where masonry is likely to be wet over long periods, the bricks should be of low sulfate content and the mortar made of sulfate-resisting cement. Porous brick or block should not be exposed to frost action.

A dramatic example of sulfate action was the sudden mysterious bending of the top 20 ft of one radial brick chimney, 140 ft high at a Westchester, N.Y., hospital. The similar companion chimney remained vertical. The

explanation was found after a study of use and exposure. Both chimneys had served well for about 25 years, one as the stack for the incinerator and the coal-fired hot water boiler, the other as the stack for three coal-fired heating boilers. Soon after the heating units were converted to fuel oil, the top of that chimney started to bend towards the northeast. The top was noticeably out of plumb, at least a foot, when a careful inspection reported no cracking of the brick or the mortar. The southwest side was growing longer.

The oil being consumed had a high sulfur content, leaving some sulfur dioxide fumes congealed on the inner face of the chimney near the top. Prevailing wet winds are from the northeast, which washed that side of

Fig. 6.3 Leaning chimney.

the chimney and concentrated water on the opposite side of the inner face. There was an internal shoulder in the brickwork 20 ft from the top and the sulfuric acid formed by the combination of the sulfur fumes and water accumulated up to that depth. The acid acted on the cement in the joints to form calcium sulfate, which takes up water readily and expands as it crystallizes. The expansion enlarged the joints in the southwest quadrant, pushing the chimney over into the wind without forming a single visible crack. The cure is not to use oil with high sulfur content and to eliminate internal shoulders. The same phenomenon of chimneys bent into the wind is almost universal (Fig. 6.3).

Masonry enclosures of roof areas that are exposed to high sun temperature or surfaced with materials having high expansion coefficients, must be detailed to give the roof freedom of movement, otherwise the masonry develops cracks, bulges, and unstable corners. Thermal change of masonry materials is not a completely reversible action. When expansion causes the formation of minute cracks or separation between brick and mortar, grit formed at the cracks as well as airborne dust partly fill the gap and on cooling the wall cannot come back to its original dimension. Every succeeding cycle thus adds to the total length and not many repetitions are needed to cause distress. Walls in excess of 100 ft in length have been known to grow, sliding on the foundation, until the ends of the building hang over the concrete foundation, which, being buried, is not affected by thermal changes to the same extent (Fig. 6.4).

Masonry must be built in separate units with isolation joints to give freedom of expansion so that cracking will not start or else be properly reinforced with steel. In an apartment house in upper Manhattan a roof garden was added and a red ceramic tile floor installed. After one summer the expansion of the tile pushed out the brick parapets that rotated about the top-story window lintels and cracked the brickwork at all returns in the wall to such an extent that the top-story tenants complained of daylight coming through the walls. An expensive wall reconstruction included providing a wide expansion filler between the tile and the parapets.

Thermal changes depend on sun exposure more than on ambient temperature. Corners of buildings have to act as hinges when one face expands more than the other and vertical cracks form at the first vertical joint. The length of wall to be considered for expansion is the developed length, not the projected dimension. An L-shaped plan of a building acts as if it were one long wall, and is in some respects, because of the hinge action, even more vulnerable than a straight wall. Vertical cracks develop in the reentrant angle made by the walls.

Fig. 6.4 Expanding building moves off its foundation.

Moisture saturation in the mortar joints of a high granite base course caused rusting of iron cramps inserted to hold the stones together. After some 25 years, the rust accumulation and expansion was sufficient to spall the granite, shearing out sections 4 in. thick. All metal imbedded in masonry should be noncorrosive. In the same building steel columns encased in brickwork rusted to the extent that laminations of rust over ½ in. thick were found when the cause of the vertical brick cracking along edges of columns was investigated.

Disregard of the effect of restraint, in other words omission of necessary expansion freedom, is the usual cause of the masonry distress seen in many buildings, more in recent structures than in older ones, which were built more slowly and with weaker components. The small amount of deformation in restrained condition that is required to cause failure may not be readily appreciated. Average values are for concrete: 0.10 and 0.01 percent, respectively, in compression and in tension; for brick: 0.20 and 0.016 percent, respectively, in compression and in tension; for stone: 0.25 and 0.007 percent, respectively, in compression and in tension; for steel: 0.13 percent; copper: 0.29 percent; and aluminum: 0.39 percent. Thermal

Fig. 6.5 Stone cheek wall moved by expansion.

change and moisture expansion can easily provide these deformations in masonry where freedom for movement is not available (Fig. 6.5).

Incompatibility between brick and block units making up a wall results in cracking along both horizontal and vertical lines in the brick facing. One of the earliest examples studied in detail was the Lincoln Park Homes development in Columbus, Ohio. Completed in 1942 the 26 two-story apartment dwellings have walls made up of face brick backed with load-bearing structural clay tile. After a few months the brick facing developed cracks which broke up the faces generally 3 ft apart vertically and 4 ft apart along horizontal lines. Cracks were pointed but reopened and new cracks also developed. The interior plaster cracked opposite to the wall cracks. Tests of the masonry materials showed high moisture expansion of the red clay tile backing, about 0.2 percent. Repairs consist of a thin cement paste with waterproofing admixture covering the exterior face, followed by a burlap bag wiping. Cracks redevelop and grouting is required about every three years.

6.4 *Joints and Cracks*

Depending on the firing temperature, clay masonry units absorb moisture in varying degrees from the air and from direct rain exposure. Water ab-

Fig. 6.6 Masonry cracking from roof expansion.

sorption always results in volume expansion, which unfortunately does not completely disappear when the water evaporates. At the same time all masonry increases in volume with a rise of temperature. If, in the expansion cycle, some fine cracks or separations develop within the masonry units or at the contact surfaces along the joints become dust filled, an equal drop in temperature will not cause a full amount of volume reduction. Minute ravelings and airborne dust enter the gaps and inhibit full closure. It does not require many cycles of thermal change to form visible cracks, usually darker than the masonry exposure and subject to frost damage if moisture gets in during the winter (Fig. 6.6).

Commonly called expansion joints, isolation separations between rather

limited lengths of masonry must be provided if cracks are not desired. The joint detail either works and protects the masonry from cracking and prevents the infiltration of air and moisture into the structure, or it does not work. The great majority of joints do not work and this is the true explanation of why walls crack in spite of joint provision. In one such building where mysterious cracking started the usual round robin of placing the blame the bottom of the joint in a two-story wall was of zero width, and the more visible top was a full 1½ in. In a large new multistory hotel with workmanlike jointing in the brickwork in accordance with the plan details a little probing proved that the expensive sealant was ¾ in. deep and backed with a very strong mortar in the rest of the 12–in. wall thickness, explaining why the brick facing was cracked.

Joint sealants must be compatible with the materials bounding the joint. The Johnson Wax Tower in Racine, Wis., has large areas glazed with 2-in.-diameter Pyrex tubing. In 1939, when the structure was enclosed, the recognized sealant was putty of lead oxide and linseed oil. It did not adhere to the Pyrex and the building leaked badly. A succession of newly developed sealants were then used without much better result. Preformed polyvinyl chloride gaskets were soon replaced by a synthetic rubber compound that lasted about three years. In 1952 a specially designed two-part sealant was pulled away by thermal changes, leaving appreciable separations along the tubes. Finally in 1958 the 133,000 lin ft of glass jointing was filled with a one-part silicone rubber by pressure injection. After about 20 years of trial and error a solution was found.

The bonding capability of the sealant to the materials in the wall must be available if a good joint is desired. The masonry, concrete, steel, or glass surfaces must be chemically clean and dry and at a proper temperature if a good joint is desired.

Masonry at roof lines, especially parapets, is too often noted in a state of distress. The two-directional shrinkages at corners, together with lifting of the corners of the roof slab, tear the masonry apart. Long-time deformations in concrete or in timber cannot economically be resisted. Only properly designed and constructed isolation details, allowing room for movement, will protect against such cracking. Compatibility of deflection between supporting members and the usually less flexible masonry loadings is necessary if longitudinal separations in the joints are to be avoided.

Compatibility, almost equality, is needed in the moisture and thermal change of the various masonry units making up a wall. An ornamental limestone facing for a library in Brooklyn was backed with clay tile to make up the main walls. The limestone showed considerable distress and the wall was removed and successfully rebuilt with a brick backing. No old

stone covered building is ever found with masonry backing of considerable lower rigidity or volume stablity than the facing material.

6.5 Weather Tightness

Ever since people have built masonry walls of earth materials, sometimes waterproofed with the dung of their domesticated animals, the purpose has been to keep the weather out of the enclosure. Thousands of years of experience and experiment have not yet solved the problem for many modern buildings. Sometimes the unwarranted use of new materials when a very low budget must be met results in serious failure. One such example was the use of lightweight concrete blocks for 8-in. walls of two school buildings in Long Island, N.Y. Economy and "functional" design left the walls naked in the classrooms and with a pastel shade cement paint exterior. With occupancy came drying and shrinkage of the blocks, to the extent that daylight was seen at many joints and the teachers could not use the seats along the walls. Litigation was started by the school board to collect the cost of repair and a waterproofing stucco covering of the walls.

Old buildings sometimes leaked, but when high-speed production methods displaced craftsmanship in the mid-1920's, weather tightness became an unusual characteristic of masonry enclosures. Just as some very-long-span bridges have permanently assigned paint repair crews, there are skyscrapers that require continuous joint and masonry caulking and repair. The tendency toward denser masonry units and stronger mortars results in minute separations along all brick or stone edges. Every square foot of brick wall has about 6.5 lin ft of joint. The walls of a building 100 × 200 ft in area 20 ft high, have over 15 miles of joint. If only 15 percent of the joints are slightly open, water penetration starts to enter the masonry. Although leakage starts with 30 percent of the joints open in the minute width of 0.005 in., which for each square foot of wall equals a slot 1 in. long by ⅛ in. wide. The necessity for absorption capability of the face brick and of the mortar is evident if the exterior surface of the wall is to act as a protection against the weather.

Cavity wall construction has been treated as a cure-all for masonry weather tightness, but is not the solution unless proper measures are taken to provide easy and continuous exit of moisture penetrating the cavity, and sufficient isolation jointing to prevent cracking of the masonry.

Curtain wall construction as a mark of modernity, rather than low initial cost has proved to be a bonanza for waterproofing maintenance specialists. Curtain walls are subject to all the forces causing trouble with

masonry walls, but more so. Since the coefficient of expansion for aluminum is 2½ times that of glass, when glass and aluminum are combined, appreciable differential movements are found even in one-story-sized panels. Wind-pressure deformation and suction effects have broken the sealant joint between large glass panels and the metal frame, have broken glass panes, and even sucked out full panels from the structural frame. Expansion and contraction, lateral movement, excessive wind pressure and suction all put added stress on the sealant. When the critical sealant is prone to hardening, shrinkage, loss of solvent or bond failure, a weathertight enclosure is not obtained. Some cases are described in Chapter 7.

6.6 Masonry Cladding

Masonry skins are placed on structural frames for weather protection of the enclosed space, for appearance, and as a protection of the structural elements against fire and against climatological change. Brick veneer covering on timber-framed houses has proven successful and of long life, if proper waterproofing is placed between the timber and the brick and if the brick wall, usually 4 in. thick, is self-sustaining on a proper foundation. Self-supporting masonry covering of steel-framed buildings became so massive as frames became taller, that the panel wall, supported by the floor constructions, became an economical necessity. The interaction of the frame and the more rigid masonry enclosure brought new problems. Any change in relative position imposed shears in the more rigid, and, in this case, also brittle, component. Normally masonry cover was not installed until the steel frame was topped out and much of the floor slab weight in place. In a number of very high steel-framed buildings, where vertical mortar joints at decorative corner bond or recess arrangements had not been provided, vertical cracks are practically continuous at all building corners, about 8 in. from the edges. These result from thermal changes on the face having the greater sun exposure, the corner tends to act as a hinge. Where quoins or ornamental pilasters were built, the vertical mortar joint provided the hinge flexibility necessary for the very small angular motion.

The vertical crack at the corners is usually close to a steel column. Moisture running down the wall penetrates the masonry, enters at the crack, and soon saturates any mortar cover on the steel column. Rusting of the steel then forms an expansive iron oxide which pushes the brick cover outward, aggravating the crack condition, and speeding up the cycle of disintegration of the masonry cover, especially where the cover on the column is only 4 in.

Elastic shortening of the steel columns from loadings applied after the wall enclosure is complete can seriously affect stone facing when no relief is provided in the horizontal joints. The common detail of lead joints at every second story for stone covering supported by steel framing seems to have been overlooked in the rush of construction starting about 1928. A monumental college building on Park Avenue in New York City built in 1940 was covered with half-story high limestone sheets, carefully set in thin high-strength mortar joints. Compression and some thermal movements from the exposure of 200 ft of wall to the west warped the face so that under the almost horizontal morning sunlight the north and south faces looked dangerous. The shadows of the small projections of each stone edge relative to its neighbor suggested much larger out-of-position displacement than had actually occurred. A number of the stones had to be removed and reset with lead joints. The somewhat earlier Pittsburgh Post Office and the New York Federal Office Building on Vesey Street had similar distress of the limestone cover.

The Veterans Hospital at Albany, N.Y., constructed in 1950, is a 10-story cruciform steel framed structure, with each of the four ends extended into an L-shaped unit. There are therefore eight end walls, about 40 ft width, two at each of the four cardinal compass points. The walls were limestone covered, with a joint at the corners. Shortly after enclosure of the building, after its first winter, cracks developed in the exterior corners, chiefly noticeable at the second-floor level but some extending the full height of the building with maximum width at about third floor level. On the east exposure all four corners were cracked; the south and west had three each, and at the north faces only the northwest corner opened up, showing the dependence on thermal exposure. None of the 16 interior corners, all of which were at 135 deg, showed any movement.

Masonry covering on concrete frames is subject to all the troubles found in steel buildings, and some others as well. By 1948 the New York City Housing Authority had 10 years of experience with brick covered reinforced concrete frames from 7 to 14 stories high, using stone or gravel concrete and solid walls of 4-in. brick backed with 6- or 8-in. tile or concrete block, supported at each floor level by suspended steel lintel angles. An inspection of nine projects consisting of over 100 buildings found corner cracks in the horizontal joint at the bottom of roof spandrel in over 50 buildings. They occurred in skeleton buildings and also with bearing walls where a continuous concrete distributing wall beam was incorporated in the roof slab. Cracks were concentrated close to the corners, but occasionally extended 10 ft along the horizontal joint. Parapets were all 12-in. solid brick and in one project numerous generally vertical cracks appeared close to all corners and at any offset in the line of the parapet. Later

project designs incorporated anchorage for the parapet wall to the roof slab with fairly close expansion joints or even complete omission of the masonry parapet to eliminate differential thermal and shrinkage movements between the concrete roof and the masonry. In the Mulford Houses Project in Yonkers, N.Y., in 1939 a 1-in. gap was provided between the face of the roof concrete and the masonry wall; all masonry cracking at roof level was eliminated.

Brick facing on tall concrete frames requires provision for differential dimension change in addition to lateral thermal length change. The delayed shrinkage and creep of the loaded columns, the elastic shortening from loads added after the walls have been built, especially where walls are started at lower floors before the framework has been topped out, impose compressive forces into the sheet of brickwork beyond the outer face of the concrete structure (Fig. 6.7). Where masonry headers connect the face brick with the backup, which is supported directly by each floor level, some of the induced compression is distributed into the floor construction. The shear value of the headers is the limit to such load relief, and in some wall failures, such headers are found sheared at the lower floors. Where

(a)

(b)

Fig. 6.7 Masonry distress from load transfer as column shrinks.

flexible metal ties are used to connect in place of headers in solid walls and always in cavity walls, no such relief is possible and the entire shrinkage effect of the columns is imposed on a 4-in. sheet of brickwork as a compression member. The steel angles, usually connected to the floor construction at each story and rigidly mortared into the brickwork, are nothing more than metal fillers in the wall.

Use of the recently popular cored brick with glazed face only worsens the problem, the brick is weaker and the face is more brittle. Use of the light-weight concrete mixes, with higher shrinkage, more creep and lower elastic modulus than normal concrete increases the amount of column shortening and higher loads are then imposed on the brick. The result is a large number of buildings with vertical corner cracks, push-outs above the steel lintel levels at lower floors, brick face spalling especially at the courses adjacent to angle supports, cracked brick along the edges of wall openings where concentration of imposed load is highest. All of these observable defects, causing unsightly exteriors, leakage of walls, falls of brick faces on the streets below, and sometimes fall of complete wall sections, can be eliminated by proper introduction of horizontal relief caulked joints underneath the continuous steel lintels at each floor, together with vertical relief joints near each corner or offset in wall face.

The Structural Clay Products Institute has issued many technical warnings since 1962 that successful use of ceramic glazed brick facing for exterior walls requires perfect control of moisture infiltration. That includes use of cavity walls with flexible anchorage to columns and floors, a vapor barrier on the warm side of the cavity, weepholes that are permanently open with ventilation holes near the top of free standing walls such as parapets, adequate continuous flashing, adequate expansion joints, not less than ¾ in. wide/100 ft of wall on each side of each corner. Considerable information has appeared about the litigations resulting from masonry defects on concrete framed buildings; examples are too numerous to enumerate, but some are discussed in Chapter 9 (Fig. 6.8).

In 1966 two incidents of large areas of brick cover falling from concrete buildings caused some awareness of deficiency in Code requirements for such construction. At a Tampa, Fla., bank building a long diagonal crack developed in the brick covering in the unpierced lowest three stories of the 10-story concrete building. The masonry bulged out as much as 12 in. in some places. The wall consisted of 4-in. brick, with header bricks bonded in every seventh course to 8-in. concrete block. The cored bricks sheared at the plane of contact with the block which was the plane of the concrete frame. One day after the crack developed, some 60 ft length of the wall facing, three stories high, collapsed into the parking lot, which fortunately had been barricaded off.

In Cincinnati, Ohio, a six-story wall-bearing apartment was placed on a

Fig. 6.8 Masonry distress at lintel from transfer of compression.

concrete-framed street-level garage. The wall consisted of a 4-in. cored face brick connected to concrete block by headers at every sixth course. Wall thickness varied from 20 in. at the second story to 12 in. at the top, but the brick layer was supported by a continuous steel lintel at the second floor above the wall-less garage. When the building was two years old the lower three stories of facing brick peeled off several wall faces, with the cored headers sheared. This happened in July and so neither accident can be blamed on frost action or on omission of metal ties, the usual explanation for such failures. The cause of this collapse was attributed to shrinkage of the concrete-block masonry backing. Storage bins and concrete tanks with brick covering have similarly lost part of the ornament, probably from the dimensional changes of the container under full and empty condition, together with thermal and moisture expansion.

6.7 Partitions

Interior masonry partitions are subject to cracking if the tiles are not dimensionally stable. The serious distress in the partitions of the Virginia

State Library, built in 1940 as a steel frame and concrete building, was explained when the plaster was removed in 1948 and the fire-clay tiles were found to be crushed. Testing of the material showed the sample tiles to have undergone a moisture expansion, averaging 0.114 percent, with the mortar unchanged. Individual tests showed considerable variation in the amount of such expansion and in the length of time required for equilibrium to be reached.

In 1966 seven of eight elevators in the 30-year-old 24-story Suffolk County Courthouse at Boston were taken out of service when the clay-tile shaft walls fell apart. The damage was blamed on vibration effects from adjacent construction, but possibly from the elevator use as well. Also in 1966 a free-standing 8-in. concrete block partition at a Columbus, Ohio, high school, used to carry bookshelves on each side, collapsed when a student was stacking books on one side. The unbalanced loading caused failure and a fatal injury.

6.8 Ornamental Screens

Sculptured masonry screens are becoming common as decorative separations of public spaces. When exposed to the exterior, they can become dangerous elements. In 1964 at a shopping center in Norfolk, Va., an 8-ft-high ornamental screen of 3-in. thick perforated elements on each side of an 8-in. air space was blown over by a gusty wind and hit a passer-by. Wind exposure is not reduced by perforations on one side and the suction of restricted orifices may even increase the wind effect over that of a solid surface.

In an eight-story concrete apartment building in New Brunswick, N.J., the outside entrance walkway at each floor level was made of 6-in.-long salt-glazed octagonal tile duct laid up as an ornamental screen between the concrete columns and floors. After a very hot day it was noticed that a number of these screen units were loose with gaps at the contact between mortar joints and the concrete. Without investigating the cause, which was the different thermal expansion of the two materials not only in magnitude but in the time effect, all open gaps were filled with a rapidly setting epoxy mortar. In the autumn, when the concrete frame cooled off, 22 of the tile panels crushed, some falling out completely. Although no exact record of location was recorded, the job diary noted that 22 panels had been recaulked following the hot day. Replacement was with a precast concrete block which had thermal characteristics similar to the concrete frame (Fig. 6.9).

Fig. 6.9 Failure of tile screen from shrinkage forces of concrete frame.

6.9 Plaster

Delayed hydration in white coat plasters, usually accompanied by expansion of the putty, is the cause of the popping and bulging seen several years after application. The plaster on the walls of the Westminster Hospital in London, Ontario, showed extensive bulging after eight years of use. A thorough study of the materials showed that an inclusion of overburnt Mg0 was hydrating and expanding the plaster body. The magnesium oxide converts to a hydroxide with a volume increase of 117 percent. The time of complete hydration depends on the temperature at which the Mg0 was made. If prepared at 800° C, Mg0 hydrates completely in three days; at 1000° C in about three months; at 1200° C in about three years. If the Mg0 is made at 1300° C, hydration is not complete after six years of exposure.

Two identical food stores were constructed in the same year in Virginia and Maryland suburbs of Washington, D.C. The main store areas had large expanse of white plastered ceilings, perforated by combination air-conditioning outlets and light fixtures. After two years one store was in good shape, but the other one was closed because of the spalling, flaking, and falling of plaster. Both had been built by the same contractors and the only variation in the plastering was in the lime, purchased from different

producers. The distressed ceiling had a high dolomite lime which caused the delayed action in the completed plaster. The entire ceiling was removed and replaced with the competitor lime of low dolomite content.

The Corps of Engineers in 1961 investigated failures of metal-lath and gypsum-plaster suspended ceilings in Army hospitals and classroom buildings. In all cases failure or hazardous conditions threatening failure have occurred as a result of corrosion of galvanized tie wires. The ties secure metal lath to furring channels and these furring channels, in turn, to runner channels. The most serious corrosion occurs in the lathing tie wires where they are embedded in the gypsum plaster and in the furring tie wires where they come in contact with the plaster. In a nationwide survey of Army buildings by division and district offices of the Corps the galvanized tie wires showed no signs of corrosion where the wires were not in contact with the gypsum plaster.

Failure of ceilings in a hospital under construction at Fort Lee, Va., alerted the Corps to the danger. Since then serious corrosion of tie wires has been found in every area of the country except the San Francisco and New England divisions. The Fort Lee ceilings failed within six months of plastering, but another ceiling in a classroom at Fort Belvoir, Va., had been in place more than 10 years before it failed. Close to Washington the Engineers found evidence of serious corrosion in a new hospital at Fort Eustis, Va., in tie wires that had been in place only three or four months. Suspect conditions also were uncovered in new hospitals at Fort Meade and Andrews Air Force Base in nearby Maryland.

Cause of the corrosion has not yet been definitely determined. Chemical analysis of the base coat of gypsum plaster in the Fort Lee Hospital showed a high percentage of chlorine, but whether sodium chloride or calcium chloride was not determined. To prevent tie-wire corrosion in new construction, the Corps changed its specifications to require stainless steel wire of nonmagnetic type. Army Engineers warn that casual inspection may not reveal the extent of the corrosion, only rust stains at the surface of the plaster, while within the plaster the wires may be completely corroded.

An interesting cause of plaster failure is found in an article in the *Canadian Field-Naturalist,* July 1961. The walls and ceiling of the heating plant at the Kingston, Ontario, penitentiary were insulated by a blown-on covering of cement and asbestos, a common plaster used as combined thermal and sound insulation. Sparrows cut trenches in the plaster and prepared roosting sites every few feet. Their activities reduced the life of the insulation and the falling debris was a nuisance and a hazard. Similar trouble has been noted in aircraft hangars. The only cure is a hard exterior coating on the insulation to prevent the birds entering the softer inviting nest locations.

7

Timber Construction, Glass, Aluminum

Timber Material; Timber Framing; Glass; Aluminum; Aluminum Conduit.

When new uses are proposed for old materials or new materials substituted in traditional assemblies, special investigation is required to determine the adequacy of strength and the compatibility of distortions. Critical stability during assembly of the components is quite different for materials having different elastic and thermal reaction. Thoughtless substitution of one material for another often brings a new set of conditions and an unsafe design.

7.1 Timber Material

It has been long recognized that timber requires protection against moisture change, yet failures due to timber disintegration are still reported. A 1960 investigation of termite attack indicated that over 75 percent of the houses on the Delmarva peninsula are infested. Old concrete form lumber left on basement walls and on crawl space piers acted as gangways for the termite tubes to reach the floor framing. Any timber located in dark and poorly ventilated spaces is attacked. Shelter tubes were built from the ground up brick walls and concrete piers as high as 4 ft to reach untreated timbers. The use of pressure-treated

lumber in vulnerable locations would have eliminated much costly repair.

Dry rot attacks the ends of beams and trusses imbedded in masonry or so covered by roofing that air circulation is prevented. In 1909 a 20-year-old six-story factory collapsed in New York City because dry rot weakened the oak columns set in cast-iron socket seats. The lobby of the Strand Theater in Pittsburgh, Pa., collapsed in 1922 when the 2 × 10 joists disintegrated in the crawl space environment from dry rot, after only eight years use.

7.2 Timber Framing

Volunteer-constructed church buildings are not as common now as they were 50 years ago, and the lack of experienced labor showed up in collapsed structures. In 1910 a church at Barberton, Ohio, fell just before completion and in 1912 the floor of a church at Harrington Park, N.J., fell during the dedication ceremony.

Erection failures of timber trusses and arches are fairly common. Lack of temporary bracing to replace the completed roof deck in holding the long slim members in equilibrium or the accidental blow from the erection equipment are the usual causes. Laminated timber arches were erected to span 100 ft for a supermarket at Hicksville, N.Y., in 1958. The ends of the arches rested on 10-ft-high steel columns and the purlins were placed. The two end arches of the five spans were anchored to deadmen, using three ½-in. steel cables as a brace until the concrete block end walls could be completed. Wood planking was being hoisted to the roof while one guy was being shifted. The arches started swaying, the other two guys snapped and the entire roof framing fell to the ground.

In 1954 a roof over a shopping center in Redwood City, Calif., had seven 105-ft span timber trusses sitting on temporary posting until the masonry supporting walls were built. The trusses failed from lateral buckling under construction loading. At Ridgefield, Conn., in 1959, five of eight 120-ft two-hinged arches had been erected for a school gymnasium and connected by the purlins. To clear space for the sixth arch, a guy wire was loosened and the entire mass fell.

One of two intersecting 308-ft span glue-laminated wood arches had been erected, but the record span fell over at the Coliseum at Richmond, Ky., in 1962. The same design was later successfully erected with the experience gained that lateral support was essential.

Ten tepee-shaped 73-ft span-laminated arches were erected on ma-

sonry walls and permanently anchored for a church in Cocoa Beach, Fla., in 1965. Only a few lateral connections were in place when a normal wind toppled the skeleton structure. A few months later the same shaped arches for a church in Des Moines, Ia., on a 48-ft span, with some steel guys and 2 × 4 cross bracing, were exposed to a sudden wind gust. Nine of the 10 spans fell in a straight line, and the end one remained upright, aided by a guy cable.

In 1965 the roof of a shopping center unit at Atlanta, Ga., was built of 50-ft span light timber trusses spaced 2 ft apart and supported on heavy timber girders to which they were toe-nailed. The end nine were braced across as an anchor and 41 trusses were being tied diagonally to this anchor. Possibly jolted by a crane, the 41 sections fell over, pulling the nails out. Wind load transmitted through the bottom chords across a building crippled the unbraced lower chord and collapsed the roof (Fig. 7.1).

A timber shell roof collapsed in Alexandria, La., in 1964 when two guys wires parted. The 240-ft dome was framed with 36 glue-laminated timber beams supported on a tension ring on 30 ft high steel columns. All the beams were in place with purlins attached, ready for the installation of a wood fibre concrete deck. With the failure of two of the four erection cables connecting the tension ring to the 10-ft-diameter compression ring, the center ring rotated and the entire roof collapsed

Fig. 7.1 Timber truss compressed by lateral wind load.

within the arena. The temporary pipe shore for the compression ring had been removed an hour before the collapse.

A covered sports arena in Vingaaker, Sweden, built in 1964, had 80-ft span specially designed steel-reinforced laminated-wood I beams. The timber roof had a slight pitch in one direction and was designed for 40-psf snow load. After two years use under a load of some 5 in. of snow, the roof collapsed, carrying with it the timber columns at one side and one brick end wall. The girders were trapezoidal in profile, from 30 to 75 in. high. The chords were laminated wood T's with stiffened plywood sheets as web members, square steel bars were fitted into grooves in both chords, and it seems that these bars separated from the wood.

The roof over a warehouse in Santa Clara, Calif., built in 1963, had two laminated timber girders, 7 × 29-in. section spanning 62 ft. After they had been completed and occupied for two weeks, a heavy rain loaded the roof with 6 in. of water and the flat surfaced roof with peripheral drains collapsed during the night. Failure was near midspan with the laminations completely separated.

The roof over a warehouse in Minneapolis, Minn., consisted of 72 hyperbolic paraboloids made of plywood and supported by center posts. Following a heavy rain in 1961, six of the spans along three walls collapsed and the walls bulged outward. Wooden cross members failed throughout the 130,000 ft^2 roof. At the east gate of the Seattle World's Fair an 80-ft-square, 60-ft-high timber hyperbolic-paraboloid shell was constructed in 1962 as the main entrance feature. The shell consisted of three layers of 1 in. planks mechanically fastened together in the field and with a 6-ft-wide strip acting as edge beam along all four sides, also field glued. Several days after the falsework was removed, the shell collapsed. The site was cleared immediately and a smaller, simpler gate structure was completed in time for the opening date.

Temporary use does not permit complete disregard of structural requirements, even if the structure is for religious purpose. A temporary meeting house, 132 × 240 ft area with a 30-ft timber truss roof, using a single 1 × 8 as a bottom chord collapsed in Lawrence, Mass., in 1917 under a heavy snow load.

7.3 Glass

Difficulties with glass used in large units and in combination with new forms of construction have indicated the necessity for more research in compatibility and connection between glass and other structural materials. For many years plate glass window insurance limited permissible

coverage with heat-absorbing paints. Black covering on parts of store windows exposed to sunshine always results in cracked glass. Thermal expansion of large glass units requires better sealants than the old form of lead oxide and oil putty. Large glass areas partly shielded by structural eyebrows to reduce heat gain in air-conditioned buildings are subject to warping strains that require special detail to avoid cracking along the shadow line.

In 1962 the Indiana State Office Building, Indianapolis, suffered some $70,000 in repair costs to stop breakage of the windows in the lobby. Similar large scale breakage occurred in large bank windows in Madison, Wisc., and in Fargo, N.D., in 1965. At the latter location, eight 10 × 14-ft windows faced south. The glass was ⅜ in. thick, smoke-gray tinted heat absorbing. Cracks developed in clear fall afternoon weather and some panes were replaced only to crack again. In the early summer panes on the east face behaved similarly. The top of the windows is shaded by concrete slab projections and the edge temperatures are somewhat reduced by the heavy masonry walls. To reduce heat radiation into the building, heavy drapes are hung close to the glass. Heat absorption distorted the glass and rough edge cuts started the cracking in the stressed faces. The large panes were replaced by three mullion-separated vertical sheets of ¼-in. heat strengthened glass.

In 1965 the Merchants Bank in Fargo, N.D., used 10 × 14-ft dark glass panes with draw curtains and suffered considerable breakage. Glass failures from wind pressure and suction in many tall buildings such as the Chicago Civic Center and the New York Chase Manhattan Tower have been previously discussed; but often the real cause is from thermal stress, imperfectly cut edges, and bending stresses from the frame connection detail. The Chicago Civic Center used bronze-tinted, heat-absorbing glass, ½-in. thick, above the twenty-fourth floor and at the corners of the building from the tenth floor up. Lower glass panes, roughly 9 ft square, are ⅜ in. thick. More than 40 panes broke in 1966, half of the thicker units and mostly on the south and west elevations where solar effects are greatest. Thicker glass can take more wind load but is more highly stressed by thermal gradient.

The breakage of glass enclosures of the clubhouse and grandstand at the Maywood Trotting Track near Chicago started almost immediately after completion in 1966. The large special glass areas are suspended by patented devices to reduce the width of obstructions from mullions and frames. Although the German method seems to have been successful elsewhere, the 8 × 25-ft panes gave so much trouble that the system was abandoned and conventional metal framing was installed. Even air-

transmitted vibration can distort and may even fail such large glass areas.

Glass for letting daylight into the enclosed Houston Astrodome was one of the features of that project. However, the sudden change from sun to shade made by the pattern of steel framing and skylights made it impossible to play league baseball, and one sports writer predicted that Houston would be the first ballpark to have a game called because of sunshine. Taking a lesson from greenhouse technique, some 700 gal of off-white acrylic paint was sprayed on the dome and the light transmission reduced to tolerable intensity—but the natural grass required replacement with synthetic green carpets.

7.4 *Aluminum*

There are many advantages in the use of aluminum, but its high thermal expansion often more than neutralizes the advantages. This was to a large extent the case with the leaky chapel at the Air Force Academy in Colorado Springs, Colo. The beautifully conceived 17-spire, aluminum-faced steel frame with vertical strips of thick multicolored glass units was completed in 1962. Use of the main level large chapel was impossible because of the high intensity of rain penetration (Fig. 7.2). The altar and pews were removed and the organ covered with plastic waterproofing until a solution to the problem of weather tightness could be found and implemented.

The walls are made of triangular facets of aluminum sheets, subdivided by aluminum girts set normal to the exterior edges of the serrated faces. The sloping sides put these girts in a sloping position downward towards the building, and all rain hitting the faces was led in concentrated streams against the caulked joints. The glass units facing the bright sunlight absorbed so much heat that cracks developed and the caulking was squeezed out. Thermal differential expansion on the opposite sides of the walls warped the frame and racked the joints. The two joint sealants, vinyl weatherstripping and polysulfide compound, as well as a replacement test section with rubber caulking compound did not absorb the dimension changes without fracture or loss of bond. Rain penetration confirmed by hose spray tests showed leaks along the joint sealers and also the strip windows.

There are 24,800 lights of 4 × 8-in. stained glass, 1 in. thick and set in subframes. A count of cracked units in 1962 showed a total of 141, of

Fig. 7.2 Leakage at aluminum panel wall jointing.

which 50 were concentrated on the east face in the three middle bays. This is less than 1 percent of the total panes, but enough to warrant some correction.

A careful study of the construction details and available joint sealants

indicated that major corrective work was necessary to provide a watertight structure. The recommendations included:

1. New cap covering the ridge.
2. Added gutters to collect rain accumulation of the upper surfaces and leaders to conduct this water past the lower surfaces, with low-voltage heating elements within the water conductors to prevent ice formation in winter.
3. A storm window covering over all the strip windows, using translucent glass set in weathertight frames covering the present caulked joint of the stained glass and providing a vertical moving column of air to reduce heat absorption.
4. All joints to be sealed with polysulfide.

Wind conditions made the repair program a slow operation. Men on the suspended scaffolds hooked to the spires could not work in heavy cross winds coming off the mountains. Successful completion, tested by high-pressure hoses played on the walls, permitted dedication of the chapel building in September 1963, after an expenditure of over a half-million dollars on repairs. The increased thickness of mullion and edge strips resulting from the applied gutters and leaders did not detract from the graceful lines of the original structure but were necessary to permit use (Fig. 7.3).

The final report by the Air Force consultants with the mission to investigate both design and construction and to agree on sufficient corrective work to stop the leakage and permit use supported the contractor's contention that he had performed all the requirements of the contract but had raised the possibility of some criticism of the design. However the consultants did state

"In the design of such an unprecedented structure, with sloping wall surfaces and the use of materials not normally combined into surfaces of this nature, complete weathertightness may not result from the basic design without some further development after the space dimension variations and thermal changes in the framework and in the combination of non-similar materials has been determined.

". . . In an unprecedented design . . . the lack of prior experience in the construction of such a building makes it most unlikely that complete success will result without some added or corrective work . . ."

Aluminum as a structural material first obtained public notice in the framework of the dirigibles. The failure of the ships taken from the Germans after World War I, however, showed that metallurgical control of the aluminum alloys was seriously needed to provide reasonably long

Fig. 7.3 Repair of walls by adding gutters and leaders.

life. The failure of the American-built rigid airship "Shenandoah" during a storm while passing over Zanesville, Ohio, on September 3, 1925, with fatalities numbering 14 men, started a thorough investigation in the aging of the duralumin alloy. Obviously the use of helium obviated the danger of fire. The design was a strengthened copy of the Zeppelin ships, after a thorough analysis to provide a factor of safety of 2 under all operating conditions. Tests of members recovered from the wreckage showed that the metal, which had been in service for only two years, had developed random patches of intercrystalline corrosion, covering only a small portion of the total area of the structural sections but seriously reducing the strength of the framework. The material had measurable embrittlement with sharp reduction of ductility and toughness. Although the change in metal characteristics was not sufficient to degrade the factor of safety to the point of failure under static and aerodynamic loadings, there was a decrease in resistance against shock and impulse. The failure of the Shenandoah marked the end of further large scale development of lighter-than-air space travel in the United States.

Aluminum alloys of more stable chemical composition were developed chiefly for planes and some of them showed up in competition with normal structural materials. With added strengths and more certain control of aging, a logical use was in bridges, especially after welding problems had been solved, at least partly. One of the early aluminum-welded bridges shown off to an engineering convention group in Canada was somewhat of a failure when it was noticed that a web member of the truss had completely parted from the chord. Thermal shrinkage had broken the weld.

In 1961 two identical aluminum 80-ft-span highway bridges were erected near Amityville, N.Y. The designs followed airplane wing technique, consisting of triangular beams 7 ft wide and 3½ ft deep with 0.081-in. rolled-flat aluminum sheet on the sides and 0.032-in. thick corrugated aluminum covered with a concrete slab for the top. A bottom plate of 0.102-in. aluminum connects the lower chords of the beams. The alloy used was 6061-T6 and connectors were cold drawn rivets of the same alloy. The side plates of the beams developed fan-shaped buckles of 4-in. radius and some 0.02 in. deep at the support points of the beams in one bridge only, within a short time after bridge opening. The center beam was first affected but soon the outer beams showed similar distress. Trucks were banned from the bridge until some reinforcing plates were added to stiffen the webs. The bridges have performed satisfactorily since then.

Corrosion and pitting of exposed aluminum surfaces in window frames and ornamental details on buildings exposed to salt air environment has been solved by the use of protective coatings on such items. In Great Britain some 13,400 prefabricated one-story aluminum-surfaced homes, constructed about 1950, with an expected life of 60 years, were found to have considerable corrosion in all extruded members after 10 years use. The sheets were rolled from scrap aluminum similar to British alloy DTD 479, which is not suitable for extrusion. For the necessary extruded members, the alloy was blended with 10 percent for sheets and with 35 percent aluminum for extrusions. Both compositions developed laminar deterioration in the direction of extrusion and at grain boundaries. Since houses are located both in rural and urban areas, it is evident that site conditions did not influence deterioration.

7.5 Aluminum Conduit

Under a variety of conditions, aluminum imbedded in concrete may corrode, causing concrete to crack and spall. Three causes of the corrosion are galvanic action between aluminum and reinforcing steel, cor-

rosion due to stray electric currents, and corrosion caused by the reaction of aluminum with alkalies present in concrete. Of these the first two—galvanic action and stray electric currents—appear to pose the greatest hazard.

Galvanic corrosion of aluminum is speeded if chlorides are present in the concrete, in which case they serve as a strong electrolyte to transfer current between the dissimilar metals. Keeping chlorides out of concrete seems a practical impossibility. They may be introduced in admixtures, water, aggregates, or other ingredients. Direct coupling of steel and aluminum may cause aluminum to corrode, even without salts to serve as an electrolyte. It does not seem that protective coatings or coverings on the aluminum can be relied upon to guarantee a trouble-free structure. Stray electric currents apparently cause the most rapid and severe corrosion of aluminum imbedded in concrete.

Alkalies in concrete cause corrosion of aluminum during the curing period, but corrosion seems to cease as the cement hydrates. Alkaline corrosion can be more severe, it appears, under conditions of alternate wetting and drying.

Because of its lightweight and ease of bending, aluminum tubing was a natural use for electric conduit when the 1959 strike made steel conduit unavailable. Code objections to aluminum because of low melting point were readily modified where necessary and the economy of use resulted in the establishment of the material as a recognized item. Sheet steel or malleable iron electrical outlet and junction boxes were used because of the lack of fireproof covering. Electrical systems combining dissimilar metals were buried in concrete floors in the normal way (Fig. 7.4). There are a few earlier cases that could have warned of probable results. In a hospital building at Smiths Falls, Ontario, built in the late 1940's, very bad spalling occurred in the concrete along the lines of conduits; calcium chloride had been added to the mix. It has been claimed that in chloride-free concrete, imbedded aluminum has good resistance to corrosion, both alone and even when coupled to steel by wiring contact. There is full agreement that corrosion comes easily when the concrete contains chlorides and the conduit, which is the ground wire of the electric circuits, is coupled to steel.

At the Luna Park Houses, a group of 20-story concrete apartments built in 1959 in Coney Island, special permission to substitute aluminum conduit was obtained only for the duration of the steel strike. Four or five stories in the mid-heights contain the substitution. At about two-year age, all the ceilings of slabs containing aluminum conduit started to crack and spall, beginning at the connection with iron outlet boxes. Continuous repairs with epoxy are still required to prevent the spalled

Fig. 7.4 Aluminum conduit in concrete slab.

pieces from dropping out. The concrete mix had no calcium chloride but the site is near the ocean. None of the floor slabs with iron conduit show such cracking.

A similar housing project in upper Manhattan, also of concrete with no admixtures but quite remote from the salt air environment, shows similar distress along the lines of aluminum conduit. The first floor of a residence hall at a university in the Bronx has a terrazzo finish placed monolithically on a reinforced-concrete slab. Aluminum conduit had been used when this building was built in 1960. In about 18 months random pattern cracks showed in the terrazzo which continued to widen and cracks in the ceiling below followed the same pattern. Concrete cuts

showed the existence of an electric conduit at each crack. Analysis indicated varying content of chlorides to a maximum of 0.1 percent (2 percent calcium chloride in the mix will show 0.2 percent chloride content). Boxes in the electric lines were cadmium-plated steel. The conduit was found corroded, after two years use, to depths of 10 to 50 mil. The aluminum-chlor-hydroxide product expands enough to crack a 2-in. concrete cover if the aluminum corrodes to a depth of 10 mil. Daily washing of the floors probably speeded up the corrosion. No possible cure could be devised and the electric conduits were replaced by ceiling-exposed steel conduit for all circuits, the imbedded conduits were cleared of wires, blown out, and filled with bentonite slurry to seal off any air contact, and the terrazzo was completely removed and replaced.

After three buildings in the District of Columbia reported trouble from ceiling cracks where conduit had deep cover, even failure of the wiring, permission for any further use of aluminum conduit was revoked in 1963. Meanwhile the Stadium in the District started to crack quite seriously along both horizontal and vertical lines of imbedded conduit and received some unwanted public criticism.

Uncoated aluminum reglets set into concrete parapets, with calcium chloride in the mix, corroded extensively and cracked the concrete after only nine weeks exposure. Coated-aluminum reglets are better, but every discontinuity in the coating permits the start of corrosion which travels to large areas under the coating.

Of course very moist environment has caused similar troubles with steel conduit. In a group of two-story concrete apartments for Navy personnel at Key West, Fla., imbedded-steel conduit rusted so badly that wires were shorted after some three years, and ceilings cracked and spalled. No trouble was found in any room equipped with an air conditioner. The concrete was very vulnerable, since the local aggregate had been washed in sea water and brackish ground water was used in the mix.

Investigations in Great Britain of the private housing built since the war, shows considerable galvanic corrosion in the aluminum supports for drain boards on porcelain sinks, at brass nipples on steel pipe, and at steel nipples on copper tanks. The general recommendation is to provide protective coating, on all copper, cadmium, zinc, lead, and aluminum covered by or in contact with concrete. Any requirement to provide a concrete without chlorides and without moisture as a guarantee against corrosion is normally impossible of performance.

8

Structural Steel

Assembly of Framework; Lateral Bracing; Bridges, Design and Erection; Cranes and Towers; Hoists and Scaffolds; Pipes and Tanks; Welding; Cables and Wires; Corrosion.

Incidents in which steel is the structural element and in which failure was caused by improper design or detail are included in Chapter 1. Natural or man-made disasters involving steel structures because they were in the path are included in Chapter 5. Under the various headings of this chapter are described the difficulties and failures that arise from the use of steel as a structural element, with the clear understanding that many of these incidents could have the same result were other structural materials used.

8.1 Assembly of Framework

The assembly of a most careful design of framework requires the awareness that the components by themselves are not necessarily stable and that at no stage in the assembly is the factor of safety equal to that of the completed structure. The nonexistence of floors and walls reduces the strength of the frame for resisting lateral forces and in the case of compression members, even for vertical forces. The usual failure occurs in one-story structures, where columns are small and limber, and erec-

tion is so rapid that temporary sway bracing is omitted from the construction program.

Typical of roof-truss erection failures is the 1937 case in Los Angeles, Calif., where 18 100-ft spans were erected for the roof of a 150 × 300-ft warehouse and all trusses were held by guys on one side only. Complete collapse came when one guy wire was loosened. In 1955 similar long spans for a roof in East Los Angeles, where tapered girders were used, failed from torsional bending, even though most of the roof purlins were in position (Fig. 8.1). As a result rulings were issued to require all tapered steel girders to be braced at not more than 30-ft intervals with braces capable of resisting a 1000-lb thrust at either top or bottom flange. Again at Long Beach, Calif., in 1956, 11 160-ft welded rigid frames twisted over before final plumbing but with the roof purlins and some monitor framing connected on one side of the spans. The other side was held by guys which required change to clear added framing. Failure occurred during such substitution of members, during a gusty wind, and 500 tons of steel fell.

In 1956 the Cleveland Auditorium roof had an erection failure after four 68-ft girders resting on columns were assembled for the roof. A

Fig. 8.1 Erection failure of unbraced long-span beams.

nudge by a crane boom with guys only on one side of the 36-in. girders may be the full explanation for the four girders and eight columns falling over. None of the fallen steel could be reused. A rigid frame of 98-ft span covering a gymnasium in Bombay, N.Y., fell and collapsed part of the steel frame in 1957, with another gymnasium in Newcastle, Ind., making similar news in 1958. A third gymnasium in Beaver, Pa., became a sidesway failure in 1959. The roof consisted of five two-hinged tied arches with a 204-ft span, and 37 ft high. The steel was erected on timber falsework bents. The underfloor horizontal ties had been installed, and with final bolting of purlins and knee braces completed, four arches fell over when a guy connected to the top chord was adjusted. In 1962 a one-story school in Atlanta, Ga., with three lines of columns supporting conventional roof beams and bar joists, fell sideways while the interior line of columns was being plumbed by the steel erectors.

When a completed roof for a one-story factory 175 × 143 ft in plan and 15 ft high collapsed in 1959 at Westbury, N.Y., considerable study was made of the inherent stability of this type of structure. The design was the same as that used in 25 other buildings located in an industrial complex and complied with the local town code. Columns were 6-in. wide flange sections, in rows 25 ft apart and spaced alternately 25 and 33 ft. Each column was anchored to the footing with two ¾-in. bolts. A spliced continuous steel girder sat on the columns; spliced continuous beams sat on top of the girders spanning 25 ft at a spacing of 8 ft. A 2-in. gypsum roof was cast on plasterboard form supported on steel bulb-tees and was covered with built up roofing.

The entire roof as a unit fell parallel to the beams, ripping the outer line of columns from their anchorages. At the column bases nuts ripped off the anchor bolts or base plates broke loose. Girder webs buckled and the beams moved laterally along the girders when the connecting bolts sheared. There was no wind during the time of failure. After some tightening up of local code controls, the building was rebuilt with diagonal bracing added at each column; knee braces had originally only been placed to the girders along two bents. Base plates were strengthened by providing steel levelling plates in place of shims for full bearing.

Erection of a prefabricated steel warehouse building in Cleveland, Ohio, in 1965 was unsuccessful, possibly because hemp rope was used to guy the framework. The building was 100 × 146 ft in plan with 24-ft spacing between 100-ft-span rigid gable frames. Without any wall or roof covering, but with purlins and some wall bracing in place, the frame collapsed over a Sunday with no observers. In 1966 a third of a pier shed at the San Francisco docks 100 × 225-ft total area, fell during erection of the truss roof, with trusses spaced at 50 ft.

Multistory steel frames have experienced some collapses, such as the 11-story Toronto building described in Chapter 5. In 1960 an almost completed steel frame for a school at Bessemer, Ala., 320 × 85 ft in plan and 72-ft-high collapsed by lateral drift. Together with failure of a school at Orlando, Fla., and a shopping center in Birmingham, Ala., the three failures resulted in strict control of foreign steel importation into the area covered by the Southern Building Code Congress, since it was reported that some uncertified foreign steel may have been used in these frames.

A steel-frame tilt-up procedure combined with hoisted lift-slab floors, using the erected frame as a support for a crane hoist has been used in Paris for low cost apartments. In 1964 one such frame assembly for a 10-story building collapsed into a mass of twisted steel as the last bent was being raised into vertical position, but before any slab had been lifted to brace the columns.

On November 1, 1966, a seven-story steel frame for a zoology building at Aberdeen University, Scotland, collapsed while men were completing the installation of high strength bolted joints. Loss of life required a court of inquiry to fix the cause of failure without being controlled by the written agreements as to responsibility for proper performance. Although such findings are not allowable in civil court actions that may follow an accident, they are widely publicized and may become precedent for future legal action. The jury ruled that the collapse was caused by wind pressure on an unstable structure. The joints between the steel columns and the horizontal steel beams were not strong enough by themselves to hold the framework rigidly in its unclad state. It found that both the consulting engineers and the steel subcontractors had made mistakes for which they must accept liability. It also criticized the lack of liaison between them and the lack of adequate supervision of obviously necessary safety factors.

The framework was quite conventional, but to provide clearance for precast concrete exterior wall panels, beam-to-column connections were limited to the webs. The engineers claimed that erection safety was not their province and the steel erector argued that although he had developed the details, which had been approved, there was no information that added strength was needed during the erection stage. Witnesses testified that the frame was already out of plumb several days before failure, which was triggered by the yield of the bolted connections on the webs of the fascia beams.

In 1966 there were erection failures of a four-story frame in Pittsburgh, Pa., and of a three-story frame in Hamilton, Ontario, a warehouse roof in Raleigh, N.C., and some 10,000 ft² of roof over the Ford Research

Center in Basildon, England, none completely explained. In 1965 some 300 tons of steel being erected in Jacksonville, Fla., for an 11-story General Services Administration contract collapsed without warning and without complete explanation.

Steel is normally erected by heavy moving equipment, with dangerously high impact loading if the boom contacts previously erected and often not completely fastened, steel. Collision accidents are quite common, usually resulting in a slight dent or repairable distortion. Some jobs are not that fortunate, such as the Landsdowne Stock Show building in Ottawa, Ontario in 1906 where six roof trusses were erected and braced with guys. A load of plank fell out of the sling and fell on a guy, pulling all trusses over. In 1930 for a four-story concrete building with steel column cores for a Brooklyn, N.Y., bakery, the 65-ft-high columns were erected full height and guyed and braced at both the third and fourth floor levels. A crane boom hit one column resulting in a slow progressive total collapse of the columns. Similarly in 1934 a school building in Bayside, N.Y., with all steel for the one- to four-story frame erected and guyed, a column was hit by a bundle of reinforcing being hoisted to a lower deck, and all the framing collapsed.

The fifth roof truss spanning 113 ft for a school auditorium at Everett, Wash., erected in 1939, hit the last preceding truss and all five fell over in slow succession. Russia has had the same problem as reported in 1961, when a crane swung a precast concrete wall panel into an eight-story steel frame with complete failure.

A 300-ft-diameter steel dome being erected in 1966 at Kingston, Jamaica, had 15 of 24 trussed ribs in place with the center compression ring on a temporary wood tower and the outer ends framed to a peripheral tension ring supported by concrete columns. Sudden collapse was blamed on failure of the center tower. Inspection of photographs reporting the failure proved that one of two cranes had been moved after the collapse of the ribs and the compression ring. The rig operators admitted moving the crane after the boom had contacted one rib and this movement with rigging entangled in the steel work, pulled ribs to the ground in a corkscrew shape (Fig. 8.2).

8.2 Lateral Bracing

The need for lateral bracing of all compression members to avoid buckling has been well-known from the early days of structural engineering. The lesson had been learned the hard way with massive collapses. In 1911 the Buffalo Porter Avenue Pump Station roof covering 100 × 360 ft

Fig. 8.2 Torsional failure of steel-framed dome.

with 23 trusses located 53 ft above the highest floor level, failed completely when the sag of the trusses pulled the wall columns inward. Some blame was cast on the steel supplier when it was determined that all steel was ⅛ in. shy of required thickness. In New York the balcony of the Orpheum Theater, located 59 ft above the main level, collapsed in 1913 because of insufficient bracing of the high columns. In 1915 at Oswego, N.Y., the steel trusses covering a 66 × 100-ft warehouse, with concrete roof being placed at lower chord level, turned over when a guy was loosened. The upper chords had no permanent bracing. Similarly, in 1917 85-ft roof trusses over the Buick showroom in Kansas City, Mo., fell over during roof construction when a gusset plate in the unbraced compression chord bent. The complete collapse of a theatre in Brooklyn, N.Y., in 1921 where the roof consisted of eight 80-ft trusses, wall-bearing at one end and sitting unconnected on a cross truss at the other end, was caused by the failure of the unbraced column at the end of the cross truss. In 1920 the 48-ft-diameter concrete dome roof of the Christian Church in Long Beach, Calif., was being built on a steel frame of trusses with no bracing for the top chords. A truss tipped over and buckled, causing the fall of the dome.

The collapse of the main entrance tower to the Fuller Brush Company building at Hartford, Conn., in 1923 resulted in the death of 10 men. The tower was a massive masonry structure projecting from the main

Fig. 8.3 Beam failure in torsion under cross wind.

building face, 28 ft square and 113 ft above the ground. The top of the tower contained a 50,000-gal water tank supported by four inclined steel columns seated on grillage in the walls at about 65 ft above the ground. The bottoms of the legs were connected by horizontal channels to resist the outward component of the loads on the legs. After the masonry walls were completed someone ordered the horizontal ties and also diagonal sway bracing removed, on the argument that they were no longer needed because the walls would resist all wind loads. The steel interfered with a new proposed use of the space below the tank. Collapse gave little warning as the tank twisted the four legs in a clockwise direction, shearing off the tank support beams.

Details must be developed to maintain stability during erection and before walls are built to provide bracing (Fig. 8.3).

In 1961 a 8000 ft^2 area of steel framing gave way in Saginaw, Mich., while a gypsum deck was being poured on permanent sheetrock formboards. Support of the bulb-tees on flexible steel joists often permits lateral displacements of roof areas in such type of construction. An engineers' committee appointed by the governor placed the blame on absence of continuous technical inspection.

An increase in floor area for exhibition space in the New York Coli-

seum was provided in 1966 by filling in the center-core open area at the top-floor level. Columns had been originally provided for the possibility of closing the open well at all levels since there had been some doubt in the minds of the feasibility planners whether 300,000 ft^2 of exhibition area was necessary. The planners were very wrong in 1955; present requirements for exhibition could make use of a million square feet. The 150-ft-wide clear opening was designed to be spanned by 26 120-ft steel trusses supported by 15-ft outriggers connected to the existing framing steel. The main floor is strong enough to act as an erection level with a crane on wood mats to protect the floor finish. When the first new truss had been positioned and bolted to its supports, it started to vibrate laterally until it fell off the support, doing some minor injury to completed trusses on the main floor. To avoid a repetition, two trusses were cross braced, lifted as a box unit, and connected to the existing floor framing for lateral stability. After they were so braced, each succeeding truss was lifted into position and held by the crane sling until lateral connections were complete.

8.3 Bridges, Design and Erection

Engineering magazines from 1875 to 1895 were as full of reports of railroad accidents and bridge failures as today's daily news reports of automobile traffic accidents. Even as late as 1905 the weekly summary of technical news includes a description of the most serious wreck of the week, usually tied in with a bridge failure. Better design procedures, better control of material production and of erection methods, together with more carefully organized maintenance, all resulting from public criticism and demand, have considerably reduced, although far from entirely eliminated, the incidence of railroad wrecks and bridge collapses. Recent reports of such nature are squeezed off the front pages of our newspapers by plane and highway accidents. The *Railway Gazette* in 1895 published a discouraging summary of iron bridge failures resulting from railway traffic, listing 502 such failures in the United States for the period 1878 to 1895 inclusive, from reports collected by C. F. Stowell, and noting that the first 251 cases occurred in 10 years while the second 251 occurred in eight years, and in the four years of 1888–1891 there had been 162 such accidents. These reports received wide publicity and attention and must have had considerable influence on the designing engineers as well as on the bridge salesmen of those years.

Complete collapse of bridge structures is not common now, but difficulties are not unknown. Cases in which wind and storm are prime

causes are described in Chapter 5. A few cases of serious trouble from improper details will illustrate the type of modern bridge failures.

In 1952 a 96-ft span of the Sullivan Square highway overpass in Boston, moved off one of its supports and the end of 17 beam stringers fell to the ground. Although all of the engineering reports do not agree as to whether the cause stemmed from the design or from the erection, the facts indicate an unwise detail at the support. The structure consisted of parallel stringers supported on three-legged plate girder bents. The end of the span that dropped was mounted on expansion rockers which were designed to be vertical under full dead load at 70° F temperature. The same bent carried an adjacent 30-ft span that had an expansion rocker at this end so that the bent was without lateral support. The concrete on the 30-ft span had been placed, but not on the 96-ft span.

Failure came a short time after completion of the concrete placing, and after failure, the free standing bent was found to be inclined away from the collapsed span, with two anchor bolts sheared off and two others bent at the base of the columns. It was agreed that any nonvertical position of the rockers would induce a thrust into the bent with bending of the columns (Fig. 8.4). Such bending of the bent would then increase the inclination of the rockers with successive steps repeated until the bent supports at the base gave way and the ends of the beams were free to fall. The disagreement in the reports was whether the initial out-of-vertical position of the rockers came from the erection in such position or from a thermal and loading change as the adjacent span was concreted. Certainly it is wiser not to have a loose link in a chain of supports.

A triangular-framed water pipe line bridge collapsed in Winnipeg, Canada, in 1957 when the pipe was first being filled. The structure was 280 ft long, carrying a 36-in. water main on shallow steel floor beams spanned between the bottom chords of two sloping Warren trusses. The single common top chord developed an S-curved deformation over some 50 ft of length adjacent to one support. A more conventional bridge design was followed for reconstruction, which added $5600 to the replacement cost of $79,500; the concrete piers and foundations were reused.

A trapezoidal piece of elevated highway structure in Chicago shifted laterally in 1966, after five years use, sufficiently to displace the bearings away from the base plates; the span dropped 3 in. to rest on the concrete pier. The 25-ft-wide section is 35 ft long on one side and 80 ft on the other. One end is fixed to a steel bent, and at the other the beams rest on sliding bearings 4 in. wide, which dropped when the supports had bent sufficiently. The lateral force required for such distortion came from an

Fig. 8.4 Displaced bridge bent after span fell.

accumulation of braking traction from vehicles on the roadway, unequal thermal changes in length of the deck because of the large discrepancy in dimension of the two sides, and possible expansion-joint locking by the filling of the space with grit and hard debris. When the span was jacked up, taking the load off the pier, the support moved somewhat towards the fixed end, indicating that some lateral movement had occurred. After raising the span to proper elevation, larger base plates were installed and shimmed tight to the pier. To avoid future separation of the two supports cable ties were added to connect the piers.

The assembly of carefully detailed steel expressway frameworks often does not produce the smooth riding pavement planned, when the

camber provided in the stringers and floor beams is not ironed out by the dead load. With slab and pavement thicknesses fitted between prefabricated steel expansion dams, sitting directly on steel floor beams, a series of scallops result. Such a rough-riding expressway in Brooklyn was closed to traffic in 1960, six months after dedication, when the asphalt surface became too rough for traffic under even restricted speeds. Correction required the removal of the asphalt, disregard of thickness and grade levels as called for by the plans, and an empirical placement of a wearing surface to connect the tops of the expansion dams, located 60 ft apart. The concrete deck had been placed with reference to the top of stringers, in a series of curved surfaces that were expected to flatten under the weight of the concrete. Possibly the rigidly welded end connections of the stringers prevented the mechanically bent stringers from giving up their camber and the hardened concrete fixed the beams in the curved position. Not providing camber in the stringers will usually result in an irregular profile with the expansion dams as high ridges. In any case the thickness of the wearing surface can be varied to correct for deck irregularities. On New York's Tappan Zee Bridge, where the concrete slab was the wearing surface, formwork for each panel had to be tailor-fitted to the surveyed steel elevations with a variation in slab thickness to provide a reasonably smooth roadway. Here too the camber did not reduce sufficiently under dead load.

Most bridge failures come during the erection stage. To omit a description of the two incidents connected with the construction of the Quebec Bridge over the St. Lawrence River would disregard some basic lessons of the structural design discipline. Generally those lessons were that major collapse can result from disregard of the safety of small details and that both compression and tension supporting members require close attention. Complete and detailed reports, both technical and popular, are easily available in the literature following the first collapse on August 29, 1907, and the second fall of the center span when almost in the closure position on September 11, 1916.

The great cantilever bridge with a clear center span of 1800 ft and 500-ft anchor spans, far beyond any precedent for such structures, was started in 1902, but it was three years before any steel of the main span was erected. At the time of collapse, the south anchor and cantilever arms were complete and about one-third of the suspended 675-ft truss span had been cantilevered out from the balanced arms. The steelwork projected 800 ft over the river from the main pier when some movements of the bottom compression chord of the south anchor arm were observed and field men were sent to New York and to Phoenixville to seek advice from the consulting engineer and from the fabricator. The

sent to the site to suspend work arrived just after the collapse, started at the bottom compression chord in the second panel ward of the south river pier. The detailed report of the Royal nission of Inquiry summarized the causes of failure as follows.

"The collapse of the Quebec Bridge resulted from the failure of the lower chords in the anchor arm near the main pier. The failure of these chords was due to their defective design.

"We do not consider that the specifications for the work were satisfactory or sufficient, the unit stresses in particular being higher than any established by past practice. The specifications were accepted without protest by all interested.

"A grave error was made in assuming the dead load of the calculations at too low a value and not afterward revising this assumption. This error was of sufficient magnitude to have required the condemnation of the bridge even if the details of the lower chords had been of sufficient strength because, if the bridge had been completed as designed, the actual stresses would have been considered greater than those permitted by the specifications."

The wreckage had been cleaned up, the design modified, and a different erection method developed for the reconstruction of the Quebec Bridge, which started in 1914. The anchor and cantilever arms were completed and a completely assembled suspended span had been floated into position for lifting to the deck level on September 11, 1916. The first five lifts of 2 ft each had been entirely successful and the barges had been removed to leave the span suspended by cables from the main framework after the third lift. While retracting the jacks for the next lift, the lower shoe at the southwest corner failed and the span slid sideways and fell into the river. The investigation covering all possible causes of failure definitely placed the blame on the failure of one casting on which the span was suspended during the lifting. This same casting had served as a support during the span assembly and a greater reaction had actually been imposed on it than the load at failure, because of the weight of erection equipment on the completed span before it was barged to the site. The suspended span was rebuilt and lifted into position successfully to complete in 1918 the Quebec Bridge now in service.

Another cantilever bridge erection failure came in 1949 at the five-span highway crossing of the Bluestone River near Hinton, W. Va. Erection of the central 278-ft span was progressing by conventional cantilever erection methods without falsework. Ten of the 12 panels had been cantilevered out from the center pier after completion of the 232-ft

and 208-ft outer spans. While the guy derrick anchored to the floor steel in the ninth panel was booming out a lightweight erection truss to connect to the next pier, the free end of the cantilever dropped 140 ft to the bed of the river, coming to rest 20 ft upstream from the original alignment. The downstream top chord ruptured and the upstream bottom chord buckled, both in the second panel. No overstress was found in the reevaluation of the design. No cause of the failure was reported.

In 1956, during the erection of a plate girder bridge for the Fresno Freeway across the tracks of the Southern Pacific Railroad in California, the second girder to be placed struck the first one already in position on the piers as one lifting crane capsized. Both girders fell on the tracks, smashing the second crane and delaying construction of the bridge by three months. Another plate girder bridge, a 180-ft span over the Brazos River near Hempstead, Tex., fell into the river during erection in 1956. The span was assembled on falsework support and failure came when an 80-ton erection traveler was setting an 85-ft section. One girder had been completed and the floor system was in place for the greater part of the span, which rested on 60-ft-high piers. Temporary supports were on piles and the south column of one bent failed when piles gave way.

In 1957, while two floating cranes were placing a section of the bridge across the Rhine River at Düsseldorf, Germany, on top of the main pier, the 118-ft, 400-ton mass rotated, one barge settled, and the section fell into the river. A complete photographic record of the failure was made by a motor driven camera taking progress pictures. As the barge listed, possibly from a leak that came below water level under the bridge loading, both legs of the derrick jackknifed. The accident came so slowly that men had sufficient warning to get out of the way; only one worker was injured.

One of the best documented investigation reports on a failure concerns the collapse of the anchor span under erection and the adjacent complete span in 1958 at the Second Narrows Bridge in Vancouver, B.C. Steel erection had started at the shore end and four 282-ft spans were completed. The 466-ft anchor span for the 1100-ft main cantilevered span was being erected with the planned use of two temporary bents on piles. The steel had progressed four panels beyond the first bent and reached the location of the second bent. At this stage the reaction on the first bent was maximum. This bent consisted of two box-section columns seated on a double tier, steel beam grillage resting on deep driven concrete-filled pipe piles. The lower grillages consisted of four built-up 36-in. deep plate girders each supported by six piles, the 24 piles being cross-tied both ways into a stable tower. On top of the two

lower grillages were placed four continuous 36 WF160 stringers with wood block separators, four sets in the full length of 42 ft, located 4 ft beyond the falsework column and at the third points in the distance between columns. The columns were 36 × 30-in. built-up sections with 5¾-in. base plates. Below these base plates and also at the contacts between the stringers and the grillage beams, pads of ¾-in. plywood had been inserted to make up irregularities in the steel faces. The webs of the four stringer beams crippled under the load, causing the steel of the anchor span to fall. Because the span was connected to the pier at the shore end, the pull placed the pier so far out of plumb that the rocker bearings of the next span lost full support and that span also fell down (Fig. 8.5).

The failure was caused by insufficient strength in the webs of the stringers, with no cross bracing and accentuated by the concentration of load resulting from the plywood packings. In addition the design calculations were in error in computing shear resistance on the basis of the entire cross section of the beam rather than that of the webs only and also making an error in the web thickness of the beams, calculations took the webs to be 1 in. thick instead of the correct figure of 0.653 in. Reconstruction required steel replacement and replacement of two concrete towers that had been pulled out of plumb. Erection procedure for the final bridge followed the original plan.

Failure of this temporary erection bent was similar to the incident in 1941 during the construction of the Charter Oaks Bridge over the Connecticut River at Hartford, where a 220-ft section of the 270-ft plate girder span fell into the river. The crippling of the grillage beam webs formed the same shape as was noted in the sudden failure of similarly unstiffened I-beam grillages under columns in a New York building in Chapter 2 (Fig. 2.1).

Failure of a temporary tubular scaffolding trestle used to support a 170-ft anchor span of the 310-ft cantilever bridge over the Manchester Ship Canal in Barton, England, in 1959 caused the fall of four 60-ton girders. Coroner's investigation disclosed that the pipe falsework was inadequate for the load, with omission of bracing members in the 65-ft high tower and the use of corroded pipe units. The scaffold buckled when the girder load was applied. Another European steel bridge failure was the 240-ft girder bridge on Italy's Highway of the Sun placed on supports 147 ft above the Lora River.

Movable bridges to erect bridges are recent developments to aid in precast beam and cast-in-place bridge decks. One such device used to speed up construction of the Essinger Throughway in Stockholm, Sweden, failed in 1965. A 150-ton steel assembly, 165 ft long, rode on

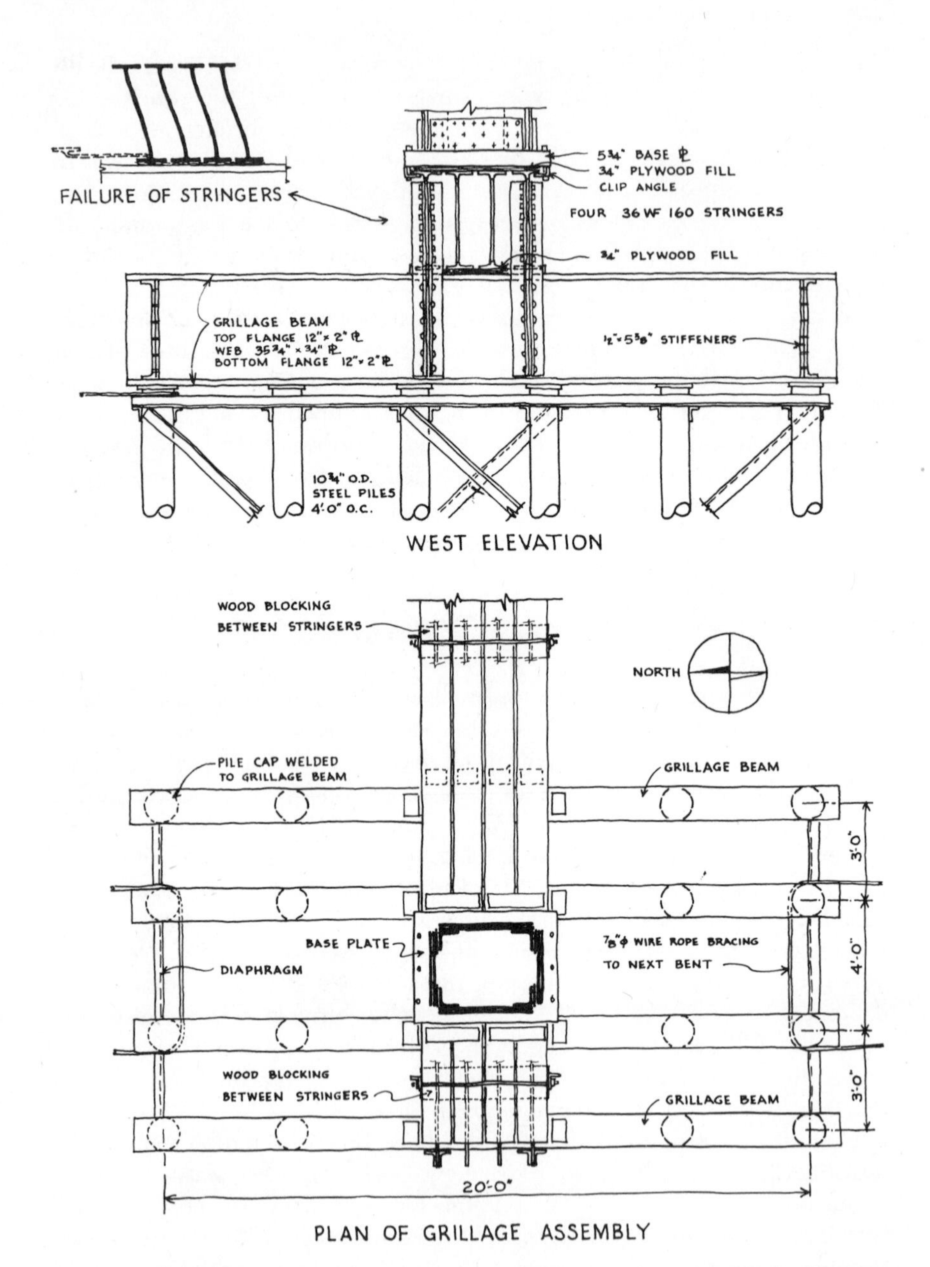

Fig. 8.5 Major failure from collapse of grillage under erection tower for bridge. (Redrawn from A. Hrennicoff, "Lessons of Collapse of Vancouver 2nd Narrows Bridge," ASCE Paper 2305, December 1959.)

the completed piers and spans over the work. It carried hydraulically operated hinged forms that swing down and out after the concrete has set, and also had conveyors for concrete and reinforcing as well as hydraulic jacks and rams for forward motion and positioning. Apparently not foolproof, a form section unlocked during positioning and caused some injuries. Three weeks later, as the device was being positioned onto a pier, the framework collapsed and resulted in more serious injuries and one fatality.

A steel-beam telescopic extension launcher for setting 27½-ton concrete girders spanning 90 ft on a Seattle, Wash., viaduct, after placing 400 girders without incident, collapsed while resetting an outside beam. The outer transfer rail anchored to the pier, slipped off and the launcher fell sideways, twisted into a distorted mass and was held from falling to the ground when the box beams contacted the piers. After some repairs and replacing the rail with a flat channel to act as a stop for side slip of the wheels, the launcher went back into service.

8.4 Cranes and Towers

Faster construction schedules for steel erection and for concrete placement have brought a market for longer and longer booms on mobile cranes, usually on wheels to permit operation on public highways. Crane boom failure was not unheard of even when a 60-ft boom was regarded as a piece of heavy equipment. With boom length extended beyond 300 ft, very careful handling of the load transfer and release is necessary to avoid sudden snap and failure. Typical of such incidents are the failure of a 125-ft boom in Burlington, Ontario, in 1958 when the load of steel placed the crane off balance and although the men jumped clear, the boom hit 12 cars parked at the job. In 1959 a 190-ft crane boom placing concrete on a 20-story Manhattan housing project while hoisting a normal 2-cy bucket of mix buckled at 38 ft above its pivot. The boom fell across the work deck and down the other side of the building, causing one fatality but very little damage to the forms.

Fear of catastrophic accident from long boom operation caused the introduction of a bill in the 1964 New York State Legislature to limit crane boom lengths to 170 ft, but the law was never enacted. In 1967 two long boom failures in densely populated areas of Manhattan again brought some pressure to enact a limit, but nothing has been done. A crane boom 200 ft long with a 40-ft jib had just completed erection of the structural steel for the Ford Foundation building on East Forty-Second Street, and the boom was being lowered to a horizontal position for

dismantling. When it reached 50 ft above the street it tilted the crane and when the operator tried to prevent it, the boom buckled, swung one way and crashed into a bus and a taxi. Fortunately only four persons were slightly injured. At East Fifty-First Street a similar crane toppled over, but the boom fell into a 40-ft deep completed cellar, with no injuries. Even the operator, who was thrown clear of the cab, was unhurt.

Short booms in recent years have been the cause of spectacular incidents. During the construction of a highway overpass in North London, England, a mobile crane with an 83-ft boom was lifting a king post to be set on top of a braced Gabbard Tower derrick frame. To avoid working during street congestion, the lift was scheduled for a Saturday morning, June 20, 1964. The 15-ton crane had a tubular steel boom of welded units fitted together with a gate section for ready assembly. The maximum capacity was rated at 5 tons on 15-ft radius by 1964 British standards, but was rated at 7½ tons by the manufacturer. It had been used safely to erect the north derrick frame and the complete southern derrick with a 6-ton king post. While lifting the north king post, which, with chain sling, weighed more than 7½ tons, the boom buckled and the king post drifted eastward across a sidewalk and fell on a bus. Seven passengers were killed and 32 injured.

The accident caused a Parliamentary investigation, with a 50-page report in great detail. It was established that the crane boom assembly had used incorrect lugs in the gate section, which was the hinge for folding the boom in transit through the streets. The lugs were designed to be rectangular plates to fit into a yoke of triangular plates. The inner lugs had been made triangular and bending stresses from the load on the boom had reached the point of failure. In addition the adapter gate section had been placed at the point of maximum moment in the boom rather than at the top of the square part of the boom, where it was supposed to be set. However, no one in the field was instructed on this point. Furthermore, the safe load indicator required for all booms was not operating, and the amber light overload warning was not received. The crane ratings are based on the equipment being on a level surface; actually it was standing on a cross slope of 1:30, which further increased the bending stress on the boom. Blame was placed on the crane manufacturer and on the contractor, who were assessed $16,800 and $11,200 respectively as their share of the cost of inquiry (Fig. 8.6).

In the record of this inquiry the investigator advises that construction accidents under the heading of "collapse or failure of a crane, derrick, winch, hoist or other appliance used in raising or lowering persons or goods" reported in Britain have increased from 78 in 1959, steadily to

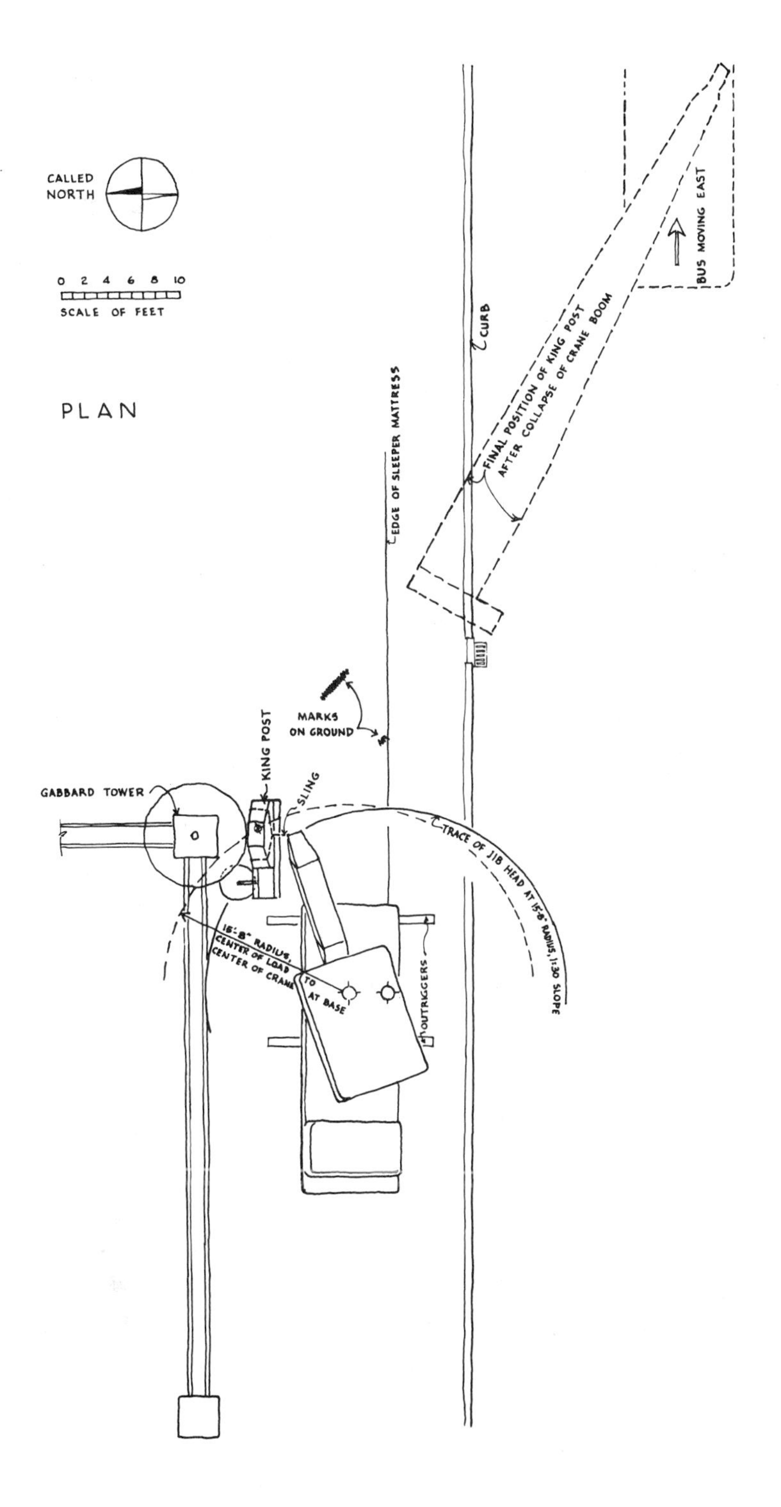
CALLED NORTH
0 2 4 6 8 10
SCALE OF FEET
PLAN
BUS MOVING EAST
CURB
FINAL POSITION OF KING POST
AFTER COLLAPSE OF CRANE BOOM
EDGE OF SLEEPER MATTRESS
MARKS
ON GROUND
KING POST
SLING
GABBARD TOWER
TRACE OF JIB HEAD AT 15'-8" RADIUS, 1:30 SLOPE
15'-8" RADIUS,
CENTER OF LOAD TO
CENTER OF CRANE AT BASE
OUTRIGGERS

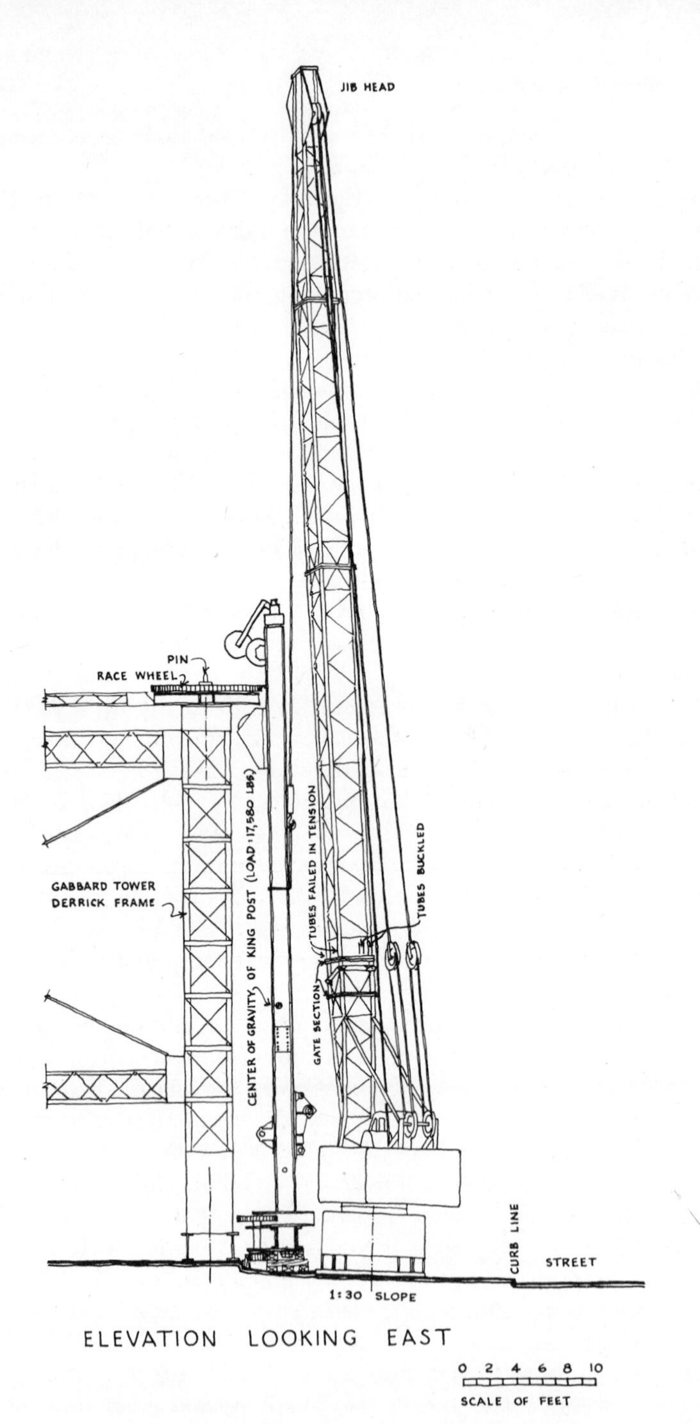

Fig. 8.6 Crane boom collapse from eccentric load on weak detail. (Redrawn from "Report of the Investigation of the Crane Accident at Brent Cross, Hendon, on 20 June 1964," presented to Parliament September 1965.)

140 in 1963, which reflect increased activity in the construction field but also indicate necessity for better control.

When a crane crashed into an Atlas Silo at Roswell, N.M., in 1961, six men were killed and 14 injured. The 20-ton crane with a 75-ft boom was used to haul men out of the 175-ft deep pit. When the crane hit the bottom the fuel tanks exploded, engulfing the entire silo in flame. Certainly the crane was not overloaded and some operational explanation is the probable cause of failure.

Failure of a 90-ft crane boom supporting 144 ft of pile driving leads at San Francisco's Golden Gateway project in 1965 put the hydraulic pile driver out of commission. The fast acting rig delivering 120 blows per minute may have transmitted vibration to the boom. At the Pan American Building in New York in 1961 a 75-ft boom truck crane lifting an 8-ton girder to the third floor whipped out of control and the load hit the boom. The impact started a 500-lb steel counterweight ball swinging, and it cut loose from the load line and flew through the framework.

Guy derricks mounted on structural framework require careful in-step operation for stability. At the Chicago Civic Center Building in 1964, the derrick dropped two stories, from the fifth to the third deck, while lifting a 40-ton column. Shortly thereafter the same derrick dropped a 17-ton plate girder, which slipped from the sling.

A 10-ton derrick mounted on the twenty-seventh floor level of the Houston Lighting and Power Co. building in 1967 went over the side of the building while lifting a 3½-ton load, tumbled over, and crashed through a walkway, smashed a truck, all without hurting anybody. The load was a precast wall panel and apparently the load was allowed to free-fall to the proper level with sudden brake application, upon which the beams supporting the crane buckled.

Tower cranes are not foolproof either. A Swiss model was mounted on the sixth floor of a concrete building under construction in San Mateo, Calif., in 1965. While hoisting concrete to the eighth floor, workmen heard some steel popping and lowered the bucket. The mast legs collapsed and the horizontal boom fell over the side of the building. In 1967 a fatal accident from a tower crane failure in British Columbia resulted in tighter regulations requiring competent supervision of erecting, servicing, and dismantling tower cranes. A climbing crane mounted on the tenth floor of a concrete building in Tampa, Fla., in 1967 had just deposited a load of reinforcement when the operator heard some steel popping. Inspection showed nothing wrong and as the 87-ft boom was positioned for another load it tipped over backwards, rotating about the 8-ton counterweight. The end of the boom speared a car in a lot across the street injuring three occupants. All 24 bolts in the turntable were sheared.

Failure of steel transmission towers is usually caused by some freak wind condition as in the cases cited in Chapter 5. In 1965 nine riggers erecting twin 200-ft-high radio towers at Bloemendal, South Africa, were killed when a cable guy pulled out of the anchorage and both masts collapsed. The evidence is that a nut on the turnbuckle of the guy cable was missing and the guy parted. An unsolved mystery was the failure of a 1262-ft TV tower at Nashville, Tenn., in 1957. The erection was complete with 12 guys in place awaiting addition of the antenna when the mast crippled at about midheight as a guy cable gave way. The steel used was a low carbon with 90,000-psi yield point and there were no joints in the structure at the level of failure. To eliminate complaints and legal action of neighboring properties, the site for reconstruction was moved to a 100-acre site in West Nashville. Redesign doubled the number of guys and used lower working stresses in the tower legs. No definite report of the source of the horizontal force causing buckling failure has been made available.

8.5 *Hoists and Scaffolds*

Vertical transportation of men and materials in multistory construction, replacing the traditional ladders and hod carriers of olden and of even fairly recent days, requires the use of various types of hoist towers, hoists, and suspended scaffolds. The bamboo scaffolding with rope lashing has been known in the Far East for centuries. The pole and plank scaffold of European countries, often with rope as the connector at all joints is being replaced rapidly by steel pipe hoist towers and vertical man-lifts. The plank ramps for pushing wheel barrow loads to the upper levels have disappeared, but the next generation may reinvent that device as a motorized escalator, although there are now many belt and chain conveyors as well as concrete and mortar pumps which elevate materials continuously.

Temporary structures are too often not designed as carefully as those for permanent use. In a three-year tall-building construction job the hoist may make as much mileage as the finished elevator over the full life of the building. A number of failures have been investigated to determine the causes and provide warnings of insufficiency in design.

In 1956 at the Federal Office Building in Edmonton, Alberta, a tubular steel hoist tower, 158 ft high and of 12 × 12-ft constant section was erected for a hoist with 3000-lb capacity. The tower was not guyed and had no attachment to the building. A gusty 26-mph wind toppled it to the ground as a twisted mass of pipe and plank.

In 1962, at the Marina City 60-story apartments in Chicago, a tempo-

rary personnel hoist was installed in an elevator shaft. When safety clamps failed to engage the guiderails and stop the hoist, the cab fell 10 floors to the bottom. Although it was cushioned by a hydraulic buffer at ground level, six workmen were seriously injured. A similar accident occurred at Houston, Tex., in 1965. Men riding on a material hoist at a 33-story building were killed when the car started to fall freely at the twentieth floor when the winding drum brake became ineffective. A shaft that applies resistance to the brake band had shifted so that only a narrow rim was in contact. The failure could have started with a missing or sheared cotter pin needed to hold the brake shaft in position.

Powered scaffolds operated from roof cranes are needed in the modern air-conditioned buildings with no sills or often no window openings, for the exterior window cleaning. Such a device was installed at the new Equitable Building in New York in 1962 following the development of equipment for window washing 10 years earlier for the Lever House. The scaffold platform was raised by wire rope operated by a rail-mounted crane on the roof and guided by aluminum mullions in the facade. Four men on the platform were killed when it fell freely from the nineteenth floor to the sidewalk. Two separate safety devices failed to operate. The scaffold cables will not run unless a deadman's button is depressed, and the motor stops if the cables slacken. All the teeth were stripped from a speed-reduction gear on the electric motor, which must have run free even after the safety devices were called on to stop the motor.

The cause of pipe scaffold failures is very simple when all the facts are available, usually after some fantastic explanations were set forth. In 1955, during a very hot spell in August, two sets of seven pipes at the eighteenth-story level of the 45-story scaffolding that connected the material hoist with the Socony-Mobil Building in New York suddenly bowed out several feet without any failure in the couplings. The planked runways were 40 ft long and the framework was connected to the building at each level as well as by diagonal guys. The top of the scaffold settled 7 in. The accepted explanation, appearing in all the local newspapers, was heat expansion of the 2½-in. pipes, although the photographs showed no bend or distortion in the adjacent hoist tower legs. The heat was the cause, but not its expansive effect. Masonry labor had quit because of the heat, but laborers continued to unload trucks of masonry partition blocks, which were hoisted to the nineteenth level and stacked on the platform until the supporting posts were overloaded. Repairs were rather simple. After adding some more cables to release the weight above the bent columns, two new pipes were inserted on the sides of each bent one and some diagonal bracing to tie the new supports to the frame (Fig. 8.7).

Fig. 8.7 Material hoist scaffolding distorted in hot weather from overload.

In a case of complete crippling of a hoist tower at the S.W. Brooklyn Incinerator Project in 1960, runways 25 ft long connected the tower to the building and were supported by three pairs of pipe columns onto the roof at the second floor level where the building sets back. During the installation of the roofing at this level at least one of the supporting legs had been removed and not replaced. When building materials were hoisted to an upper level and stacked on the runway, the horizontal

level collapsed and pulled the tower over until it draped the building wall. Among the rubbish on the low roof level, which only suffered one punch through by a 1½-in. pipe brace, were found 25 palettes with an average of 500-lb loads of brick, tile, and mortar materials.

During a resurfacing and strengthening of the interior face of the 293-ft high concrete cooling tower at the West Ham Power Station in England, the pipe scaffolding erected within the tower collapsed in 1965. Several months' work was required to cut away the tangled mass of scaffolding. In the investigation the designer of the frame said that the erectors had "strayed" from his design, and that he had not authorized the variations made by the erectors. The erectors, in turn, testified that the design was impossible to follow as certain permanent parts of the cooling tower were not allowed for in the design, and the alterations made would not have caused the collapse. Future litigations in civil court will probably further obscure the true cause of failure.

8.6 Pipes and Tanks

Stress conditions in steel used for pipes and tanks seem to be easily determined because of the certainty of the loadings, yet there have been serious failures. Because these uses are not usually classed as construction, only a few incidents are included, leaving the subject to the texts on metallurgy. The failure of an 18.5-ft-diameter steel culvert in Montana in 1965 was blamed on hydrogen embrittlement of the steel. Hydrogen inclusion under high-stress concentration causes loss of ductility and cracks that propagate to the point of structural failure.

The culvert was carefully built with 86 ft of earth cover. Metal was ⅜-in. thick and a culvert ring was made of 12 corrugated plates longitudinally bolted with eight bolts per foot, two in each crest and valley. Bolts were ¾ in. of heat treated alloy approximately 50 percent stronger than A325 steel. Failure was noted in the full length of 328 ft in at least one seam of each section, compressing the rings to 16 ft and resulting in steps in the alignment as much as 5 ft. Even though the bolt steel was stronger than the A325 metal used for the sheets, the metal had a lower level of stress at which embrittlement could start. Galvanic action in the wet conditions liberated atomic hydrogen that caused bolt failure. Removal of large parts of the culvert were required and replacement was with larger bolts of A325 steel.

Liquid-gas steel tank containers expose the metal to very low temperature while under maximum stress. A large tank for this use developed a crack in the outer shell and after more than a year of use the tank

exploded with injury to over 200 people and substantial property damage. The steel had become highly brittle when exposed to the minus 250 degree temperatures. In the court litigation the designers were held liable for specifying a steel that changed in strength under the intended use.

The molasses tank disaster in Boston Harbor in May 1919, among other serious damage, put the elevated railroad out of service for several weeks. The 90-ft-diameter riveted-steel-plate tank was filled to a height of 49 ft when the lowest ring plates failed in tension. The ½-in. plates were stressed to 26,400 psi based on gross section, and with almost half the section eliminated by rivet holes, failure came at a stress of about 50,000 psi. An identical tank in Havana harbor failed in 1925.

Two identical steel rectangular bulk-oil storage tanks failed early one morning in 1948, one in Linden, N.J., and the other some 40 miles away in the eastern part of the Bronx, N.Y. The stored oil was # 6 fuel requiring steam to keep it fluid. Temperatures were close to zero and during the night the moisture leaving the tanks froze up the vents with a solid ice closure. At early morning, when oil was being withdrawn for distribution, the reduced internal air pressure was not balanced through the vents and the roofs collapsed. The two failures were reported within

Fig. 8.8 Steel fuel tank collapse from unbalanced barometric pressure.

half an hour. At each location bulldozers were called to push aside the sticky mass of some 250,000 gal of congealed heavy oil (Fig. 8.8).

The failures of steel pipe in natural gas lines is described in Chapter 5 under the heading of Explosions.

8.7 *Welding*

Welding techniques have developed a long way from the blacksmith and forge days, but all new techniques bring new problems. And welding problems are often miscalculated with tragic results. Bridge type structures require special attention because of the usually heavy plates and sections. The investigation by P. P. Bijlaard of the collapse of the Vierendeel welded span at Hasselt, Belgium, in 1937 pointed out that steel under certain high-stress concentration and low temperatures will loose its ductility and brittle cracks will form. Such a state of affairs exists in welded thick plates where shrinkage in all directions is significantly impeded. The bridge collapse followed fractures during cold weather of the heavy plates of normal structural steel metallurgy.

In 1958, during the assembly of the second Carquinez Bridge in San Francisco Bay, Calif., welded sections were found to have minor fractures, but enough to cause question as to suitability. The use of 90,000-psi yield point steel was chosen to reduce the weight of the new bridge. Defects were discovered by magnaflux testing of the filet welds in H sections used as tension members, both in the A242 and T-1 steels. A total of 56 members required correction, removing the defective weld metal with a carbon arc torch and replacing the weld manually. Exposed, flame-cut edges of plates had been prepared carefully for welding by removal of hardened metal, sharp corners, and abrupt surface irregularities. The detailed study of procedures and of materials resulted in safer methods for welding such heavy members.

The failure of the Kings Bridge in Melbourne, Australia, in 1962 contributed further information on proper welding methods. The bridge over the Yarra River and a railroad consists of two parallel 2300-ft-long one-way bridges with spans up to 160 ft. Like the Duplessis Bridge near Three Rivers, Quebec, failure came on a cold day. Cracks started in tension flanges at welds connecting different size plates. Both bridges used steel not considered the best for welding; a rimmed steel with carbon content as high as 0.4 percent at Duplessis and a semikilled steel with 0.23 percent maximum carbon and 1.80 percent manganese at Kings. The failure at Melbourne came more than a year after completion, when a 38-ton crane crossed the bridge on a day with 30°F

temperatures. The bridge sagged 18 in. and wide cracks opened in the deck for a distance of about 100 ft. Investigators reported poor control of fabrication with traces of the red lead shop coat found in the cracks of the welds, improper steel for welding with considerable variation in metallurgy of samples taken, and unawareness of the difficulty in proper production. The project was a design-construct contract with no outside supervision of the performance.

During the fabrication and erection of the Gateway Arch at St. Louis, Mo., in 1964, wrinkles developed in the stainless steel faces from distortions caused by welding the vertical joints. When one considers the difficulty of hanging large mirrors to avoid distorted images, it is not surprising that the wrinkles were magnified by sunlight reflection.

A 300-ft-high office building on Fenchurch Street in London was designed with a slip-formed concrete core to support welded-steel girders cantilevering out in four directions. Hangers attached to the ends would then act as the support for 24 floor levels. During fabrication some brittle fractures were found in the 14 × 16 WF sections weighing 264 lb/ft, and they were rejected. Replacement steel was also rejected because of rolling defects. The third shipment also failed to pass tests of potential brittle fracture. The designers then changed to prestressed concrete girders. The specifications had called for a low alloy steel of 55,000-psi yield point. The entire building had been designed as a steel frame and all the floors remained as designed, with a peculiar combination of concrete beams as the major support (Fig. 8.9).

The competition of reinforced concrete in Britain, as indicated by the steel tonnage used for structural shapes in buildings as against that for reinforcement forced the steel industry in 1962 to advertise a new steel product of higher strength. In the period 1955–1958 about the same tonnage was used for shapes and reinforcement. The trend then went in favor of concrete and in 1962 only 37 percent of the steel was used in shapes. The British Iron and Steel Federation then announced "The New Standard Structural Steel" as in effect the beginning of a new era in steel construction. The need was for a steel of higher strength but also weldable, but the two qualities tend to quarrel with each other. The reaction in the British press was that the new specification steel did have greater strength but was too tricky to weld, with resulting internal stresses that make the metal crack suddenly. Plates can be controlled by rolling and heat-treating, but sections are more prone to brittleness. Most of the heavy beams successfully used with this steel were welded from plates. The steel industry replied that these heavy rolled sections are listed as columns and not beams. But the handbook of the new sections included tables of safe distributed loads for these sections act-

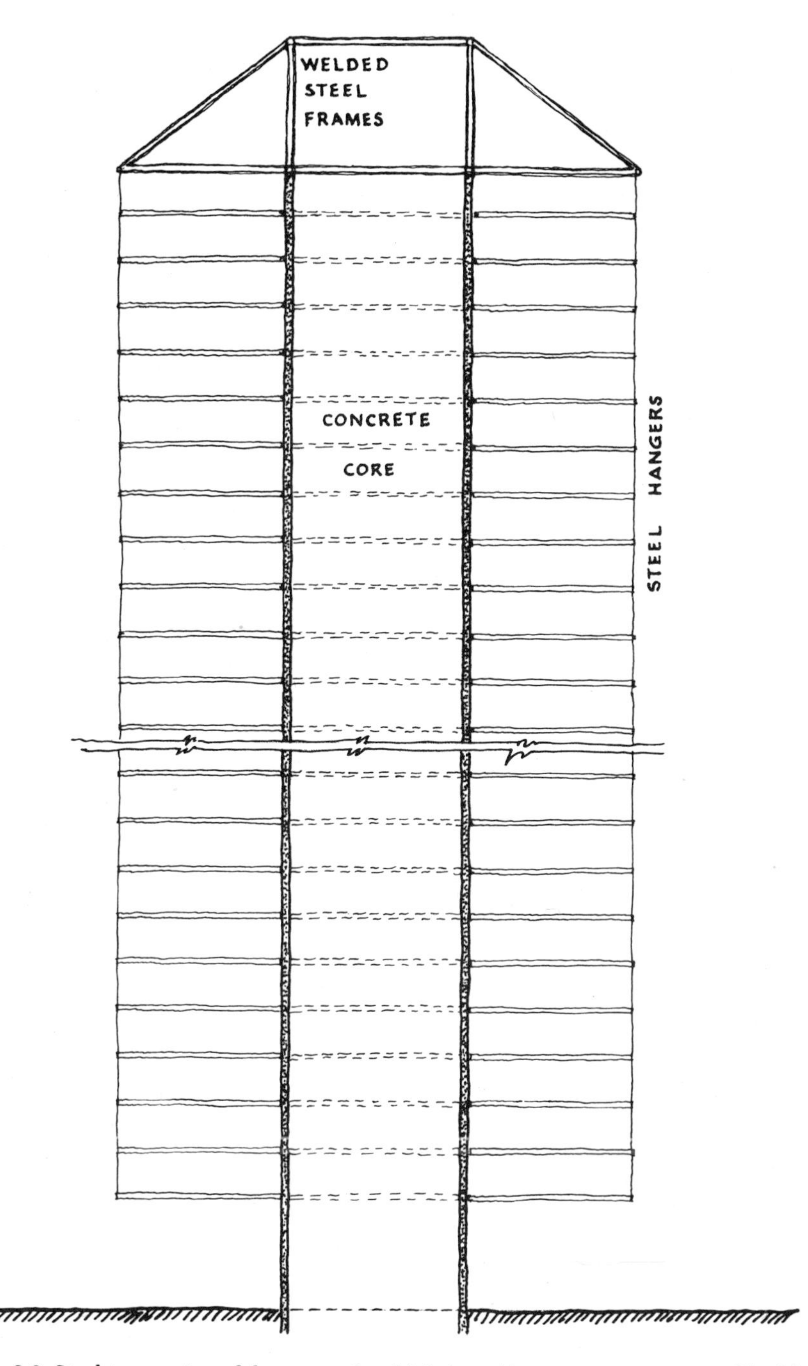

Fig. 8.9 Steel truss rejected because of weld failure (diagrammatic section of building). [From the Sunday *Times* (London), July 4, 1965.]

ing as beams. As a corrective measure the steel mills opened a new plant to heat-treat rolled sections, making brittle fracture much less likely.

For a recently built New York City tall building, rolled jumbo sections of similar size were specified in three-story lengths, for columns with welded splices around the edges of milled faces. The sections were found to be out of square and often with the webs not exactly in the center of the flanges. Considerable corrective welding and packing out was required to eliminate overhanging column flange edges at the splice joints. In a subsequent project, to avoid this difficulty, all column sections came with a shop-welded butt plate on top of the milled length and the upper column was then welded to the plate. Thereby all irregularities were hidden from view and the plate was sufficiently thick to distribute the load from one section to the lower one.

At a roof for a factory extension in Los Angeles, in 1960, nine tapered shop-welded I beams spanned 96 ft and were supported on 10 × 10 timber columns. As the column cap clip angles were tightened, the girders buckled and fell down. No purlins had been placed but the exterior wall and columns were pulled inwards as the steel girders bowed until the center rested on the floor. A welded joint at the connection of rod bracing to the column clip angles had been field tacked but not completed and this omission was blamed for the failure. The long girders twisted 90 deg and such rotation could have been resisted by connected roof purlins or other form of bracing to make the top flanges laterally stable.

Field welding requires some form of inspection to prove the sufficiency of the connections. Several nondestructive test procedures are available to supersede visual inspection and the acid etch test. The latter is supposed to expose slag inclusions and nonhomogeneous weld metal filler. At the 43-story all-welded Wells Fargo building in San Francisco, Calif., with rigid connections for the seismic resistant rigid frame, visual inspection was first checked by a cobalt-60 photographic device that was found impractical in checking the root welds. An ultrasonic reflectoscope was then brought and in the lower floors weld rejections were very high, as much as 45 percent. The full penetration, single V butt welds up to 2½ in. thick were made with backup strips at the roots. Horizontal flange welds were made with automatic gas metal-arc equipment, and vertical web joints of beams to connector plates manually shop-welded to the columns, using shielded metal-arcs. Defects uncovered were mainly slag inclusion at the face of the backup strip, with some porosity in the weld metal. Change of electrode to E7018 with carbon dioxide shielding reduced the percentage of rejections to about 2 percent of the welds. One complaint was the presence of mill

scale on the rolled A36 sections, requiring a higher melting point to fuse the base metal.

Two 80-ft high-riser pipes, 16 ft in diameter, connecting the 24-ft diameter penstock to surge tanks, were welded at the Garrison Dam in North Dakota in 1955. The risers are made of plates varying from 1½ in. at the penstock to ⅝ in. at the tanks. The complex shape of the connected risers and the reinforcement of the penstock pipe required a complicated sequence of assembly. The various pieces acted as wedges and levers as the welding expanded the steel in all directions. Cracks developed both during welding and after completion and were up to 3 ft in length. An attempt to relieve the stress buildup by cutting out large areas of the penstock until the interior connections were made and then replacing the closures did not solve the problem. Further consultation modified the plan to consider the steel merely a liner, even if cracked, and the lower 43 ft of the risers together with 62 ft of the penstock were encased in a block of 7000 cy of concrete. Remedial costs were $1.5 million together with a four-month delay in power production.

High-steel chimney stacks have their own problems. When welded they are prone to collapse in time of low temperatures due to brittle fracture and also to vibrate under low wind velocities more than is found with riveted stacks. The rivet heads and lap splice projections spoil the smooth formation of eddies and also the energy absorption of a riveted joint, from its inherent flexibility, resists vibration buildup.

Brittle fracture of some steels at low temperature was correlated with notch effects and high local stress concentration by the 1957 report of the National Research Council Committee on Ship Structural Design, the result of 13 years research instigated by the brittle fractures in welded ships of World War II. Elimination of sharp corners at openings reduced failures in Liberty Ships and Tankers from 7 to ½ per 100 ship years. The improvements consisted of changes at hatch openings, riveted shell seams, and some crack arrestors. Some rather minor revision in design and details, together with more rigid control of welding enforced in 1947, resulted in complete elimination of such ship losses up to 1955. The cost of the research program was $3.5 million, about half the cost of a merchant ship without cargo, certainly a profitable investment.

8.8 Cables and Wires

Hoist cable failure has been largely eliminated, especially in elevators, by rigid periodic inspection and replacement when the first indication of excessive wear or frayed wires is noted.

High-strength wire as used in prestressed concrete work is covered in Chapter 11. Similar wire has been used in parallel wire-suspension cables, long before the popular acceptance of prestressed concrete. Two cases of serious failure both came in 1929, when the completed cables for the 1850-ft Ambassador Bridge at Detroit and for the 1200-ft Mount Hope Bridge near Providence, R.I., had to be removed and replaced with wire of a different metallurgical composition. The decision to demolish came when breakage was observed in wires where they curved over the strand shoes in the anchorages, at the top of the towers, and at the suspender points of the main cables. The Mount Hope Bridge was completely erected with considerable formwork and reinforcement in position for concreting. At Detroit only about one-third of the main span steel work had been erected. Demolition was an unprecedented job, almost a mirror image of the erection procedure, except that the cables were cut up into short pieces for manual removal and scrapping. Dismantling was more rapid than assembly and most of the structural framing was salvaged for re-use after new cables were spun. The wire in the cables was a heat-treated product of high yield point and was replaced by the material traditionally used in suspension bridges in the United States.

An attempt to save cost on steel hoist slings for steel erection in Toronto by a Canadian contractor who used nylon rope, resulted in a stiff fine when the sling broke and the falling steel killed a laborer. The Construction Safety Act requires that every reasonable precaution be taken to ensure safety of the workers.

8.9 *Corrosion*

Steel exposed to moisture and oxygen will rust and be continuously reduced in section. A 44-year-old roof in Brooklyn, N.Y., collapsed in 1965. The building started as a movie house and was converted to a supermarket without altering the roof. Five steel bowstring trusses projected above the roof, which was a concrete slab encasing the flat bottom chords. Roofing was pitch-pocketed around the lower ends of the vertical and diagonal web members; the rest of the trusses were painted. Moisture filtered down the contact between the steel and the pitch and seriously corroded the invisible steel. The painted steel was in good condition. The entire roof collapsed with a crash and air shock that blew out the street windows. Luckily the 50 occupants escaped with only a few minor injuries.

Steel buried in masonry must be guarded against moisture attack,

Fig. 8.10 Granite fracture from rust formation on iron cramp.

because the rust formation will push out parts of the encasement and crack the face of the walls (Fig. 8.10). Steel buried in the ground is subject to corrosive attack, depending on the soil condition, and there is one case on record where an underground piping system was largely destroyed by corrosion in less than 12 months. The school board owner made claim for negligence in failing to properly test the soil and to provide cathodic protection in the existing very corrosive soil conditions, and obtained a $35,000 settlement of the claim.

Corrosion is speeded up by stray electric currents, especially dc currents, by microcurrents developed within the body of the metal under high stress at the contacts between slag inclusion and base metal, and at contacts between metals of dissimilar electrode potential in the presence of any moisture. Failures of this nature are of greater interest in the products manufacture industry than in construction. These so called stress corrosion incidents, however, often affect the construction industry, as mentioned before in discussing steel weldments and in Chapter 11 in the case of curved prestressed strands. Similar cracking in the leak detector and high-pressure systems of the Homogeneous Reactor Experiment No. 2 at Oak Ridge, Tenn., delayed completion of the installation.

9

Concrete as Material

Concrete Mix; Surface Reaction; Internal Reaction; Shrinkage; Expansion and Plastic Change; Surface Disintegration; Fire Effect; Rusting of Imbedded Steel.

Concrete as a structural material differs from all others because it does not come ready-made to the site and is so prone to misbehaving if any one of the many controls is either disregarded or improperly implemented. A witty description of the necessary precautions appeared in the *Engineers Journal of Ireland* in 1966 under an appropriate heading.

"The Lady is for Mixing"

"She requires coddling and wooing; she requires all manner of treatment and courtship before disclosing her qualities; she gives off warmth; she shrinks; she is fickle; she is the subject of books and books. She is a good mixer and she requires constant supervision.

"Above all—given proper consideration she ages gracefully.

"Concrete is a lady.

"The make-up of any lady may involve the artful use of creams, lotions, powders and perfumes. They can impart qualities of mystery, attraction, allure.

"Milady concrete, we are told, also requires her pills and potions. Admixture may not be a poetic word, but that is how these ingredients are described.

"The choice of admixture depends, one feels, on an equally rational

decision. Regrettably, however, milady concrete must rely on her progenitor for the choice of that which will make her shrink or expand, retard or accelerate, be strong, resilient, dustproof, water repellent, plastic—even lovable.

"The formulae which confer these infinitely desired properties, are, as with the cosmetics, closely guarded secrets. Progenitors of concrete, be they architects or engineers, may be told in strict confidence part at least of the secret, but only on request.

"So concrete is a lady but . . .

"She may be a bitch betimes, but if she has quality she requires little addition or embellishment."

9.1 Concrete Mix

No collapse can be attributed solely to an improper concrete mix or to noncompliance with the proportions set by a controlled mix design procedure. Practically every report on failure indicates, however, that test cylinders or cores of the hardened concrete are below design strength. When the first Joint Committee Code for Reinforced Concrete was issued for discussion in 1932, the writer said in reference to the section on design mix controlled concrete that he "would like to raise the question of what should be done in the case of a building designed and constructed on the assumption that the concrete would test 2000 lb in 28 days and the tests actually showed that the concrete was much weaker."

The answer to this question has been the specified additional curing, the taking of cores for later tests in the hope that the concrete strengths indicated will be sufficient, and, finally, the making of full-scale load tests to prove that the design included too high a factor of safety and therefore the weaker concrete provided was sufficient. The true answer is a little more integrity in the furnishing of concrete with the contracted aggregates, cement, and water, plus or minus other ingredients.

In his own experience the writer has come across some embarrassing situations, probably typical of the experience of other engineers. At one site where concrete was under extremely careful control and test, the mix contained Type-I cement and a field added air-entraining agent. During cement shortages in July 1955, a carload of Type IA was brought in and used. The result was a concrete with a density of 135 lb/ft^3 and a strength of less than 2100 psi at 28 days. The design mix was for 3000 psi and some 20,000 yd^3 of concrete had been produced and tested satisfactorily. The low-strength concrete came near the end of the

job and 170 yd^3 of it was delivered on two separate days. The error was not discovered until some intricate imbedded steel items were encased. Acceptance of the job was delayed for over a year, after removal and replacement of the concrete. Of course this trouble, assuming that the use of the Type-IA cement was an accident and without knowledge, could have been avoided if the test cylinders had been weighed as soon as made. In that way a mix error can be picked up early and the concrete removed before too much cost is involved. The writer used this simple field check on mix as delivered at the New York Coliseum job and found it a rapid check on plant deviation from design mix, which did occur even with plant inspection.

In two building operations designed by the writer and under simultaneous construction some 30 miles apart in Texas, the concrete design mixes being identical, the cement brand the same, and the aggregates from one source but with separate mixing plants, the 28-day tests were consistently higher at one job. The average difference was over 500 psi, not explicable from any variation in slump, temperature, or testing procedures.

On the other hand, on two similar housing projects in New York City on opposite sides of a main road, with concrete coming from a single plant and the trucks routed at the entrance to either project as agreed on by the two contractors, with different testing laboratories, one project showed not a single cylinder below the required 3000 psi, and the other had consistent low tests.

One can only conclude from such experience that the cylinder test data are indicative but not conclusive, and no simple correlation between failures and mix production is to be expected.

Improper aggregates have caused orders to demolish concrete already placed because of low indicated strengths. Much publicity was given to the case of Public School 90 Brooklyn in 1963 when 28-day cylinder tests indicated 40 percent below the 3500-psi design strength of the lightweight aggregate concrete in some of the tests. When the city ordered the entire concrete removed, some ⅓ of the total required for the building, the contractor asked for determination of strength on the basis of the New York City Building Code, namely, (a) core tests to check the cylinder tests, (b) analysis of the design to see if the indicated concrete strengths were sufficient for the intended loadings, and (c) static load tests under 150 percent of the live load. Some 14,000 ft^2 of floor were involved and the concrete was a lightweight mix, the first use of such aggregate in New York City schools.

Much of the concreting had been performed in cold weather and an inspection of the mixing plant, correlated with analysis of the concrete

cores, indicated that the aggregate had been piled without protection, contained much ice and was completely uncontrolled as far as separation of sizes was concerned. The core tests confirmed the wide variation in concrete strength, although low results were not found in some of the suspected areas and were found in areas where the cylinders indicated sufficient strength. Under a program of 26 load tests in the areas where low-strength concrete was suspected and tests carefully monitored, every test passed the requirements of the code. Technically this was sufficient for approval of the completed work, but the Board of Education refused to accept the results as conclusive and insisted on the removal of about half of the originally rejected areas. Reconstruction and completion were carried out with a modified concrete mix design with better control of the aggregate gradation and temperature control, after a nine-month delay.

Special concrete mix used in the piers and connecting bridge seat girder for a landscaped footbridge connecting the additional east campus with Columbia University's center mall failed to test to the desired 3750-psi strength and the 80 yd^3 of concrete was demolished. With a specially prepared crushed red granite aggregate, the laboratory tests indicated that 566 lb of Type-I cement would provide the desired strength. Field tests ran 20 percent low and were confirmed by tests on cores. The work was rebuilt and completed with a new mix formula requiring 680 lb of cement and removing all minus #4 screen fines from the crushed rock aggregate. There was a delay of some four months in 1963 to revise the mix requirements.

In 1965 some 64 yd^3 of concrete slab, beams, and columns placed in the floor of the A. J. Loos Fieldhouse, Dallas School District, were ordered removed when the cylinder tests did not come to the required 3000-psi strength.

Improper aggregates in a concrete mix can be the cause of much trouble in appearance and even use of the facility. Even the watchdog agencies and committees of Congress have become aware that "permitting the use of materials that were known to cause serious pavement defects" will reduce the usable life of a concrete runway. The quote is from the discussion at a budget request to reconstruct the 1959 Selfridge Air Force Base runways, which had seriously raveled by 1964.

Aggregates used in a concrete mix must be compatible with the other ingredients. In 1947–1948 an investigation was made of abnormal cracking in concrete highway structures in Georgia and Alabama. An analysis of the concrete records compared with actual performance, checked by extensive laboratory tests of aggregates and cements, showed that the defects seemed limited to concrete mix in which natural siliceous aggregates from Montgomery, Ala., were used with three ce-

ments having alkali content (percent of alkali as equivalent Na_2O) of more than 0.6 percent. Investigation covered the history of 294 bridge structures, all at least five years old. The cracking from the "Alkali-Silica" reaction seldom starts at a younger age. Defects were observed in piers, pier caps, wing-walls, railings, and curbs. The work was corrected by adding concrete covering after removal of the cracked and loose concrete and on less affected surface by paint sealers. All future work was performed under a specification restricting alkalies to 0.6 percent in cements, which has been effective in control of such failures.

Another careful analysis of concrete mix failure resulted in the determination that adding a water-reducing admixture to a cement with low SO_3 content will cause a delayed set for several days. In January 1960 the concrete placed in the arch and sidewalls of a railroad tunnel in rock, around the Oroville Dam reservoir in California, refused to set normally. When the forms were stripped, most of the concrete in the arch back to the steel tunnel bracing came down with the form. The mix remained plastic for over five days.

Aggregates had been carefully prepared and controlled, and the mix contained five bags of Type II cement per cubic yard. Slump at the 7½-cy mixer was held to 5½ in. with 2½ percent air entrained by a resin agent, and a water-reducing admixture was also added in solution to provide ¼ lb of sodium ligno-sulfonate solids per bag of cement. To test the effect of the admixtures, after two 48-ft lengths of tunnel behaved the same way, a third section was placed with a mix containing no admixtures and an additional ½ bag of cement per cubic yard. This set up in the expected time period and forms were removed in 10 hours.

Not wishing to give up the advantages of the admixtures in the pumping of the concrete, the contractor brought in a new shipment of Type II cement and placed the next section with the original mix. It was successful, but the following day's placement again came down with the forms. Further work was with a different brand of cement and no slow set was then noted. The three sections of slow-set concrete were trimmed out and after becoming hard were fixed by shotcrete when one month old. The chemical analysis of the various materials showed conclusively than an SO_3 content below 2 percent in the cement made it incompatible with the lignin-base admixture. The extremely sensitive effect of very small percentages of some admixtures must be carefully considered before permitting their use. Such analysis must keep in mind that accurate control of all other ingredients in a concrete mix is desirable but seldom available. This is true even in precast work (Fig. 9.1).

Scaling of a newly completed pavement near Deming, N.M., in 1965 was mysterious because it was confined to some 30 areas, each about 3 ft

Fig. 9.1 Precast elements made without control.

square, along 2000 ft of the two lanes then paved. A careful analysis of all ingredients could point to only one possible explanation. The cement was shipped to the site in bulk, using railroad cars that had previously hauled potash, which could have contaminated some of the cement.

To avoid normal shrinkage of large concrete masses, especially when placed by tremie in water-filled forms, the procedure of first placing the aggregate and then injecting the mortar, which displaces the water, has been developed. At least two incidents of nonsuccess are reported. The center pier for a movable railroad bridge connecting Staten Island and New Jersey was founded on a caisson pier that was so filled. Upon completion and dewatering, the concrete was found to be discontinuous

and of insufficient strength. A new foundation support for the pier was then provided by inserting drilled-in caisson piles, socketed into the rock, right through the grouted preplaced aggregate.

The second incident, coming about the same time, was the discovery in 1958 that part of the steel sheetpile enclosure of the south anchorage of the Mackinac Bridge was not completely backed by the intruded aggregate placed in that cofferdam. The sheetpiling had been left as protection against ice and scour. Some 260 lin ft of the face was found loose, the sheetpiles were removed and replaced by an outer layer with the space to the concrete filled with new concrete to form a solid mass. It must be pointed out that this procedure for placing concrete in difficult positions, usually under water, is usually satisfactory and economical. The repairs of a large cavity in the Arizona spillway tunnel of the Hoover Dam and the rehabilitation of Barker Dam by adding a layer to the upstream face, were both performed by the prepact method of concrete placing.

Concrete strength gain of a mix depends on the temperatures of the mix as well as of the outer air, although ultimate value of strengths is not as much affected. Frost affect is covered in Chapter 10. In extremely hot climate especially with low humidity, such as exists in the California desert areas, proper control of the concrete mix can only be obtained during night time placing. Concrete placed in 1962 at three buildings in Athens, Ga.—a dormitory, a pharmacy school building, and a sports coliseum—failed to gain the expected strength, apparently because of the hot weather during the construction period.

9.2 *Surface Reaction*

The hardening of cement in concrete or mortar is not a true chemical reaction and the resulting crystallization of the amorphous cement is subject to alteration when in contact with many corrosive materials. Disregard of this possibility has resulted in much surface and some deeply formed disintegration of concrete, with unsightly surfaces and expensive corrective work.

Recognition of this danger came as early as 1880, when Professor Prazier of the University of Aberdeen, Scotland, investigated the causes of the previous concrete in the graving dock at Aberdeen Harbor. He warned that overcalcined cement used in concrete exposed to sea water will absorb magnesia, an ion exchange phenomenon, resulting in expansive surface disintegration which in turn exposes more surface to the same action. Similar trouble was encountered at many harbor works in

England, and M. Vicat came to the same conclusion in France, in the study of disintegration of the concrete blocks made of lime and artificial pozzuolana at Marseilles, Rochefort, Algiers, and Cherbourg, after only a few years exposure to the sea.

Normal chemical action on concrete from salt spray, ice abrasion, tide, and weather is found on many waterfront structures. After about 21 years of use, the concrete piles and caps of the trestle bents of the James River Bridge at Newport News, Va., required a $1.4 million repair and replacement job in 1955. Some 70 percent of the 2500 piles were found in need of repair. All the piles were therefore jacketed over the tidal range and the steel sleeves form left as covering. In the period 1947–1952 about 60 percent of the piles not affected at the first inspection were seriously damaged (Fig. 9.2).

Similar deterioration was found in the precast concrete piles driven in 1932 near Ocean City, N.J., where by 1957 the 22-in. dimension had become reduced to 12 in. Some 750 piles were jacketed with pumped colloidal grout within a steel sleeve. Protection against the natural aging influences of salt water aided by the physical action of tide and wind has resulted in the recognition that especially dense concretes are necessary and that moisture tight surfacing is economically advisable.

The chemical reaction of sea water on concrete, followed by rapid disintegration, has been researched for many years. In 1917 Rudolph J. Wig of the Bureau of Standards and Lewis R. Ferguson of the Portland Cement Association reported on personal inspections of 130 structures where concrete was exposed to sea water. Locations included all coasts of the United States, some in British Columbia, Panama, and Cuba. Quoting from their report:

"It is not the intention to demonstrate that concrete cannot be used for marine structures. On the contrary, it is the opinion of the authors that it is a most economical material to be used if it is made so as to successfully resist sea water action, and it is their opinion that it can be made so. The articles will, however, illustrate in an emphatic manner the rapid deterioration which results from the use of too little or too much water in mixing concrete, the origin and development of deterioration from normal construction seams which are common to most concrete work, the protecting influence of adequate fender systems or granite facings, the need for radical modification in the design of reinforced concrete marine structures, and the conditions under which sea water can be used in mixing the concrete without detrimental effect.

"It is a fact that plain concrete, if exposed to physical abrasion, will very soon commence to disintegrate, due to chemical action. The evi-

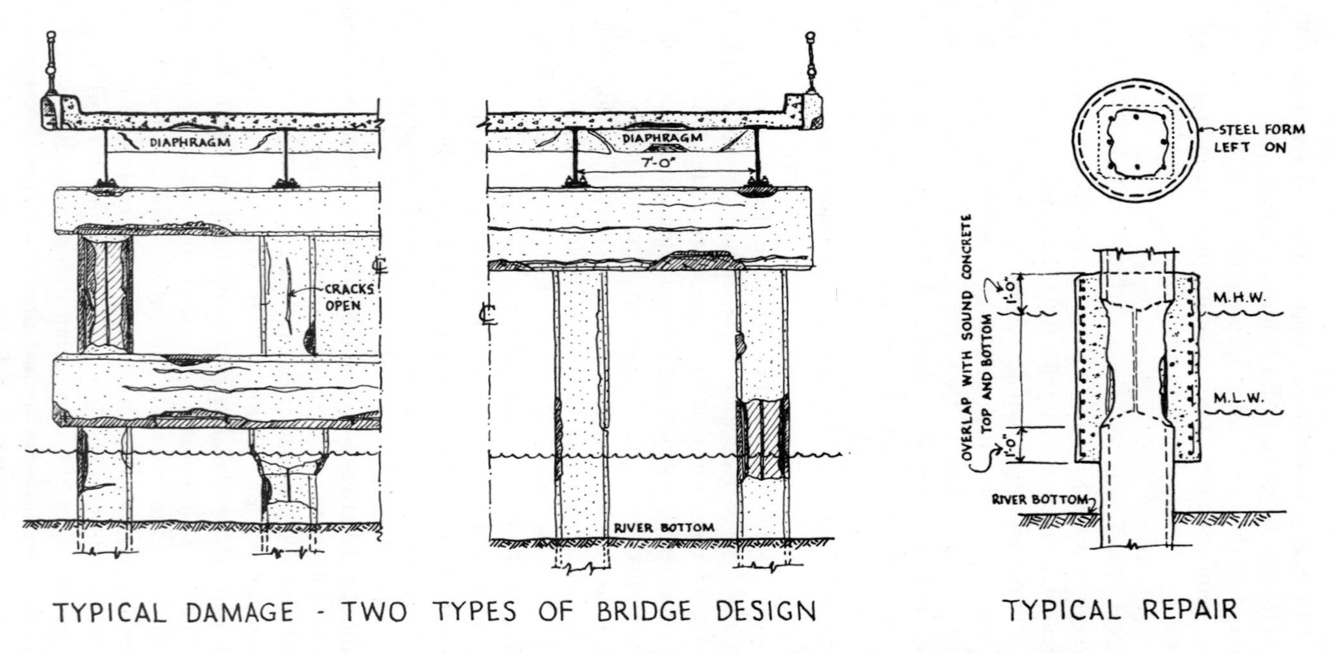

Fig. 9.2 Concrete disintegration in tidal water. (Redrawn from article by George Vaccaro, "Salt, Tide Damage to Bridges Erased by Novel Facelifting," *Engineering News Record*, May 3, 1956.)

dence collected by the authors shows that reinforced concrete will, in practically any climate, begin in a short time to fail above sea water due to the absorption of the sea water or penetration of sea air bearing chlorine to the reinforcement. This reinforcement soon rusts and the concrete cracks."

Rapid chemical attack was found on surfaces tooled off for artistic effect, at sharp edges and at seams in the concrete placing. Any protective coating of steel, stone masonry with tight jointing, or wood sheeting inhibits such action. Disintegration by chemical action is slow unless there is frost action or continued mechanical abrasion, removing the decomposed cement products as they are formed. Reinforcement imbedded in first-class concrete to a depth of 2 in. in accordance with existing theory and practice is not perfectly protected against corrosion by sea water action. Cracking of the concrete in sea water always starts above the high-water line and seldom is the corrosion found below low-water line. Electrolysis is not the only cause, as was shown by the cracking of concrete piles built in 1911 in Long Beach, Calif., harbor, which had remained as cast because the pier had not been built. The authors of the report conclude that concrete work to be exposed to sea water must be considered as requiring special attention and entirely free of imperfections normally acceptable in concrete production.

Experiences reported by J. J. Yates of concrete piers for railroad bridges in the vicinity of the lower New York Harbor, covering the years 1900–1917 indicate difficulties all caused by poor concrete placing. The center piers of the Hackensack and Passaic River Bridges of the Central R. R. of New Jersey were constructed in 1870 of Rosendale cement. The concrete was placed with circular wrought-iron caissons left in place. By 1902 the caissons had completely disappeared to a depth of several feet below water, and the concrete had been eaten away to a depth of about 2 ft between high and low water. Repairs were made by filling concrete within new steel bands around each pier.

Between 1903 and 1905 seven concrete piers were constructed for the drawbridge over Newark Bay within timber cofferdams, in the dry. By 1909 five of the piers showed disintegration up to 20-in. depth, but none beyond any point 1 ft below mean low water. Repairs by gunite lasted about four years. The action was not general over the entire surface but confined to local areas, and two piers were completely free of defects. The bridge over the Shrewsbury River, N.J., built in 1913 had 10 piers, two of which showed signs of failure about a year after they were built. Concrete had been placed by tremie and alternate layers of concrete and a soft putty containing salt crystals was found on the removal of these two piers. Laitance accumulation of 2–3 in. in 15-ft depth of concrete

placing was common when placing concrete using gravel aggregate in sea water. With crushed stone aggregate, however, the laitance layer was twice as thick. Complete removal of these deposits is necessary to avoid weak seams for the beginning of sea water action.

The durability and precautions necessary to obtain it in concrete in coastal structures is completely covered in the Beach Erosion Board Technical Memorandum No. 96, June 1957, and practical experience is described in the discussions and paper by William G. Atwood and A. A. Johnson on "Disintegration of Cement in Sea Water," *ASCE Transactions*, 1924, and in many recent articles in the technical press.

Bursting of precast concrete beams in the lumber drying kilns at Muskegon, Mich., in 1956 was caused by the expansive forces from the rusting of the reinforcement after the concrete had absorbed moisture including lignite from the green lumber. Although factory cast and of dense concrete, the usual cover on the steel in the 14-in. web beams is less than 1 in. and not sufficient to stop air and moisture attacking the steel and not thick enough to resist the expansive force of the rusting. Actually enough concrete spalled off to permit shear failures in some of the beams especially with slippage of the stressed corroded reinforcement. Concrete exposed to the high humidities of special uses should be waterproofed with either bituminous or plastic skins and as a second precaution, concrete cover over the reinforcing should be at least 2 in.

An investigation of concrete deterioration of a foundation for a silo in Copenhagen Harbor in 1957 indicated sulfate attack by the ground water with aggressive carbon dioxide, also a primary agent of disintegration. Excessive formation of secondary chemical deposits were found in the concrete exposed to ground water.

Excessive use of calcium chloride for snow removal, together with the leaching of tannic acid from the soil adjacent to new concrete curbs caused a complete disintegration during the first winter of the curbs in a plant built in 1947 in Syracuse. In some lengths there was no sign of curb above the asphalt pavement after the snows melted in the spring. The remainder of the concrete was saturated with a tannic acid compound largely activated by the calcium chloride.

Tannic acid wastes in the Providence, R.I., harbor had penetrated the adjacent soils to such concentration that the concrete footings for a housing development built below ground water level during pumping of the footing pits, and flooded after the concrete looked hard, were affected sufficiently to require removal and replacement. Most of the concrete remained soft and friable for more than two weeks after placing. Reconstruction was successful when the pits were kept free of water for about a month after placing.

At the Yonkers Raceway in New York State, the top deck of the garage

built in 1955 was covered with a ¾-in. application of tar macadam as a wearing surface and waterproofing protection for the lower levels. The deck slopes uniformly at a 1½ percent rate to trough drains spaced some 250 ft apart. After eight years use in summer weather only, the racing season was continued into the winter months. Some rapid method of snow removal was necessary to provide for the peak-load parking. Calcium chloride could not be used because it might be carried over to the stable area and it is injurious to horses' hoofs, so common salt was used in copious quantities. By the third winter the salt had filtered through construction joints of the tar surface, near the low area, penetrated through the 3-in. concrete of the waffle slab, and covered the entire ceiling area with a complete coat of white salt powder. It had also completely softened the concrete so that a man stepped right through a pan area and was kept from falling only by the mesh layer that had been placed in the top of the slab. Every dome area was sounded, and all hollow or soft indications were noted to be removed. All loose surface topping was cut out, soft concrete replaced, and the exposed concrete sealed with asphalt and recovered. Granular salt had been found piled up several inches thick on top of the deck when the extent of damage was being surveyed. Snow blowers were recommended as the safer method of keeping the deck available for winter parking (Fig. 9.3).

After some sad experience with the specified separate 4-in.-thick concrete wearing courses on the 200 bridges of the New York State Thruway, some after two years of service, the N.Y. Department of Public Works modified its standards in 1959 to provide asphaltic concrete wearing course and open grating drains to remove the brine from the direct salt application for snow removal. The new standards will permit a concrete surface if cast monolithically with the structural slab. Most of the bridge decks have been resurfaced or surface sealed to protect against the chemical action of future salting.

The de-icing problem for pavements is not yet completely solved. The old faithful rock salt and calcium chloride effectively remove snow and ice, and if used with moderation will usually not harm good normal concrete, and even less good air-entrained concrete. New agents have come on the market, but any that contain either ammonium sulfate or ammonium nitrate will actively attack any kind of concrete.

Application of plaster or paint on concrete surfaces for interior finish has been a troublesome item. For many years after direct application of finish plaster on concrete ceilings became an acceptable practice, a patented gypsum product was sold with guarantee of adhesion. So many claims for noncompliance arose that the product was taken off the market. Plastic spray coatings were then developed to act as bonding

Fig. 9.3 Effect of salt de-icing on concrete deck.
Upper—extent of concrete softening.
Lower—salt penetration to ceiling below.

agent between concrete and white plaster. Although very successful in most instances, the permanence depending on a continuous supply of moisture vapor to maintain a colloidal gel, there have been a number of failures apparently from the dehydration of the concrete and of the plaster.

The action of lactic acid on cement is so rapid and complete that the use of concrete floors in process plants where lactic acid can form is most inadvisable. These include milk pasteurizing plants, cheese manufacturing plants, breweries, and slaughterhouses. Even the joints between the packing house brick used for floor surfacing, unless made of special cements or sulfur compounds, will deteriorate within a year. When complete sanitary control is required, even a surface defect in the joints permits the accumulation of milk or other wastes that soon alter to lactic acid and cause further chemical reaction with the cement.

Investigation of complaints of floor leaks in the Bush Terminal structures in Brooklyn showed that circular wood vats with the sides projecting below the vat floor, used for the cooking of candied fruit peels, had permitted the sugar syrup to eat into the concrete floor and form in some cases a complete ring of disintegrated soft concrete through the entire floor thickness. Sugar or molasses mixed into wet concrete will inhibit the setting of the cement and have been used as a means of preventing the hardening of concrete that is to be removed, as in formwork failure incidents.

Cinder concrete, which was formerly used to a much larger extent in the New York and Boston areas, is especially vulnerable to chemical action. With even small traces of sulfur present, electrolytic deterioration is rapid on all metallic imbedded items. In a floor covering several ammonia compressors and carrying a large bottle-washing machine in a New York milk plant, the combination of some milk slop and some ammonia absorption completely disintegrated all of the welded wire mesh reinforcement. All the steel reinforcement and conduit exposed showed so much deterioration that the owner accepted the writer's recommendation and abandoned the building. Repair costs during occupancy would have far exceeded the replacement cost of a new plant.

The action of teredo and limnoria borers on timber is well known. Some bacteria, mollusks, and similar animal life can deteriorate concrete from the chemical wastes reacting with the cement or aggregates and sometimes from actual physical erosion to make shelters below the surface. In 1922 the concrete encasement of some wood piles in the Los Angeles harbor were found to contain boring mollusks that had penetrated from 1 to 3 in. of the 3-in. mortar coating. At that time they had been exposed to sea water for 14 years; similar conditions were found on

both sides of the channel for a length of about half a mile. Borers were found from the mud line to about 20 in. higher. None were found higher than 4 ft above mean low tide. The aggregate used was a sea sand from a beach where every variety of native rock contain perforations made by Pholas mollusks, so it is assumed that they had been mixed in with the sand in the covering operation. Adjacent concrete with stone aggregate was not attacked. The jacket mortar contained as high as 10 borers per square foot of exposure.

From Poland comes a report of concrete disintegration by bacteria identified as Merulius lacrymans and Poria vaporaria in some deep tunnels for the Warsaw subway in 1952. The concrete had been placed against fresh pine boards which contained the fungi bacteria. Although no nutrient was found in the concrete, the movement towards light and food had caused considerable damage to the concrete from the secretion of organic acids and the erosion of the affected concrete.

Sulfur-producing bacteria usually found in sewers are the usual instigator in the concrete disintegration. The hydrogen sulfide emitted reacts with the cement in the concrete unless a protective coating is applied before the sewer is put into service. The most promising of the recently developed coatings is a polyvinylchloride plastic, which seems effective even when only 0.06 in. thick.

9.3 Internal Reaction

The proper method for combining ingredients for the production of sound concrete that will have a desired strength is fairly well established, but often the desired life or result is not obtained. All objects tend to disintegrate; in the case of concrete the inclusion of reactive ingredients or imbedded materials helps the natural process. The proper concrete mix becomes chemically inert after the cement reaction is complete. Certain chemicals react badly with the binder that holds concrete together, and they include the entire family of sugars, lignites, resins, sulfur vapor, salts, and many others. The subject of reactive rock in concrete has been completely covered by the work of various Highway Research Stations and of the Bureau of Reclamation and needs no elaboration. Yet often these warnings are not heard in the general construction field.

Internal chemical changes within the concrete resulting in cracks and fractures often accompanied by crazing and spalling of the surfaces, can occur under certain cement and aggregate combinations. An alkali reac-

tion occurs between cements containing a high content of alkali and aggregates with soluble silica, resulting in the formation of the expansive gel formed by alkaline hydroxide with water and silica.

The expansive action of magnesia in plaster is described in Chapter 6.

The use of sugar to inhibit the setting of cement is well known. An accidental and undesired use occurred in a grade crossing elimination project in New York, where when the forms of a large pier were stripped, a soft brown area was exposed that had not set. The explanation was pinned to the dropping of several containers of coffee, brought to the site for the men who were placing and vibrating the concrete, into the pier form. The blend of sweetened coffee and concrete had to be taken out and replaced with regular concrete.

Cinder concrete was the standard in the New York area and elsewhere as long as good, clean, completely combusted anthracite clinker was available. However, many jobs used bituminous cinders and the incomplete combustion left free sulfur that corroded the reinforcement. Cracks and spalls were common, especially along the moist seashore. An example of the resulting troubles is the failure of the roof over the kitchen of the Atlantic City Hotel in 1921, then 15 years old.

Often the cement chemistry can be modified or an additive used to mask the reactive relation between the cement and the aggregate. In 1966 at the Jupid Dam on the Parana River, Brazil, at the beginning of the concrete placing for a 3151-ft-long concrete section to serve as spillway, powerhouse, and navigational lock, swelling and cracking was noted. Its cause was found in a chemical reaction between the alkali in the cement and the basalt aggregate. Because it was impractical to control the alkali level of the cement to a point below 6 percent and freight costs of imported pozzolans, which would serve the same purpose by diluting the alkali in the cement, were very high, it was decided to manufacture a pozzolan at the site. Local clay is calcined at 700°C in a plant built for the production of a good grade of pozzolan used as a 10 percent replacement of the cement with little loss in strength and no change in total cost. In addition to elimination of the swelling and cracking, advantages included a reduction in heat developed during setting, tighter concrete, and improved resistance to the sulfates in the stored water.

In the Oslo area of Norway a concrete-eating rock is found, locally called "aluminum slate," but actually a metamorphic compression of sulfates and carbon with water-laid silts. When exposed to air and water, the oxidation products are lime, iron oxide, and sulfuric acid. The mass increases to 3½ times the original volume. When concrete is placed against a fresh exposure, walls and floors heave and the acid

reduces the concrete to a crumbled mass of aggregate. The sulfate-resistant cements are less attacked and watertight film covering of the rock will inhibit reaction. But the jet-black rock is avoided where possible.

Aggregate coming from rocks that have expansive qualities, usually classed as dolomites, have been found to cause much of the distress in viaduct bridges when growth in length closes the expansion joints, and also buckling of concrete roads in high temperatures and also in the pattern cracking of monolithic finish. These rocks are found to contain at least 50 percent dolomite in the total carbonite content, an insoluble intrusion of clay minerals that may be reactive and an apparent separate crystal structure of the dolomite in a fine calcite matrix. Much damage to concrete has resulted from the use of such rock especially in the southern Appalachian area and in the vicinity of Kingston, Ontario; standard tests for acceptability of aggregate do not reject these materials, nor do pozzolans and chemical admixtures prevent the expansion.

Sea water may be an economical necessity for concrete mixing in locations where fresh water is not available. It serves well in the making of "shell concrete" as used in many tropical islands, where crushed shells and beach sand are blended and packed into forms to slowly harden into a natural concrete without the addition of cement. When salt water is used in reinforced concrete, especially in tropical areas where high temperatures and humidity are prevalent, expansion and rusting of the reinforcement will rapidly destroy the structure. This was the warning given as early as 1916 in the Bureau of Public Works Bulletin when all engineers in the Philippines were advised that use of salt water in concrete structures is dangerous and the use of beach sands and gravels should be permitted only after thorough washing with fresh water.

Duff A. Abrams in 1924 reported that laboratory samples made with sea water gave higher strengths at three and seven days than equivalent samples with fresh water. However, at 25 days, the sea water concrete fell to 80–88 percent of the normal mix, with a trend to equality at greater age. But in spite of these tests, he also stated that it was unwise to use sea water, especially in the tropics, on account of the danger of corrosion of reinforcement.

Examination of concrete structures along the Florida Keys where sea or brackish water was used indicates considerable disintegration. A report in 1948 of the concrete placed during 1941–1945 by the Civil Engineer Corps of the U.S. Navy in the Pacific Islands, using sea or brackish ground water, provides mixed results. At Midway Island the concrete around underground steel-tank fuel storage was found good

with no deterioration. At Wake Island the concrete basketball courts and Quonset hut floors were in poor condition with surface spalling and wide cracking. At Johnston Island the concrete recreation halls and bases around Quonset huts were in poor condition and easily crumbled by slight impact. Also inspected were some airplane warm-up areas of 6-in. pavement, underground hospitals and a communication center, gun emplacements and several utility buildings, all of which were in good condition. However it is pointed out that in most of the cases brackish water of low salinity, 57 grains of sodium chloride per gallon as compared to 1638 grains in sea water, was used and that the high water content of the aggregate further reduced the salinity of the mix.

9.4 Shrinkage, Expansion, and Plastic Change

Although a very few examples of actual collapse can be blamed on temperature and shrinkage loadings, considerable cracking and spalling results in concrete structures from lack of freedom to modify dimensions with temperature changes and from aging. Unfortunately, due to the thermal insulation afforded by the concrete to the inner volumes and to the nonuniform rate of shrinkages in elements of different thickness and shape, the dimension changes are not entirely linear. Rotational displacements result and seriously affect brittle masonry surfacing, windows, and door frames. Although the addition of uniformly distributed and continuous reinforcement is recognized as a proper resistance against thermal changes, the apparent distress from shrinkage is not affected by such steel. Actually shrinkage distress is more likely with larger amounts of reinforcement because the steel will not shrink and internal bond failures may result.

Temperature-restraint reinforcement is always specified for floor and roof slabs, but it is seldom required for beams and girders. Even the minimum 0.25 percent requirement, if applied to all concrete sections, would prevent totally unreinforced surfaces with possible shear failure. The older literature suggests a minimum of 0.3 percent of the cross section as necessary to prevent cracking from shrinkage and temperature. This amount has not always prevented the formation of cracks and Dreyer (1933) stated that the experience of the Pacific Gas and Electric Company indicated a percentage of 0.65 as necessary.

The differential shrinkage between thick and thin reinforced concrete members in intimate contact can cause an eccentric pull on the members and it is the usual reason for the cracks forming in slabs connecting to heavy girders. R. F. Blanks presented some data (1951) on this subject

and showed the results of tests by Pickett in measuring the coefficient of shrinkage at age of nine months for various thicknesses.

1½ in.	thick:	720 millionths
3	thick:	700
6	thick:	580
12	thick:	330

A report on Drying Shrinkage by Bryant Mather as Chairman of a Task Group for the Highway Research Board Committee on Durability of Concrete, appeared in *Highway Research News,* 26–29, March 1963. Drying shrinkage is there defined as the reduction in volume resulting from a loss of water from the concrete after hardening. Some of this happens at an early age with consequent development of plastic shrinkage cracking, forming internal weaknesses which later become open cracks. When volume changes are restrained, stresses are produced when later shrinkages occur, which may cause distress and failure. A complete review of existing reports indicates a lack of knowledge of the causes and cures for this important phenomenon.

Empirical experience in recent years indicates that the magnitude of the delayed shrinkage, at ages of about two years, is more critical in concretes with lightweight aggregates than in normal concrete. This is usually blamed on the saturated state of the aggregate when proportioned in the mix. Attempts to correct this condition by mixing dry lightweight aggregates result in a concrete that is "bony" and very difficult to place and finish properly so that the field is forced to add water to the mix, with resultant low strengths and increased delayed shrinkages.

The amount of overstress due to creep in long beams, 40 m and over, considered significant by G. Lefaudeux in *Le Genie Civil,* **137,** (Paris), July 15, 1960, was discussed by R. Chambaud, Henry Lossier, A. Couard, and A. Guerrin without full agreement, but the problem should be resolved if long spans are to be considered a safe use for concrete. Similar discussions have recently appeared in other foreign journals.

Transfer of loading, when columns in multistory frames are affected by delayed shrinkage usually after the building is completed and in use, has cracked and spalled rigid brick and stone coverings. This has been found more often when lightweight concrete mix was used for the columns combined with a glazed brick unit laid in tight mortar joints. The very first high-rise building with lightweight concrete in New York, covered with white enameled brick, developed vertical cracks at the corners and considerable spalling of brick surfaces at two year age. Several more buildings in the 20-story range and some less than 10 stories show the same distress. See Chapter 6.

By 1962 the then 10-year-old Veterans Hospital in Boston had so much brick facing falling from its 14 stories that the grounds had been roped off to avoid accidents and it was decided to remove the entire facing and replace with a metal curtain wall at an estimated cost of over $4 million.

Repair and legal costs on the several buildings investigated in the New York area have not been large, but the continuing maintenance costs are causing serious questions as to the suitability of the material for tall frames. A suggested preventive measure has been installed in several buildings. The joints in the masonry just below the continuous steel lintels are filled with a flexible caulking and premoulded sheet to allow each story height to take up the column shrinkage without imposing extra load on the masonry covering. Experience with this detail is only a few years old, but there seems to be no indication of distress in the facing of these buildings; one is now six years old.

Freedom from the effects of delayed shrinkages in assembled precast units as well as in cast-in-place rigid framework must be considered. When a design is not carried to completion with details consistently analyzed or when construction procedures knowingly or unknowingly introduce factors that have not been considered in the design, unexpected things happen. Cracks, spalls, disintegration at connections, deflections and deformations appear, apparently without cause. Just as an experienced foundation engineer can follow the pattern of cracks in a structure to localize the deficiency of foundation support, so will a careful three-dimensional trace of structural distress indicate a funnel surrounding the trouble. The axis of the funnel points to the deficiency which originated the distress.

The amount of long-time shrinkage of an actual concrete structure is dependent on the many factors of type of mix, temperatures during placing, and climatological conditions. One detailed record is available of the dimensional changes in a seven-story flat-slab building, 275 × 131-ft plan, with 9-in. slabs spanning in a 21 × 22-ft column array. It was built in 1939 and instrumental measurements of vertical and horizontal change were made for 16 years. After that time the shrinkage vertically was 985 millionths or about 0.1 percent, and horizontally was 460 millionths, equivalent to 0.55 in./100 ft. The horizontal change was equivalent to the effect of a temperature drop of 100°F. The vertical effect was twice that found horizontally, explained by the horizontal restraint from the floors acting as diaphragms braced by the columns. The vertical measurement is also an accumulation of plastic flow (creep) in the columns with the drying shrinkage.

All of these length changes cause cracks in the concrete. Three classes

of cracks are normally recognized in concrete: (a) those resulting from elastic strain and usually of hairline width, not considered injurious to stability, (b) shrinkage cracks from drying out of the paste, and (c) corrosion cracks from volume change of aggregate or reinforcement. The latter two types are more serious than the first.

Shrinkage of concrete floor slabs is often exhibited as a short diagonal crack across the corners of the building, sometimes quite close to the column and even spalling off the inner corner of the column above the floor level. In at least two large warehouse buildings such cracks have been noted wide enough to allow light to penetrate from one floor to another. In the standard details for such building designs, common knowledge in the 1920's but now seldom followed, a set of corner bars parallel to the diagonals of the building was always placed in the top of the slabs, hooked down into the edge beams, at all floor areas bounding an exterior edge or an interior open shaft. The extra cost of such a detail is warranted as a precaution against corner cracking.

Serious cracking in the walls of the Ice Harbor Lock, near Pasco, Wash., after five years use as a navigation control in the Snake River, resulting from a combination of shrinkage and stress changes with each emptying of the lock, required a large repair by introducing post-tensioned tendons into carefully drilled holes. By 1965 cracks had opened, similarly in the two walls at the downstream gate. The walls are 60 ft thick and 70 ft long and act as guides for the 91-ft high-lift gate. To avoid similar shrinkage in two other locks under construction, John Day and the Lower Monumental locks will both be post-tensioned to provide the necessary tensile resistances.

Expansion of concrete with age, an accumulation of growth from successive thermal change and internal reactions of aggregates can cause shape change and major cracking. During the late 1950's the Texas Highway Department constructed numerous anchorage systems at the terminals of concrete pavement adjacent to structures to restrain pavement growth. By 1963, 22 of the 152 anchor units inspected had failed. The extent of growth exceeded the capacity of the possible transfer of expansive stress to the soil, and the soil sheared along the bottom of the anchor lugs placed under the slabs.

The 710-ft long highway bridge built in 1948 at Massillon, Ohio, was closed to traffic in December 1964. The deck is a concrete-filled steel grid on steel floorbeams and stringers resting on concrete piers and abutments. Several years earlier the concrete sidewalks buckled during hot weather and the expansion joints were almost closed. The joints were cut out to provide thermal freedom. The deck had similarly broken up at several places, an abutment wall cracked and ripped out the fixed

bearing anchor bolts. Expansion in one 210-ft span exceeded 3 in. The damage to the deck of the structure in 15 years of expansive growth of the concrete, possibly resulting from the use of an expansive aggregate, caused abandonment and demolition. A new reinforced concrete deck with some strengthening of the steel framework was then placed on the original piers.

Expansion growth of the parapet wall on the Northern Boulevard overpass at Corona, New York City, crossing the Grand Central Parkway buckled the face outward some 18 in. by September 1956 and traffic was diverted until emergency repairs could be made.

The 163-ft high concrete arch Matilija Dam in Ventura County, Calif., after 16 years of expansive reaction between aggregate and cement, which caused interior swelling and exterior cracking, was ordered demolished in 1965. Further study modified the order to reduction of capacity of the reservoir by half. Two notches 140 ft long and 30 ft deep were cut in the 630-ft-long crest. It was hoped to use the reduced facility another 15 years before further action would be necessary to compensate for the weakening of the structure.

9.5 *Surface Disintegration*

A frequent item of observed concrete distress is the surface disintegration of pavements, curbs, and sidewalks. Whether the result of traffic action on a poorly surfaced concrete, aided by the chemical and abrasive additives to control snow and ice coverage, the fact that there are so many good concrete surfaces of many years life, shows that some detail of combination or of construction has been improperly introduced. In 1958 the New York State Thruway Authority spent about a million dollars in the repair of concrete wearing surface on some 80 to 90 bridges. The design consisted of a 4-in. concrete wearing surface on a waterproofed bridge slab. The top slabs in four years use, had sometimes heaved and in other areas completely eroded.

With the failure of wearing courses of more than 200 bridges on the New York State Thruway after an average of two years of service, the Federal Highway Administrator suggested that New York State's standard design be modified in 1959. The standard was changed to a 2½-in. bituminous wearing surface and by 1961 all Thruway bridge decks had been sealed with a slurry of coal tar pitch emulsion overlaid with 1½-in. asphaltic concrete. An extensive field study of protective coatings to stop surface deterioration of the concrete was carried on for several years. As a result by 1963, it was determined that a three-coat application of a

mixture of linseed oil and mineral spirits was the most economically sufficient protection, with a fourth coat applied after five years use. With a cost of 10½ cents/yd^2 for the four-coat treatment (1965 costs) it seemed prudent insurance against the high cost of concrete repair and replacement to require such treatment for all exposed concrete. Newly placed concrete must be completely dried out before linseed oil is applied, otherwise the surface is not of uniform color.

This rediscovery of a technique to control dusting and raveling of concrete floors, common in industrial and commercial construction in the 1920's when wearing surfaces were given two mop coats of raw linseed oil, is an example of changing times and fashions in a period of two generations. To repeat a common expression, "the old-fashioned way is not so bad after all."

The problem of maintaining good riding surfaces on bridge decks with the demand of the public that bare pavements be provided under all snow and ice conditions exists in more than half of the 50 states. The 50–50 mixture of boiled linseed oil and mineral spirits is now used in most areas as a sealant on the pavements and often also on curbs, railings, sidewalks, and pier tops. Many states have budgets of several hundred thousand dollars each year for bridge deck repairs that warrant protective expenditures in all new work. The far northern states, where winter traffic does not insist on full removal of snow cover, have relatively few bridge deck problems. But in heavily traveled main arteries such as the Interstate 80 over the Sierra Nevada in California, with common freeze-thaw cycles of temperatures, about 90 percent of the structures required resurfacing within five years of use. Although the de-icing chemicals are usually blamed for the extent and rapidity of surface damage, they are seldom the cause of the beginning of the disintegration. As a retardant of the action, higher air contents in the concrete up to 8 percent, lower slump and reduced finishing operations are found beneficial.

Several major bridge-deck repairs have been found necessary after deterioration rapidly progressed to a dangerous condition. Built in 1942 before the acceptance of air entrainment in exposed concrete, the Detroit Industrial Expressway, with a bituminous wearing surface, disintegrated so badly by 1957 that complete removal and replacement of areas of slab and tops of concrete beams were required. The repairs were made under traffic volumes of 15,000 vehicles daily using the bridges during the removal of the inferior concrete and replacement with high early strength mix after some repair and strengthening of the reinforcement. A number of these concrete bridges were too far gone for repair and were demolished in 1962, to be replaced by steel spans.

Built in 1952, by 1959 the concrete bridge slabs of the New Jersey

Turnpike crossing the Passaic River had also required repairs costing almost a million dollars, and finally the complete removal and replacement of the concrete deck for some 4800 lane feet. In the same geographical and climatic areas the concrete pavement of the Pulaski Highway Bridges, after 25 years use with very little maintenance, wore too smooth from the heavy traffic and required a surface coating ("retreading") to give it the necessary traction.

Before replacement of the slab on the Passaic River Bridges, a complete investigation of signs of distress, physical and chemical tests of the concrete in place and correlation between conditions of individual slabs and construction reports led to the following findings:

"1. Although concrete cylinders during construction tested at 3000 psi, the cores taken at seven years age only averaged 2500 psi.

"2. Concrete was more porous and less dense than anticipated.

"3. Readily discernible are the variations in appearance and texture of the concrete at different locations.

"4. The reinforcing steel is closer to the surface than the 1½-in. cover intended.

"5. Aggregate tests and cement content more than complied with the original specifications.

"6. There is no correlation between the variations in surface texture or appearance and the incidence of failure.

"7. Earliest and most severe failures occurred where the bars were close to the surface, but many failures did occur where the steel was covered by 1½ inches of concrete, and damage resulted from the expansive forces of the rusting steel."

In the reconstruction the slab thickness was increased from 6¾ to 7½ in. and the steel was covered by 1¾ in. and fixed in position by welding to tie bars, which were welded to the steel beams. Special care was taken to eliminate plastic cracking.

Production of turnpikes financed by investment bonds is set at fast schedules to save interest costs. Sometimes the schedules are too fast for the production of a good product. To repeat the writer's statement made several times in this connection, just as it takes time to make good bread, it takes time to produce good concrete. A rush job, no matter what the specifications say or how rigid the intended supervision and inspection, cannot result in a good job. Perhaps the announcements of records in volume of concrete placed in a day should be replaced by more pride in the quality of the job produced.

Correlation between the occurrence of surface scaling and the method of concrete installation was established by the investigation of the

bridge deck for the St. Catharine, Ontario, Garden City Skyway built in 1962–1963. The roadway is six lanes wide and covers a series of simple steel spans from 100 to 200 ft for a total length of 7082 ft between abutments. The design was based on composite action between the slab and the girders. To provide proper bond between the concrete and the shear connectors after all dead load was in place, a retarding agent admixture was used. The first choice was an organic acid admixture, some 4 fluid ounces per bag of cement. Concrete was placed in hot weather, up to 86°F with humidities varying from 39 to 89 percent. Four days after concreting the first bay of the bridge, the surface was noted to be flaking, generally along bleeding paths near the surface and in isolated areas of deterioration. The condition worsened with age and up to a 3/16-in. layer flaked off. Different methods of curing did not stop

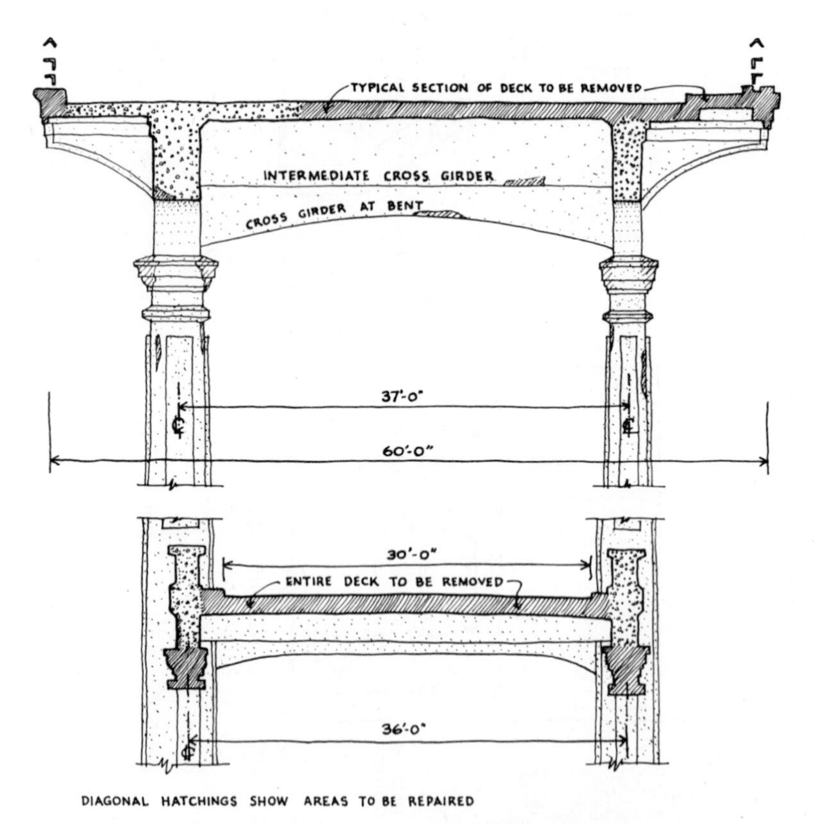

Fig. 9.4 Salt penetration through concrete deck. (Redrawn from article by H. K. Glidden, "Surgery Restores Old Viaduct," *Roads and Streets,* October, 1965.)

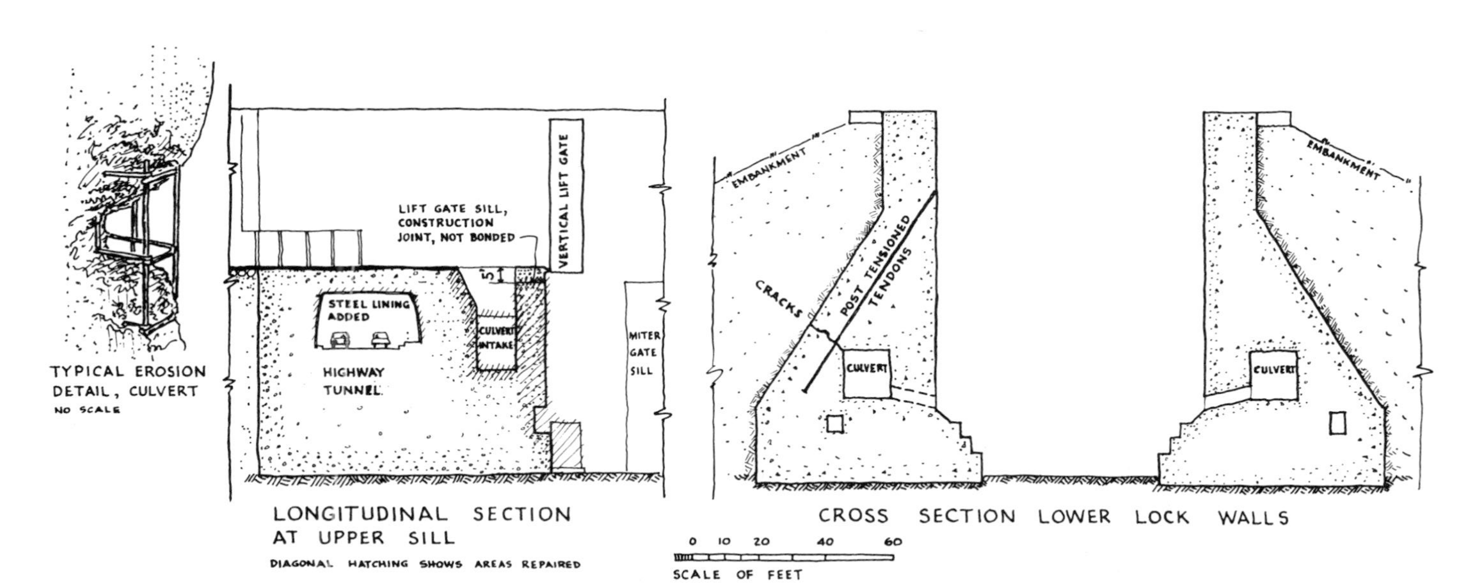

Fig. 9.5 Concrete distress in St. Lawrence Seaway lock. (Redrawn from *Engineering News Record*, August 31, 1967.)

the trouble. The retarder was changed to a lignosulfonate which caused a greater water reduction than the organic acid admixture and provided about 5 hr of retardation of setting. The earlier mixes had continued bleeding from 10 to 14 hr. There was no scaling after the admixture had been changed. Some 60,000 ft^2 of slab placed with the first admixture was corrected by cutting off the scaled finish with a bump cutter leaving a grooved surface texture. Cost of cutting was approximately 12 cents/ft^2 of surface.

How long will a concrete bridge structure serve its purpose if deterioration starts? The two-deck level Twelfth Street Viaduct in Kansas City, Mo., was 50 years old in 1960 before it was considered essential to remove and replace about half of the top deck, the full lower deck, the expansion girder bents, and considerable edges, corners, and ornaments that were reconstituted by gunite (Fig. 9.4). In the same year the 65-year-old five-span Melan arch over the Kansas River at Topeka, Kan., collapsed when one pier crumbled away. The pier split, opening up the 1-in.-thick mortar coating, which had been placed over a 7-in. shell of concrete within which was a lean mix of broken rock and sand.

The St. Lawrence Seaway has many concrete structures all exposed to the same climate and the same waters. Yet the Eisenhower Locks have suffered a series of concrete deterioration, while the adjacent Snell Lock and many other structures have remained immune to surface disintegration and to cracking. Completed in 1958 the first major difficulty came in 1962 when a 500-ton concrete sill broke loose from the floor of the lock. The sill block was anchored again to the base slab by drilling holes from 4 to 10 ft deep with diamond-studded bits and grouting steel dowels (Fig. 9.5).

Concrete defects then developed in the lower walls of the lock chamber in the form of surface disintegration and cracking. The sill incident was blamed on a poor horizontal joint permitting a 5-ft-thick concrete sill to lift 3 in. and move downstream 5 in. Deterioration to a depth of several inches was found at the end of the 1962 season along the front edge of both miter gate recesses near the floor level. Also at the culvert connections to the lock where flow velocities are high, the concrete was starting to break up. In 1964 widespread and deep deterioration was found in the 14 × 16-ft tunnels within the lock walls, which served as filling and emptying culverts. Depth of erosion of the concrete at the upstream ends of the culverts was as much as 10 in. Internal deterioration and surface cracking was common in both culverts. None of these defects was found in the Snell Locks, of similar design but of different concrete mix, since at the Eisenhower Locks 25 percent of the cement content was natural cement. Other structures with the same natural

cement content show no troubles, however. Repair and reconstruction of the defective concrete was then planned as a four-year program, during the winter closures of the Seaway.

Then in 1966 cracks were found in both the Eisenhower and Snell Locks starting at the upper corners of the culverts and running to the sloping rear faces of the lock walls. These defects are being corrected by long post-tensioned tendons in holes drilled from the front face of the walls crossing the cracks. A two-lane vehicular tunnel carries a highway under the Eisenhower Lock with a 6-ft concrete roof slab also forming the floor of the waterway. Although core tests show this concrete to be of 6500 psi strength, a steel frame and plate liner is being placed in the tunnel as a precaution against concrete deterioration. The five-year repair job of presently known weaknesses requires a $13 million budget and as yet in spite of extensive laboratory research no cause has been determined for the failures in the concrete.

Scarred and gouged-out concrete wall faces in the locks of the New York State Barge Canal have been long-time correction problems. For some 25 years up to 1950 the badly eroded areas were resurfaced with wrought-iron plates backed by cement grout. High material costs then caused a change in the method, substituting intrusion grouted preplaced aggregate. The work can only be done in winter, but one lock required $130,000 of concrete repair where the wrought iron covering was estimated to cost $190,000. No fender system had been provided in the original design.

Concrete dams holding reservoirs with widely fluctuating water surface that is frozen most of the winter months are prone to serious surface disintegration in the zone on the upstream face exposed to many cycles of freezing and thawing. Before 1918 the Southern California Edison Co. provided protection to such deteriorated faces by earth filling covered with an asphalt blanket. It was then found more economical to use a minimum 2-in. layer of pneumatically applied mortar after a careful cleaning of the surface and chipping to sound concrete. The Huntington Lake reservoir dams were so covered, but by 1955 further mortar covering was not possible since the concrete under the mortar was disintegrating. The faces were then covered with a welded skin of 10-gage steel sheet, similar to the 12-gage black iron facing at one arch of the Florence Lake Multiple Arch Dam placed in 1942. Very little maintenance has been necessary on the steel protected facings. Walkways and horizontal surfaces are repaired as needed with pneumatic mortar or a 4-in. layer of no-slump concrete, properly cured and then coated with linseed oil, which has been found a good protection for all types of concrete repairs.

More recently, since 1960, epoxy resin bonding films are placed on

the cleaned-up concrete faces of dams, spillways, and aprons where erosion has removed the skin and some concrete. Corrective replacement is then with very dry concrete mixes. Proper temperatures are necessary for the epoxy curing. Such repairs were made in 11 of the 12 50-ft-wide bays at Rocky Reach Dam where eroded pockets were as deep as 6 ft. At Priest Rapids Dam all 22 bays of the 900 × 80-ft apron and the lower 18 ft of the spillway ogee required surface repair after only two years use. To obtain proper control, large areas were covered with draped tarpaulins and space heated.

The large athletic stadia probably have more concrete exposure surface than any other structure. Disintegration is very common, starting as crazing and continuing as spalling when the reinforcement is attacked. A number of the very large stadia built during the 1920's, such as those at the Universities of Pittsburgh, Illinois, Ohio, Chattanooga, Chicago, and Northwestern were laboriously covered with double applications of Minwax to seal the concrete surface.

White's "Fundamental Causes of the Disintegration of Concrete" (1925) was then the recognized authority on the subject and his explanation was often quoted:

"Water is the cause of both the life and death of concrete,
"Reaction of cement with water gives it strength,
"Removal of the water (drying) causes it to shrink and crack,
"Restoration of the water causes it to swell,
"Alternate expansion and contraction, due to changes in moisture is the greatest underlying cause of the destruction of concrete structures,
"If concrete is to be permanent, its moisture content must remain relatively constant."

Disregard of these warnings in many stadia caused the start of disintegration with high annual maintenance. Some have been lifetime projects for the waterproofing industry.

Concrete runways with tolerances permissible for slow moving planes are not satisfactory for high speed jet operation. Typical of the many repairs necessary because of curling at joints as well as surface erosion was the 1965 repair in the 10,000-ft runway at Wright-Patterson Air Force Base near Dayton, Ohio. Included in a $1 million contract was the replacement of 6 percent of the total surface, 58 percent of all joints, and patching of 51,000 popout cavities. Joints were 28 ft apart and had been made in 1947 by replacing a steel plate form by a rigid expansion joint filler. The concrete along these joints had been overworked to smooth out the plastic concrete across the joint. All poor concrete was cut out within saw lines and replaced using an epoxy grout after thorough

cleaning of the exposed sound concrete. Where concrete is replaced on the epoxy bonding film, completion is limited to the pot life time of the epoxy. If the epoxy has set before the concrete is placed, the film acts as a bond break and the repair will soon come loose.

Resurfacing of deteriorated roadways by the use of a gang rotary saw is a rapid correction for lightly eroded pavements. In 1964 the Golden Gate Bridge in San Francisco suffered some weathering in some 800 ft of the south upgrade. Repairs by patching failed to withstand the traffic and a rotating drum carrying 120 diamond-impregnated saw blades shaved the surface, removing bumps and providing a skid-resistant texture. The grinding was in 24-in.-wide strips with equipment that could operate in a normal traffic lane.

The durability of reinforced concrete in buildings is the subject of Special Report No. 25 (1956) of the Building Research Station of Great Britain. The introductory section sums up the reasons for such a study.

"Many reinforced concrete buildings have required very little maintenance over a life of several decades (some indeed for more than fifty years) but others designed and built by the same people, and submitted to similar conditions of exposure, have in the same period reached a state where either extensive and costly repairs must be undertaken, or the building will shortly have to be written off as a dangerous structure since, though structural collapse may not be imminent, the spalling off by pieces large enough to injure anyone on whom they might fall becomes an ever-increasing hazard.

"Based on inspection of present condition and record of repairs in previous years of concrete office buildings, warehouses, factories and granaries, built between 1899 and 1940, the surprising conclusion is that the older the construction the less necessity for repairs during the usable life of the building. The common defects that tend to shorten the life of a reinforced concrete building all stem from a simple cause. The enemy of durability is nearly always water, the vulnerable element is the steel reinforcement, and trouble is certain if the concrete fails to keep them apart."

Design details must provide such protection and construction must not reduce the thickness of cover nor the imperviousness of the concrete. Similar conclusions are reached by the record of a British engineer who had analyzed causes of damage to 254 reinforced-concrete structures. No less than 227 suffered from insufficient cover, porous concrete, or both together. Chemical attack by gas works atmosphere at 32, by liquid or solid chemicals at 11, and by locomotive smoke blast at three structures, made up the second most common cause. Only one each was caused by

stress failure, by frost, and by temperature cracks alone. Compared to conditions in the United States, Britain is not so seriously affected by frost or by unsound aggregates, and mortar patching is more likely to remain as permanent protection due to the smaller ranges in humidity and temperature.

9.6 Fire Effect

Concrete is rated as a fireproofing cover of steel, 2 in. being considered sufficient for a 4 hr protection of normal fire heat. Yet intense fires have been known to cause serious damage to concrete structures. Before 1918 five damaging fires in the Far Rockaway warehouse, in the Norfolk tobacco warehouse, in Montreal, in Baltimore, and the Edison Building fire in Jersey City, led to serious investigation of the true value as a fire protector. Cinder concrete was found to be of little value if all of the unburnt coal was not removed, and after the Montreal fire stricter specifications were set on the presence of unburnt coal in permissible cinders. The Edison fire indicated that gravel concrete was of less value than crushed stone mix since the gravel normally has different coefficients of expansion along different axes and under intense heat will pop out of the concrete. Godfrey also noted that in the Edison fire the tied longitudinal bars in columns forced the concrete cover off and the columns failed.

During the winter of 1959 in an almost completed Marine Transfer Station for the filling of barges with incinerator ash in the East Bronx, N.Y., a fire started by some heating unit and fed by the timber piling and bumpers of the scow slips completely gutted the structure. Considerable damage was found in the concrete pile caps, columns, and floor slabs, although the basic frame was structurally sound. Corrective work consisted of complete removal of all spalled and charred concrete surfaces, application of a spray film of polysulfide epoxy adhesive and immediate (after a 2-hr air exposure) replacement of the missing concrete placed within forms for vertical surfaces. The adhesive was applied at the rate of 1 gal to 110 to 120 ft^2 of exposed concrete. A complete and watertight connection to the parent concrete resulted with no separations during the following winter with a number of freezing and thawing cycles. In damaged areas where watertightness was not necessary replacement was by gunite, with the damaged concrete chipped out enough to expose the reinforcing rods.

9.7 *Rusting of Imbedded Steel*

The deleterious effect of rust formation, resulting from the expansive forces generated by the oxidation of iron has been covered in many of the examples of failure described in both Chapters 9 and 10. There are some common errors that increase the effect both in extent and in rapidity of action. Rust formation is a chemical change of iron requiring the presence of water and oxygen. It is speeded up by electrical phenomena such as stray currents and internally generated electric potentials. The bibliography on corrosion of reinforcing steel in concrete is mainly a listing of cases where the mix contained free electrolytes, especially chlorine. Whether introduced by an admixture or in the water, chlorine content must be avoided and if it cannot be avoided, special precautions to produce a dense concrete must be taken. Recent promotion of galvanized reinforcement as a protection against rusting is not to be taken as a cure-all; it does not guard against electrolytic action especially in damp environment and must be considered as a temporary safeguard, good only until the zinc coating is dissolved.

The corrosion of iron imbedded in concrete was studied in depth by the Zurich Institute for Testing Materials and in 1923 the report by Bruno Zschokke lists three precautions, which are still valid:

"1. The concrete should not be too meager, but should contain the best percentage of cement, so as to make it impermeable to air. Thus, the proportion of calcium hydrate, which prevents oxidation, is increased, and the concrete is also more impermeable to carbonic acid, which neutralizes the lime.

"2. There must be no substances in the concrete such as locomotive cinders, often containing sulfur, which exert a chemical action on the iron.

"3. The layer of concrete covering the iron should be of sufficient thickness properly to cover it, and should not crack under pressure or through shrinkage."

Corrosion in marine atmospheres, especially in tropical climates is a constant source of trouble, and electrolytic action starts readily even when the mix water is free of salts. Corrosion of prestressed steel is so critical that any admixture containing chlorides is prohibited in modern work. Tests with the use of special cements such as ASTM Type IS indicate no great advantage over the usual types. Corrosion studies in

France in 1963 indicated that the causes in two special cases were the formation of a lime sulfate and in the other incident, of the reduction of the calcium sulfate by microbes into a hydrogen sulfide, both chemicals readily reacting with iron. Tests in the U.S.S.R. reported in 1958 indicate that if an additive of 2 percent sodium nitrate is used, the corrosion is reduced by 10 to 50 percent even where 2 percent calcium chloride is used. Australian research in 1961 reported that using stannous chloride, in place of calcium chloride, will accelerate the rate of setting and strength-gain of concrete and will simultaneously inhibit corrosion of reinforcement.

Chemical substances other than the chlorides will also rapidly oxidize the imbedded steel, as was shown in 1956 by the breaking up of precast beams exposed to the warm moist air in the lumber drying kilns at Muskegon, Mich. The rust formation in the steel bars started slippage from reduction in bond resistance and shear failures occurred near the ends of the beams. Beams with serious distress were replaced and all concrete was given a seal coat of two applications of asphaltic waterproofing.

Power plants and installations requiring complete grounding of all reinforcement produce a huge galvanic cell which aggravates corrosion of the underground structures. Old elevated railroad structure foundations were found split open by the corrosion of imbedded bolts and attention, unfortunately in complete isolation and separation by every minute facet of the concrete industry. Much greater value from the investment would result from a coordinated effort.

10

Concrete Structures

Design and Detail; Shear and Torsion; Compression Failure; Erection Difficulty; Frost; Temperature Change; Deformation and Cracking; Precast Assembly; Tanks and Silos, Space Frames.

Compared to masonry and timber, reinforced concrete is a relatively new construction material, and, like steel, has had to go through a period of empirical development to prove its feasibility. In spite of the intensive research in the field of concrete structures many of the succession of new developments must likewise depend on actual performance, and such trial-and-error procedures naturally have their share of failures. If they are singular in occurrence, the failures can be forgiven, although someone may have been hurt by the incident. But making the same mistake twice is inexcusable; its cause is stupidity or lack of communication.

10.1 Design and Detail

Some examples of concrete structures involved in failures because of shortcomings in design or detail are included in Chapter 1, where the error was not a consequence of the choice of material.

Regulation of concrete design procedures came from a series of spectacular collapses that resulted in taking away from the designer the full freedom of action that as a "master builder" was considered his right

and privilege. One such case was the reinforced-concrete viaduct built by the Celestial Globe Company in 1900 over the Avenue Suffren in Paris so that visitors could attend their attraction without leaving the enclosure of the Exposition grounds. The deck was 375 ft long, 16.5 ft wide and some 15 ft above the street and consisted of 30 ft spans, with concrete girders on 12-in. square concrete columns. The slab was 6 in. concrete, two-way reinforced and supported on light steel beams resting on the girders. Longitudinal stressed cables bounded the walkway as hand railings. Originally designed as a straight line, two axial deviations were introduced to avoid removal of some trees. The resulting oblique forces practically kicked the deck off the piers and when the bridge fell a number of people were trapped underneath. The Board of Experts and the Court pronounced the consulting engineer and the contractor guilty of "the most culpable negligence."

Hennebique designed a five-story hotel for Basel, Switzerland, which failed on August 21, 1901. The city building department had given the designer full control of the work. The commission of investigation found that the columns had been designed for full stress on both steel and concrete, so that in view of the possible bending and eccentric loadings, there was only a small factor of safety. Also the steel in the columns was not continuous but ran only from story to story. Although this was a serious deficiency, the actual failure was attributed to the cellar piers, which were designed as masonry but had not been built first. Timber temporary bents were provided while the concrete was being built and apparently partially removed while the masonry piers were being built, leaving the columns above unsupported. The experts listed this as the chief cause but also as additional causes (a) insufficient dimensions of columns in the first story, together with lack of control of the engineer-contractors over the construction, (b) use of improper material (unwashed sand and gravel), (c) careless construction of the concrete work (bottoms of the columns were less dense than the rest), (d) lack of tests of the cement and concrete, (e) lack of organization among the various subcontractors, (f) haste in construction. This report could be used without a single change in explaining recent similar incidents.

Deficiency in the basic design, such as amount of reinforcement at points of maximum moment, or dimensions of concrete section to provide compressive and shear resistances under normal loadings, is an extremely rare cause of failure. One case that was caught just before placing concrete in roof beams spanning 64 ft over a school auditorium in Yonkers, N.Y. (1925), has been described in Chapter 1. P. Rogers in a discussion of the writer's ACI paper in 1957 reported a similar case in which, because of computation error, less than one-third of the necessary

reinforcement was provided in 54-ft concrete girders spanning over a school gymnasium. The error was discovered after the concrete was cast, but the formwork was still in place. Rogers compensated for the designer's error by adding Stresstel bars along the stems of the beams, anchored against steel plates on the outside of the spandrel beams. By applying a tension of 100,000 psi to the bars, a partial prestress design, the mild steel reinforcement was then only stressed to 20,000 psi, which was satisfactory.

Many incidents stemming from similar causes are rapidly buried and no information becomes available. The May 10, 1962 issue of *Engineering News-Record* has a small note on a cracked folded plate roof at Fort Sill, Okla., which collapsed under a load test. In the August 2, 1962 issue there is a note on the collapse of a seven-story lift slab structure in Forth Worth, Tex., before completion. The information lost by not properly recording the incidents could have been of great value and could well eliminate future error of similar nature.

Sometimes the design error is made in the field when excess loads are imposed on structures. In 1924 at Buffalo, N.Y., the roof of a 10-story concrete building, then six weeks old (poured in January), fell when loaded with materials for the coping and roofing.

In at least one case a failure reported as being caused by design error needed some later rectification. The failure of the partially prestressed Vierendeel trusses in a Lodi, Calif., warehouse roof was attributed to improper design in 1952. Later information introduced a new cause—one truss in lifting had fallen on the others when a crane overturned and the completion schedule did not permit time to make more than surface repairs to the "few slight fissures" in the trusses hit by the falling weight.

The failure of the columns in the Chicago Grant Park Garage in 1963 must be classed as a design error, not so much in the concrete design as in the lack of provision of a possible loading condition. Frozen broken buried water lines, due to unusual cold winter weather, built up large ice volumes and introduced uplift forces which were not expected and could not be resisted. Thirteen lower columns in the two-story flat slab structure sheared as the roof and upper floor resisted upward movement (Fig. 10.1).

For simplicity in detailing and rod installation, it is often expedient to use the same size and spacing of bars for a roof slab as the typical floor of a multistory building. This tendency may explain why in a New York school building in 1953, the main reinforcement over an interior support of the floors was incorrectly copied from the design notes and the much lighter reinforcement of the roof design was shown and used.

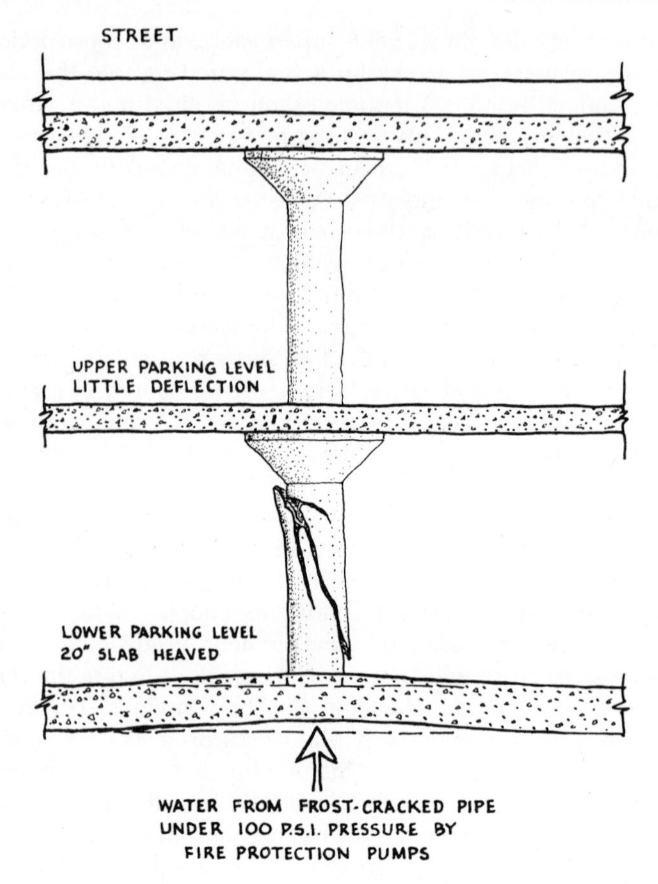

Fig. 10.1 Failure of columns from frost upheaval. (From *Chicago Daily News,* February 12, 1963.)

Although this was only one of several noncompliances discovered in the investigation, it may have had a considerable influence on the resulting large slab deflections found after the forms were removed. As a correction, a 4-in.-thick cap slab was introduced after shoring the deflected slab and wedging it upward and adding high-strength steel-wire mesh welded to studs shot into the concrete over part of the deflected span and the complete (shorter) adjacent span. A considerable portion of the unwanted deflection was eliminated and a full-scale load test showed good composite action of the unusual post-stressing procedure.

With growing experience in all types of reinforced concrete designs becoming common in all the controls through which a job passes—

design, estimating, detailing, field supervision, and construction—the possibility of a gross error in design actually getting into the construction stage is indeed remote. Yet some examples cited under other headings indicate that the possibility still exists. Concrete designs need more than a numerical checking; a cursory examination of the design drawings by an experienced engineer or constructor will always show up gross errors in design. It is the small items of underdesign, in the category of secondary stresses from volume or shape change, that, as in structural steel design, are the troublesome difficulties. An arbitrary assumption of the location of points of inflection, a procedure quite popular up to 25 years ago when the free support moments were split as desired into positive and negative factors, often resulted in cracks since the structure refused to accept the assumption if not in compliance with actual relative stiffness of the contiguous members. The reduction in mathematical processes introduced by the Hardy Cross moment distribution procedure eliminated much of this type of local failure, but only when realistic stiffness ratios are determined.

There has been a gradual but persistent tendency for design drawings to show maximum reinforcement conditions and leave to the detailers the decision of where to stop bars, development of laps and splices, and other decisions which should be part of the design. Errors that creep in are easily overlooked in the checking of shop drawings. The European practice, as well as that of many public works projects here, is to detail all the rods on the design sheets. This practice has merit, in spite of the shortage of technical design help, since the overall labor required is actually reduced, even though it means that an office will do fewer jobs in a year.

Does any design office leave to the contractor the job of determining actual dimensions of the concrete work? An editorial in *Engineering News-Record,* December 1, 1955, questioning whether the standards for detailing could be blamed for some of the "concrete structures that have fared badly" brought forth vehement denial of any correlation. It is important that details follow a recognized standard and that everyone use the same symbols and notation. In slabs with straight steel in top and bottom, it has been found expedient and economical to prepare separate detail sheets for the top steel and for the bottom steel. Since the layers are separately installed, this procedure eliminates all possible errors or misunderstanding.

In a recently constructed university building in Chicago a carefully detailed architectural exposed concrete roof construction had some indefinite requirements for the reinforcement. The vendor interpreted the necessary lengths and submitted shop details that were approved and

followed in the construction. As a result, there were gaps in the continuity of steel in one direction of the two-way system of ribs, and concrete shrinkages were not resisted. Cracks developed in the ribs at these points and considerable corrective work in exposing the ends of the bars and welding in the missing links, followed by a replacement of the concrete covering, was necessary. The ceiling height made the repair cost more expensive and the loss of the space for several months was quite serious.

During construction of the top level of a four-story underground garage, designed to be a landscaped plaza, it was decided to strengthen the deck so that trucks could use the roadway. Forms were in place and were used unchanged and only the reinforcement was increased. The rush of the change left some weak details and when the gravel and soil cover was being placed, the small bulldozer unit broke through the slab and fell to the cellar floor as the successive slab levels failed. Failure was in shear with no warning (Fig. 10.2).

The details at expansion joints need careful attention. A number of warnings that sliding supports must have freedom to slide, otherwise the brackets or supports will spall and crack, are found in the literature. H. D. Loring, at the 1923 Convention of ACI, described the proper

Fig. 10.2 Collapse of floors from load application.

details and H. Lossier in his book on *La Pathologie du Beton Arme* warns of the necessity for antifriction bearing plates. There are very few expansion joints, however, that work properly and show no distress. Many attempts to provide sliding expansion joints by placing concrete beams on piers with a troweled surface or even a paper-separation layer have proven unsuccessful.

When reinforced concrete was first used in the construction of wall-bearing mill buildings, it was customary to require a full static load testing of the floors. Under such a loading a five-story brass factory in Trenton, N.J., 300 ft long and 50 ft wide, designed for a live load of 250 psf, failed in 1903. The center support of the transverse beams was a concrete girder, carried by a series of columns, but their ends rested in pockets of the brick walls. A load of 225 tons on a 30 × 30-ft area of the second floor, the walls below being 20 in. thick was successfully carried. A load of 180 tons on a 40 × 16-ft area of the third floor, the walls below being 16 in. with 4 in. of brick covering the ends of the concrete, caused a collapse of this floor which also broke the first floor. It had been built by a reputable contractor; investigation showed good concrete and no failure in columns or footings. The walls were pierced by windows covering almost half the length. Failure resulted from lack of proper bearing in the wall, the cross beams pulled out of the sockets when they were deflected by the loading.

The year 1903 was an unfortunate one in the field of reinforced concrete. A three-story factory in Corning, N.Y., also failed under its load test. Before the test the architects had notified the contractor's bonding company that they expected failure since the wire mesh reinforcement had not followed the plans and actually in some areas had been almost entirely omitted. Their prediction was accurate. Under a 200 psf load over two areas of 5 ft^2 on the third floor, the entire rear half of the building, floors and walls, collapsed.

Also in the same year a floor in a concrete building in Petrograd, Russia, collapsed and a commission investigated the causes and made its report. Owing to some dissatisfaction with the conclusions, a second commission was appointed and conclusively disproved the first report, giving as the principal causes of failure:

1. The design, both system used and computations, was incorrect.
2. Poor cement and concrete was used.
3. Lack of superintendence.

In 1903 there was a second collapse of a grain elevator of 15 circular bins in Duluth, Minn. The same design had failed in 1900. Careless workmanship on the front section of the fourth floor of the Milwaukee,

Wisc., warehouse was blamed for its collapse, the fall taking with it all the floors below. A form failure in Pittsburgh, Pa., collapsed the roof slab that dropped clear to the basement of the nine-story apartment. It is a wonder that the concrete industry survived after the experiences of 1903.

Cantilevers seem very vulnerable, apparently because workmen disbelieve the plan requirement that the main reinforcement belongs in the top of the concrete member. The roof canopy at the Ocala, Fla., Courthouse failed in 1963 and the causes were that top reinforcing was laid too deep on the cantilevered slab, some bars were not anchored to the spandrel beams, and the canopy was loaded with about 2 tons of roofing material. In 1966 a 9½-ft cantilever canopy failed at the Dallas, Tex., Trade Market, shearing out completely from the spandrel beam. Short balcony cantilevers in a west coast steel frame apartment deflected too much and steel beam brackets were added for stability (Fig. 10.3).

Omission of some reinforcement is seldom the cause of failure, but it can exhibit itself as an unexplainable deflection. In a three-story concrete waffle slab classroom building at a military installation in Virginia, the ceilings showed large deflections in every room. The first explana-

Fig. 10.3 Repair of balconies to control sag.

tion was the students' marching in step to class, but when this was discontinued the deflections still increased and the building was vacated. Electromagnetic nondestructive probing surveys indicated a lack of rods in the center ribs in each ceiling area and when the concrete was cut away no rods were found. Since the construction had been performed under close supervision and every form had been inspected and approved, the omission of the steel could not be explained. To make the building usable, a complete frame of steel beams and columns was incorporated into the structure as a support for the concrete slabs. Further inquiry finally resulted in the true explanation. The reinforcing bars had been delivered to the job and placed, two in the bottom of each rib form. Inspection was made and recorded. The electrician then came on the deck to install lighting fixture outlet boxes and conduit. The boxes could only be placed in the bottom of a rib by removing the bars, which he did. Not being permitted to place steel, as that was the task of the iron workers, he left the bars on top of the pans used to form the waffles. The electrical work was then inspected and the concreting done under the guidance of another inspector. The "surplus" rods were thrown off the deck; the same thing happened at every panel on each of the three floors and the roof; perhaps a case of too much inspection by the book, without any knowledge of what was being built.

Edward Godfrey in an open letter addressed to W. A. Seater, published by *Concrete* (Detroit), February 1921, and in a paper presented before the British Institution of Civil Engineers, February 22, 1923, lists a number of concrete building failures which he attributed to design deficiency, some of which are described later.

From the presented information, the most repeated mistake in the early period of concrete design was to neglect a proper transfer of load from floor to column. In the 1905 (first) edition of Taylor and Thompson's *Concrete, Plain and Reinforced*, page 320, there is a section headed "Prevention of Diagonal Cracks in Beams," in which recommendations for shear, diagonal tension, and bond resistance are clearly outlined. It is indeed strange to see how little protection against failure has been added by the voluminous engineering and research commentaries in the field of concrete design to the basic simple rules set up in 1905 by Frederick W. Taylor, mechanical engineer, and Sanford E. Thompson, bachelor of science.

10.2 Shear and Torsion

The most serious type of failure in concrete structures is failure due to a deficiency in shear resistance, for then there is no warning. Shear fail-

ures are preceded by little, if any, deflections and even with little cracking. Probably more has been argued about shear resistance, and there is probably greater variation in the codes of various countries on shear resistance design, than any other facet of reinforced-concrete design. The basic unknown is whether and to what extent reinforcement acts together and simultaneously with the concrete in resisting shear, and how the torsional stress induced by nonsymmetrical loading influences the maximum shearing stress to be resisted. Unless the contractor has failed to follow the plans and specifications, all shear failures are caused either by deficiencies in the available knowledge of what requires resistance, or by the lack of proper application of available data to the problem.

Under the heading "A Few Examples of False Reasoning," H. Lossier in his *The Pathology and Therapeutics of Reinforced Concrete*, Dunod (Paris) 1955 (available in English translation from the National Research Council of Canada, TT-1008, 1962), states:

"Some engineers assign to the stirrups only the excess of the shearing stress, or more exactly the tensile stress, which the concrete is able to balance. In reality, however, vertical stirrups really become effective only after the concrete has cracked. They must therefore be capable of completely replacing the concrete, resisting, by tension, all of the tensile stresses of the concrete acting over the entire area occupied by a given crack."

American practice, as described in the successive editions of our codes, has not agreed with this argument, and there are countless examples of sufficient designs to disprove the argument. But of course we may be overoptimistic in our appraisal of the factor of safety, or else concrete can be trusted to resist some tensile forces at the same time that the reinforcing is doing the major part of the same job.

Several shear failure reports are found in the early concrete literature. In 1911, when the roof of a three-story flat-slab building was loaded with cinders, shear failure collapsed the deck, and the Indianapolis case became the subject of long discussion on how to design flat slabs.

Flat Plate Floors

Flat plates, the successor to flat slabs, are still failing in shear when all the rules are not followed, even though well-known. The *New York Times* of October 4, 1956, under the usual noninformation headlines reports from Jackson, Mich., Oct. 3 (AP), "A four story office building being constructed to withstand the shock of an atomic blast collapsed with a roar today, killing at least six workmen and injuring fifteen

others." In March 1957 the editor of *Engineering News-Record* refers to this collapse and states: "No explanation has yet been made of the crash, which brought down several partly complete floors, but a special commission that investigated the mishap on the orders of Gov. G. Mennen Williams has yet to report its findings." When the decision to reconstruct was reached in March 1957, no official report had yet been issued.

The building was cross-shaped in plan. While the fourth floor was being concreted (an area of about 72 × 144 ft, roughly three bays wide by six bays long), the 10-in. flat plate fell vertically and the impact took down all the floors into the cellar. The first and second floors were several weeks old, the third floor was 20 days old and had been reshored to the second, and was carrying the formwork of the fourth. After the collapse almost all of the columns remained standing full height with the top-story forms attached but with hardly a single piece of reinforcement projecting from the free standing columns at any floor level. Column sizes were 25 in.^{2} below the first floor and reduced to 23 and 20 in.^{2} above.

Some of the columns had 10 × 14-in. duct openings along two adjacent faces, which of course prevented running any slab steel through the columns. The design called for a square spider around each interior column within the slab thickness, but how these assemblies could be placed within the zone of high shear and still permit the duct openings is not clear. The lack of load transfer resistance from slab to column indicates the advisability of renewing the older concept of punching shear investigation as part of any plate design. In these days of air-conditioned buildings in which the concrete frame becomes merely an enclosure of mechanical equipment, equally important is the close coordination of structural requirements with the holes, sleeves, and loads required by the mechanical designs. Flat plate floors are feasible and economical only if uninterrupted continuity at columns is provided. No one would agree to a design for a plate footing with openings at the face of the columns. The structural requirements for a flat slab or plate floor are the same as for a footing.

A similar partial failure occurred in New Jersey in 1961, when a shortage of proper reshoring during the construction of a high-rise concrete apartment caused the overloading of the ninth floor and the slab sheared completely around three columns. Failure probably started at one column where a mechanical duct opening occupied the full width of the column face (as in the Jackson, Mich., case) and no slab rods could be carried through the column. With one support missing, the load transfer to the adjacent columns overloaded their capacity, and the slab

Fig. 10.4 Shear failure in flat plate: (*a*) top view; (*b*) bottom view.

sheared there as well. The cost included the demolition of both the ninth- and tenth-floor slabs and reconstruction plus the value of the time delay in completion. Cause was the attempted saving of some properly placed shores under the seven-day old slab, which was asked to carry a full deck of wet concrete, with shear resistance at one column seriously reduced below design expectations (Fig. 10.4).

Two Canadian school buildings with 6-in. flat plate floors and 12-in. circular columns showed abnormal deflections of the slabs and a typical crack pattern on the top of the slab at each column of a circle slightly larger in diameter than the column and four radial cracks along the diagonals of the slab panels. These incipient shear failures were restrained by timber shoring until permanent concrete column heads, 28 in. diameter and 11½ in. deep were attached to each column. A ¾-in. bite was cut out of the column perimeter for the depth of the cap, #4 spiral loops were placed and the caps were gunited in place. A total of 140 column caps were added at each school to increase the shear resistance, which was inadequate because of low placed steel and heavier partitions than assumed in design.

A concrete plaza deck, acting as a roof of a garage in continuous use for almost three years, collapsed in New York City on December 23, 1962. There was no warning, although two garage employees, parking cars for the night, heard some cracking and left the area before the 16-in.-deep grid flat slab with some 4 ft of earth cover crashed on top of the parked cars. Actually about half of a symmetrical entrance plaza, an area roughly 45 × 50 ft failed, the other half remained in place, apparently in sound condition. Panels were 24 × 32 ft and supported on 14 × 30-in. columns.

Failure was a clean punching shear, with almost vertical fall of the trees and little effect beyond the shear cut. In spite of the long experience of the builders, contractors, and supervisors in similar construction, the indicated (not too clearly, it is true) concrete caps extending 12 in. beyond the column faces and 10 in. deep had been omitted at all of the nine columns in the failure area, and also at all of the nine columns in the symmetrical area that did not fail (Fig. 10.5). The only difference in conditions was a stopped-up underground drainage in the failed area that resulted in a frozen saturated earth cover, the other half being well drained. The difference in the dead weight of the soils was the difference between failure and stability. Computations of shear at the columns showed a Factor of Safety of 1.05 for the remaining slab. Concrete used was lightweight, and in the failed area a remarkable lack of bond adhesion was noted between the rods and the concrete. The exposed bars came out clean and the impressions in the concrete seemed dusty and only roughly formed. The slab was reconstructed with column heads

Fig. 10.5 Deck failure in shear because of omitted column caps.

and new columns; the other half was also considerably strengthened by new girders and column jackets.

In a three-story building in Denver in 1961, when shear deflections in the range of 3 in. were found on the first floor, a system of steel beams and posts were added and the cause was never made public.

The Air Force Warehouses

The great magnitude of thermal and shrinkage stresses as compared to live load stresses is shown by failures of the warehouse roof rigid frames at Wilkins AFB, Shelby, Ohio, on August 17, 1955 and at Robins AFB, Macon, Ga., on September 5, 1956. Both failures as well as the experimental and theoretical studies made in the attempt to evaluate the forces which could cause such failures and the details of resistances which must be provided opened up a healthy discussion in the technical press. At least it became apparent that the real causes are not quantitatively known. Similar warehouses were constructed in many other locations, all being 400 ft wide and of various lengths but always in 200-ft units with complete separations between units. The rigid frame bents are on 33-ft spacing, and consist of six 67-ft continuous spans, the third one from one end having an expansion joint (Fig. 10.6).

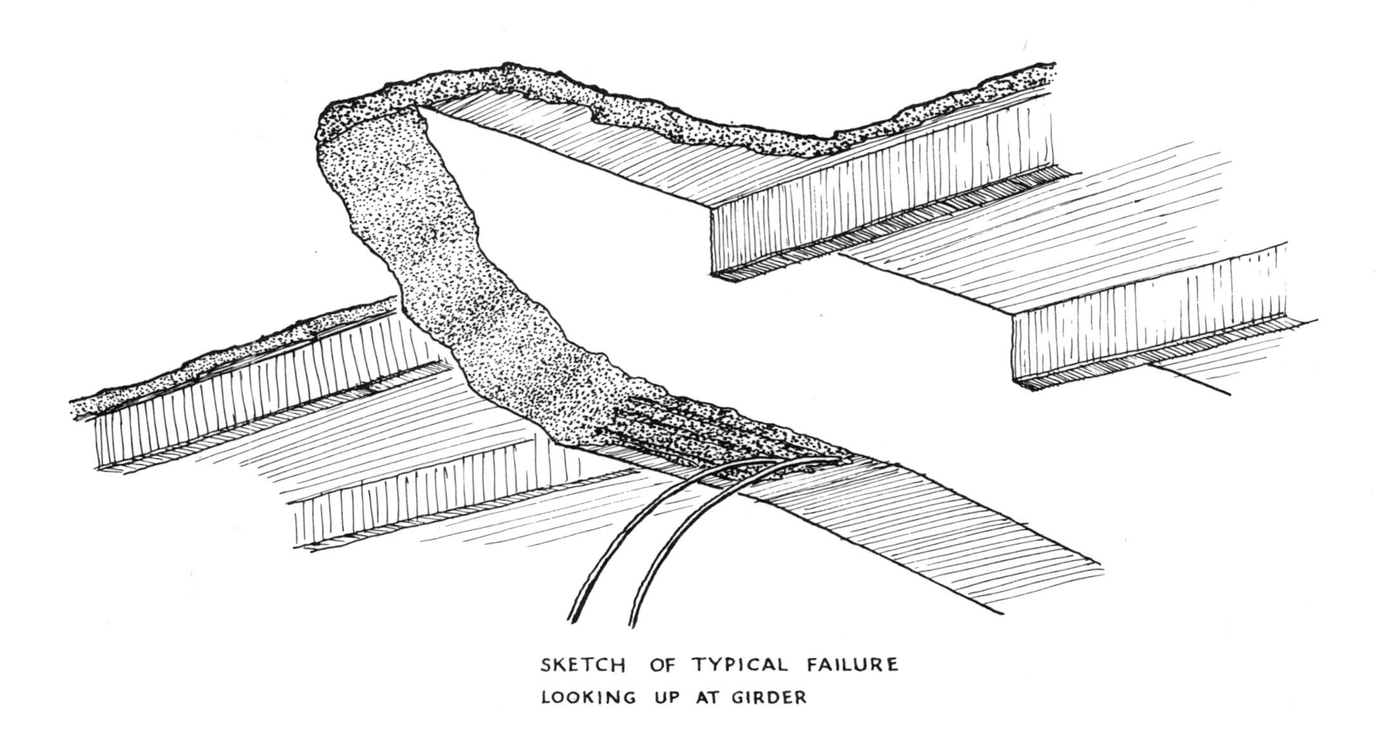

SKETCH OF TYPICAL FAILURE
LOOKING UP AT GIRDER

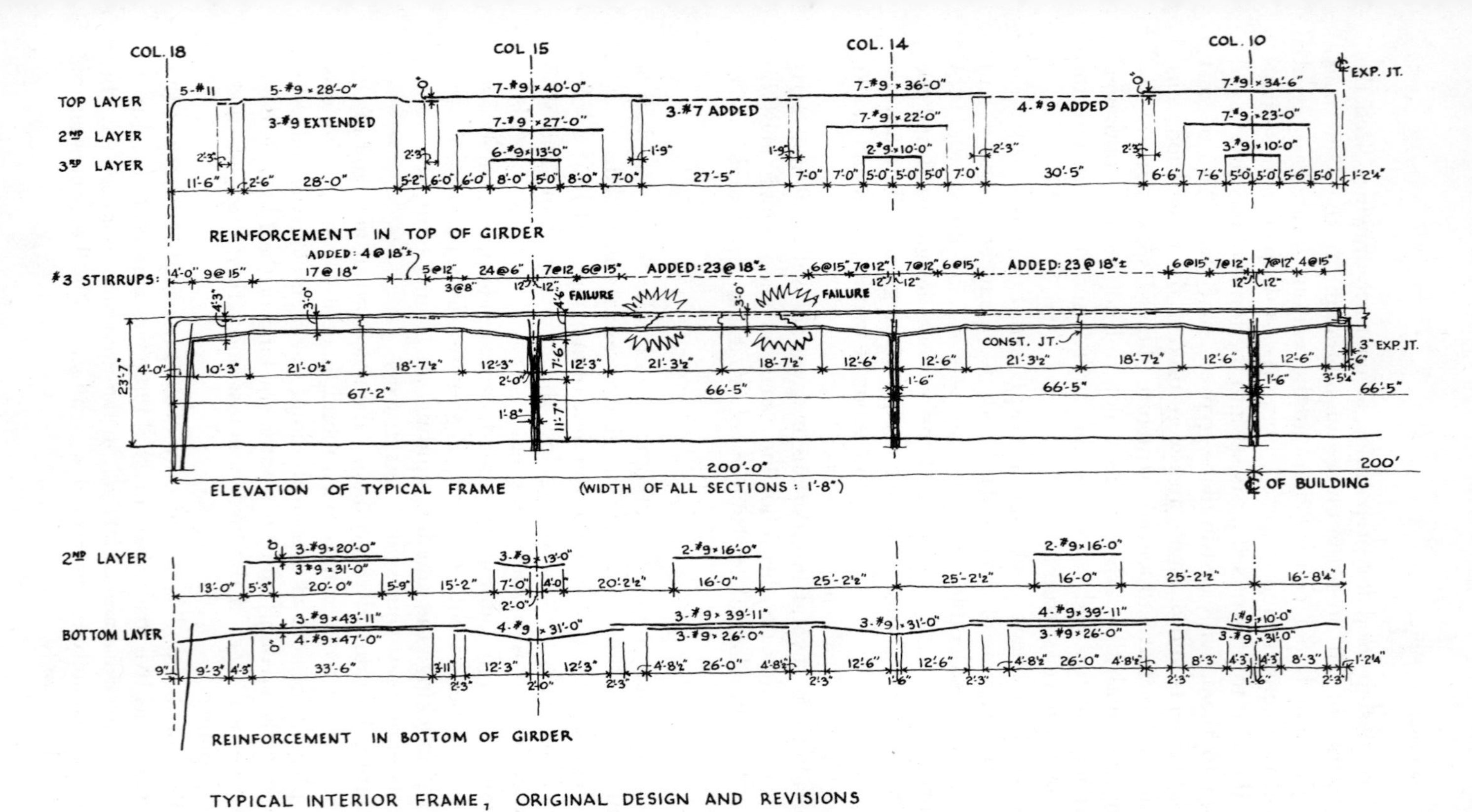

Fig. 10.6 Warehouse rigid frames. (Redrawn from Boyd G. Anderson, "Rigid Frame Failures," *A.C.I. Journal*, **28**, No. 7, January 1957. Title No. 53–34.)

Standard design used for preparation of contract drawings fixed the main girders but left open the choice of subframing in the 33-ft spans and of the actual roof construction. In the various locations a great variety of roof construction was used, from precast deep channel sections to span the 33 ft to prestressed joists set into the forms of the girders and anchored to them, with cast-in-place gypsum roof slabs. Also a great variety of cements and of aggregates were used in the several locations. Yet some girders in almost every structure showed some signs of distress, more so with the early designs which did not require continuous reinforcement in top and bottom of the girders and more so with the designs having fewer stirrups.

No amount of check computations explained the causes; after the failures, modifications of the standard code were suggested and in part approved. A considerable test program was carried on at the PCA laboratory, model beams reproduced the various designs used and the tests to failure proved that under the design loadings, no failure or even cracking was possible. Yet failures did occur at two locations and cracks at many, with practically no live loads.

Observations at several installations in geographically different parts of the country indicate the following pertinent data which must be considered in any logical explanation of what happened and how to avoid it:

1. The expansion joints do not operate, and the 400-ft length is a single unit as far as thermal and shrinkage stresses are concerned.

2. Both collapses came at about the date of maximum net heat absorption gain, the night loss becoming less than the day gain. The 1955 failure was after a most unusual hot spell.

3. Identical designs built at the same location and approximately over the same time period with the same aggregates do not show the same amount of cracking; the smaller building area units built at a much slower rate of progress shows much less distress.

4. Rotational strains of the columns about a vertical axis and of the girders at cracked sections have been found. Except where caulking was used in place of cement mortar in the joint between columns and exterior block walls, spalling and horizontal cracking of the blocks is common. The vertical faces of the girders across one serious crack are not a continuous plane.

5. Failure of the models would only occur under loadings below the total of dead plus live load when an axial force was applied to the girder model. With an axial force of 200 psi on the minimum cross section, the ultimate load capacity of the model was substantially the equivalent of vertical dead load only.

6. Although some cracks appeared during the construction period, most of the distress was not found until the frames were almost two years old.

The final reports and conclusions are available. Certainly, some lessons can be learned:

1. Just as it takes time to make good bread, it takes time to make good concrete; a rush job is vulnerable.
2. Expansion joints only serve their purpose when they operate and frozen joint details are worse than no joints.
3. Lateral restraint of all members is assumed in our basic designs; such restraint must be provided, especially at the compression faces of continuous structures, and that comes at the bottom of the girder where it sits on a column.
4. If torsional strain can absorb expansion with less internal work than axial expansion of the girder plus bending of the supporting columns, then the girders will deflect laterally in a sine curve pattern and rotate the columns; cracks must open if the girder is not braced or reinforced to compensate for such torsional strains.

In 1956 a bridge buckling in Queens, N.Y., was reported with a similar lateral distortion of the facia wall and parapet on a rigid frame overpass. The distortion provided the extra length required by thermal expansion with less internal work than an axial elongation of the unbraced parapet.

Design must take care of both external and internal loadings. Whether such change is brittleness in dried-out timber, opening up of microcracks of metals from vibration or temperature change or shrinkage and plastic flow in concrete, materials that change with age introduce internal stress conditions that must be considered in the design. Disregard of such changes will bring failure, either as collapse or as unsightly distress. This lesson is being learned in the field of aviation; it applies to all structures.

After the first indications of distress, an evaluation of the standard AF Warehouse design showed that negative moment steel had been stopped too close to the support, if unbalanced live load was assumed. The immediate revision was continuous top steel and an extension of stirrups with some indicated reduction but far from elimination of cracks. The Wilkins AFB structure did not have continuous bars and the diagonal shear failure left the ends of the bars exposed after pulling them out of the suspended part of the span. No stirrups were crossed by the failure surface. This design was in full conformity to the 1951 ACI Building Code. About 4800 ft^2 of roof fell. Here the roof joists had been made up

of 8 × 8½ × 17$^5/_{16}$ in. deep machine-made concrete blocks tied together with tensioned bridge strands into 33-ft-long beams. Beams were set into the girder forms on 8-ft-3-in. centers, and negative reinforcement ran over the beams within troughs cast in the beams. The troughs were filled with the girder concreting. Some 20 similar projects were under construction and all of them were suspended until the restudy of the design indicated advisability of providing more top steel, addition of web reinforcement, staggering of bar splices and an improved expansion joint. New designs were made available to all projects within two months of the first failure.

An immediate recommendation was made to modify the ACI Code requirements on the portion of shear to be resisted by steel, either stirrups or bent bars. The 1907 earliest standards allotted shear resistance to the concrete for the full carrying capacity and called for steel only for any excess shear stress. In 1916 European codes required the total shear where in excess of the concrete value, to be resisted by steel with an abrupt stoppage at the point where the shear stress was balanced by the concrete value. Actual practice then was to provide bent bars and stirrups over the entire span, and this was the European code as of 1925. Yet in 1924 the Joint Committee permitted the 1907 standards, and with the almost universal elimination of bent rods because of high labor costs, stirrups were only provided over the range of excess shear stress and only for that excess. Especially in rectangular continuous beams, the code was recognized as insufficient and most designers called for at least two continuous bars in the top for the full length of the beam with stirrups spaced usually 12 in. apart to make up the cage of steel. This precaution was not followed in the standard warehouse designs. Almost immediate action resulted in the revision of the ACI Code Section 801 d., the new requirements calls for shear reinforcement to carry at least two-thirds of the total shear or all of the excess wherever the shear stress exceeds the allowable value for the concrete alone.

Similar free-standing multispan frames had given trouble in previous years, usually at low shear areas and near the normal points of contraflexure. Temperature and shrinkage loadings can shift the point of contraflexure by considerable distance. A rigid frame bridge model tested at the University of Illinois in 1938 showed such failure. A two-span bakery roof of this design failed in 1954. A hangar built in 1943 consisting of nine rectangular frames each of seven 130-ft spans showed similar diagonal cracking. In letter discussions of the Coronada rigid frame failure (1955) there are several warnings of the danger in unreinforced diagonal sections and also in the heavy concentration of longitudinal bars in narrow sections, another item which resulted in weakening

in the AF Warehouse girders, exhibited by spalling and cracking in the bottom surfaces. The exterior spans had seven #9 bars in a width of 20 in.

An extensive program of analytical and experimental investigation was instigated by the several Defense agencies interested. The most important was started in October 1955 at the PCA Laboratory in Skokie, Ill., and consisted of testing to failure of 13 one-third scale models of the second span in the typical frame. The first seven beams were tested to failure under flexural loading only and the last six were subjected to a combined flexure and axial tension. Only the latter beams failed under flexural loadings in at all similar manner to the actual frames. The tests also proved the value of a suggested correction, adding exterior steel bands on the completed frames to provide the necessary shear resistance.

Inspections indicated that those designed for lower roof loadings in warmer climates showed more cracking, and the accumulation of temperature absorption seemed to have some effect. The almost daily observations at Rollins AFB near Macon, Ga., showed widening of cracks in August, with one crack becoming ½ in. wide by September 4. Finally some 6000 ft^2 of the roof, including two adjacent girders of the longer arm at an expansion joint, failed at 3:05 a.m. on September 5, 1956. This area was in a warehouse 1600 × 400 ft, which had been in use for 18 months. A similar design in a building 400 × 400 ft at the same base, constructed as a separate contract which had a very slow progress schedule, showed practically not a single crack. It seems that the extent of shrinkage with resulting axial tensions may be somewhat related to the speed of concreting or to the extent of each separate pour.

The Rollins design had the same shape as in the failed Wilkins frames, but the top steel was continuous and nominal stirrups had been provided. In both instances however, the expansion joints were inoperative. All warehouse frames that showed cracks of consistent shape and larger than hairline were reinforced by adding exterior stirrups of band steel stressed into shape by mechanical means. It was found necessary to tap the straps manually during the tightening operation so that unit stress in all four legs of the band would be fairly equal. The cost of the repair procedure plus the cost of temporary shoring of cracked girders, labor of moving and replacing stored goods and final cleanup was only a few percent of the value of the warehouses, but ran into several millions of dollars.

Shear failures at the hinged base of 17 concrete rigid frames in Austria in 1938 was caused by insufficient reinforcement above the hinge bearing. There was a complete break of the concrete section, about 26 $in.^2$

and the two stirrups at the section of the break had reached the yield stress of the steel.

Failure of the Coronado, Calif., 67.5-ft concrete rigid frame in 1954, under its own weight after erection was a failure in shear along an unreinforced irregular path formed in the gap between spliced bars. A change in detail of the reinforcement splice eliminated any possible shear surface not crossing the reinforcement. In the copious discussion after the failure was reported it was noted that every lapped bar introduces shearing stress in the concrete within the splice length. The intensity of the shear stress depends on the intensity of the stress in the bars, and the minimum splice lengths specified in codes may not be safe if splices are permitted at locations of high tensile or compressive steel stress.

When the top reinforcement of the cantilevers for a two-column highway viaduct bent was extended 40 diameters beyond the face of the column, the designer felt that the code was complied with. But upon completion of the roadway for the Harlem River Drive in New York, 1960, truck loading caused the cantilevers to crack off. The crack made an "end run" around the top steel and curved to the bottom of the cantilever. Correction and repair was by adding two poststressed rods to each bent, anchored into heavy billets on the exposed ends of the cantilevers and then hidden from view as grouted corbels on the sides of the girders.

Nature has a habit of finding the weakest path of least resistance for propagating cracks and in the case of low shear resistance, it almost always means failure. The designer and the detailer must guard against the possibility of such weak paths.

10.3 Compression Failure

The strongest factor in concrete is its compression strength and if the concrete mix can provide the desired strength, there is no reason to suspect the columns as a cause of failure. Actually very few failures have been attributed to columns, although distortion from eccentric loading and vertical change in length from temperature, shrinkage, and creep are largely connected with column action.

In the early days of reinforced concrete, columns were designed at a lower factor of safety than is recent practice, and certainly seldom was any bending stress considered. In Basel, Switzerland, a five-story hotel collapsed in 1901 because of column failure. About 10 years later there were several column induced collapses in the United States. In 1909 during a load test of the completed reservoir roof at Annapolis, Md., the

9 × 11-in. columns failed. In 1910 the Hencke Building in (
Ohio, collapsed with little warning when almost complete. 7
the four-story building was then 12 days old and some timb
still in place on the third floor. Failure started in the third story co
umns, but the investigating committee also found the architect negligent for his lack of supervision and technical control, and the contractor negligent for careless construction, for a 20 percent deficiency in cement provided for the concrete used and for lack of coordination with the architect. In addition Edward Godfrey and C.A.P. Turner criticized the adequacy of the design.

The failure of the Stark-Lyman Building in Cedar Rapids, Ia., in 1913 was generally blamed on the columns. In 1914 the U.S. Custom House built at La Ceiba, Honduras, collapsed when the 12 × 12-in. columns supporting 20 × 21-ft bays proved inadequate. Disregard of bending stresses induced in the columns of a flat slab design may be the explanation for such column failures in 1916 at the Havre Holding Co. and at the Roxbury, Mass., School Buildings and in 1920 at the Braunstein-Blatt buildings in Atlantic City, N.J.

Two recent foreign cases of complete building collapse caused by overload of columns are the eight-story apartment buildings on the Italian Riviera in 1965 and the 15-story structure in Piracicaba, Brazil in 1964. The latter was designed as a 10-story apartment and commercial building and the owner, over the protests of the structural engineer, insisted on adding five more stories. The concrete was completed to roof level when one wing collapsed and killed 40 workmen. Designed as a narrow building of one long span and two cantilevers, the two wall columns broke into rubble and the floor slabs folded over to close the end of the building with a huge concrete venetian blind.

A common compression failure comes during the lifting of long precast elements with two sloping cable connections from a single boom. The horizontal components of the lifting forces compress the casting, and buckling failure has resulted.

10.4 Erection Difficulty

Unless the formwork scaffolding gives way in the cast-in-site designs or unless there is a weak link in the program of erection procedures, concrete bridges seldom fail. The destruction of three spans of the Maracaibo Bridge in 1964 by the impact blow of a heavily loaded oil tanker and the repetitive local demolitions of the Lake Pontchartrain Causeway by barges are cases of collisions described in Chapter 5.

A mysterious bridge failure in 1966 was the sudden collapse of the

center-span four-lane crossing of the River Nethe, nine miles from Liege, Belgium. The fall came during a foggy night and eight cars plunged into the river or were smashed on the broken bridge. The bridge is of similar design to one along the King Baudouin Highway, with 166-ft-long center spans and 90-ft side spans. After the center span dropped into the river, one side span was pulled down into a sloping position.

The precast sectional bridge across Washington State's Hood Canal from December 1958 to January 1960 had a set of mishaps not caused entirely by weather conditions. This highway bridge, 6470 ft long, consists of 23 concrete pontoons with an elevated roadway structure. Individual pontoons are 50 ft wide and 360 ft long and are joined together with a steel-transition hinged span at each shore to adjust for an 18-ft tidal range. Except for the two center pontoons that can be retracted to provide a ship passageway, each unit is anchored with two 550-ton concrete blocks.

The pontoons with their sections of roadway were fabricated at a casting yard 35 miles away. In December 1958 two pontoons sank in the channel leading from the yard. One unit was complete with the deck and became completely submerged; the other pontoon had only the columns constructed and sank on a 22 deg slope, with the columns projecting above water. The sunken pontoons, each representing a half-million dollars expenditure, were salvaged by building walls from the steel panel forms attached to the sides into a floating cofferdam on the totally submerged unit and by winch lifting from barges for the other unit. Cost of salvage was about 75 percent of the amounts already spent on the two units. The real cause of the sinking was discovered a year later when a third pontoon started to sink and it was found that wood blocks plugging holes at one end of the pontoon had worked loose and allowed water to flow into the internal compartments. The holes were near water line and are used to hold anchor lines that string the pontoons together into a floating bridge.

The sunken pontoons had to be salvaged or removed since they blocked entrance to the graving dock where the fabrication yard was located. To increase stability of the floating units, the roadbed then was lightened by the use of lightweight aggregate concrete and this permitted addition of ballast inside the pontoons. In December 1959 two pontoons collided and the end wall of one was punctured by the projecting shear keys of the other, and repairs were necessary before they could be joined together. The storm in January 1960 caught the bridge while the grout in the joints was not fully hardened, and deformations of the prestressed bottom slabs and normal concrete top slabs became cracks. Strengthening of the design was then agreed upon,

chiefly by adding 24 posttensioned cables, each of 40 ⅜-in. wires, through the pontoons and by replacing the portland cement with a fast-setting epoxy in the joint fillers. Each joint required 100 ft^3 of grout, making this job one of the largest applications of epoxy in the construction industry.

Demolition of a concrete bridge is sometimes more critical than erection, especially where the load of the demolishing equipment is on the bridge. In 1958 a 150-ft span concrete arch bridge at Topeka, Kan., was being wrecked by an iron ball slung from a crawler crane on the bridge deck near one abutment. The operator was quite surprised when the bridge and the crane fell into the river together as the two concrete arches sheared near the abutment.

The omission of necessary reinforcement is an evident cause of failure whether in the case of the Wing and Bostwick Department Store in Corning, N.Y., where the rear bays at the roof and third floor collapsed in 1903 because bars were omitted from the concrete joists, or the Peoples Tabernacle House in New York, N.Y., in 1905, a six-story concrete frame with tile and rib floors where the fourth and fifth floors collapsed for lack of steel. The latter job had been designed and supervised by the pastor.

The lifting of a prefabricated structural unit, whether a precast pile, the suspended span of a cantilever bridge, a precast or prestressed or plain reinforced beam, or a lift slab section can only be safely accomplished if erection stresses do not go beyond the yield points and the lifting equipment provides sufficient continuous and uniform support. Failures in this class are always caused by inadequacies of small details or nonuniform action of the lifting procedure. Small horizontal forces can easily upset the neutral equilibrium existing during an upward lifting of a large mass unrestrained laterally.

Concrete placed by tremie in a pier form set in tidal waters was started at high tide and all of the cement and fines were pulled out by the receding tide (Fig. 10.7). When the form was removed, there was no pier.

Lift Slabs

Wide acceptance of the recently developed technique known as "lift slab" has been delayed somewhat by several such erection failures. There were three successive failures on three separate attempts to complete a 175,000-lb lift-slab roof-slab section at Ojus (near Miami), Fla., in 1952. Only the last was attributed to any structural design deficiency. The first failure occurred when a 5-in. pipe stub, supporting a jack on

Fig. 10.7 Pier concrete pulled out by receding tide.

top of a permanent column buckled and dropped the jack while eccentricity in lifting was being adjusted. The second failure was triggered by stripping of defective threads on the inserted flange supporting the slab. The third failure was a shearing in the edge of the concrete cap above the column, the computed unit shear was 93 psi over the circular concrete section. The final and successful design included some extensions of reinforcement and additional steel shear resistance.

During a demonstration erection of a lift slab in San Mateo, Calif., in 1954, while guys were being adjusted to counteract a 3-in. leaning of the columns, the 65 × 70-ft slab unit drifted laterally 15 ft and fell. In addition to six workmen, four of the 60 spectators were injured and were awarded damages of over $100,000 as a result of litigation in a California court. The verdict was against the architect, general contractor, and lift-slab contractor. Fortunately most of the visitors were cleared off the slab. Failure was partly explained as the result of eccentric loading when the observers were concentrated on one side of the slab.

During a wind storm with possible gusts of 35 to 50 mph—normal late spring weather in Cleveland—an eight-tier lift-slab garage tower drifted laterally about 7–8 ft out of plumb. It was the west tower of two. It was completely raised but none of the slab collars had been welded com-

pletely to the columns. The east tower had the second and third floors in place with the rest of the slabs temporarily held within the next two stories, and remained plumb. Emergency action stopped further movement. The *Cleveland Plain Dealer* reported the next morning April 8, 1956:

"As darkness fell, engineers apparently had won their frantic struggle to prevent their structure from crushing the two-story building next door. Cables attached to four winch trucks, telephone poles and other objects had stopped the building from listing more than the 20 degrees it had sagged earlier in the day."

Each floor slab weighed 90 tons and the accident occurred shortly after the top slab had been lifted into position. Within a month design procedures were completed for righting the 91 × 21-ft eight-story frame and its 10 columns. The complete success of the righting operation speaks well of the original design, but limber frames of such assembly require special bracing precautions.

In 1961 the lift slab building for TWA in Kansas City exhibited some unexpected slab deflections and in 1962 a seven-story assembly, nearly completed, suddenly and without any witnesses on a Sunday afternoon, collapsed in Fort Worth, Tex.

10.5 Frost

Concrete can be safely installed in freezing weather if precautions and protection equipment are prepared and available before work is started. Witness the great number of successful winter operations in Canada, where contractors have learned to live with the cold weather and protect against it rather than hope for the thaw that does not come when needed.

There have been some serious concrete failures caused entirely by lack of proper methods and protection. The collapse of a concrete bath house at Atlantic City, N.J., in March 1906, was attributed almost entirely to frost penetration into the wet concrete, resulting in separations between beams and columns and between slab and girders.

A most spectacular frost damage collapse occurred when the seven stories of a proposed eight-story concrete hotel building at Benton Harbor, Mich., collapsed on January 28, 1924. Most of the concrete in the debris could be just shoveled away; only a few of the lower story columns and first floor beams remained intact. Concreting was in December and January. Although no one was injured in this collapse, every

detail of specification compliance was checked and, as far as could be determined, the ingredients and the mix were correct, the aggregates were thawed before mixing, and the water in which some calcium chloride was dissolved was heated. Tarpaulins were available for one story only. Temperature records showed some below freezing days in December and almost no day without freezing in January. Compressive strengths of five samples taken from the debris ran from 247 to 718 psi, not enough to hold the dead weight. The concrete had not set.

In 1925 the four-story concrete John Evans Hotel, Evanston, Ill., was shored completely to the roof and collapsed when the cellar shores were removed. The concrete mix was not of the best—dirty sand had been used—but failure was chiefly caused by the lack of expected strength in the frozen concrete. In 1927 a 12-story steel frame apartment in Buffalo, N.Y., with concrete floor slabs suffered a partial collapse when the shores were removed from several spans that had frozen and the concrete below the ninth floor fell to the basement. The ninth floor slab was properly set and remained in place.

A common frost damage results from the practice of placing wooden mud-sills on frozen ground which softens from the drip and the frost protection heaters, causing unwanted sags in the supported structure. An example of a less common frost damage occurred in a substructure contract where completed piers with anchor bolts imbedded were left exposed during the winter. Bolts had been placed in oversize sleeves to permit some adjustments to fit steel base plates. Water filled the sleeves, froze, and cracked off the edges of many piers. Such open sleeves, even if only for future railings, should be covered or filled with nonfreezing materials.

Damage to precast beams after posttensioning from the freezing of the grout while sealing the cable sheaths has been reported at a number of locations. One recommended procedure, from Scandinavian sources, to avoid this damage is to mix alcohol into the grout to act as an antifreeze. The expansive force of the freezing grout acting across unreinforced faces of concrete can completely shatter the highly compressed casting.

The formation of growing ice crystals, exerting pressure of several tons per ft^2 intensity, is a cause of many other troubles. The continually expanding cracks in exposures to water during below freezing temperatures results in surface cracking and spalling of roads and walls in any available material. Frost penetration of the subgrade on which concrete footings are placed will heave the footings if moisture is available to form ice crystals. Spring thaws will then bring settlement of the structure.

An interesting situation of this nature occurred in Watertown, N.Y.,

an area of very cold winters. In the fall footings and piers for an apartment house were completed up to grade and the work was backfilled and left idle over the winter. With the next construction season the timber sills and framing were set on the piers and work was well advanced when considerable distortion was observed. A study of conditions showed that the frost action in the fills around the piers had lifted the undowelled piers off the footings, some by as much as 12 in., and had held them up during the construction work until the ground thawed. Repair work required the exposure of all piers, grouting solid to the footings and a releveling of the entire floor construction. Similar upheaval of concrete block walls backfilled to grade and left during a winter in Detroit in 1938 was observed when work started again, requiring an expensive reconstruction.

The advisability of providing an impervious surface on concrete with water exposures is often seen in dams and bridges as well as buildings. A typical example of the havoc raised by a series of freezing and thawing cycles during a long winter was the Indian Lake Dam near Brussels Point, Ohio, in 1959. The entire spillway surface required replacement after chipping all loose concrete.

As reported by Abdun-nur and Mielanz, finishes on slabs placed in cool weather where frost can be expected before the concrete is set, are vulnerable to rapid disintegration. At the Platteville, Colo., school in 1957, such conditions existed and the tile floor finish was found bumpy. Exposure and examination of cores in the concrete showed that the top ½–1¼ in. had been frozen before hardening, was severely fractured and continuously disintegrated under traffic. After setting, the internal cracks permitted formation of ice crystals that did further damage to the top layer of concrete and formed the wavy surface.

The failure to clean snow and ice accumulation out of the forms is a common error, in the expectation that the concrete mix will melt the snow and clear the space. Since heated liquids rise, the warm concrete mix water will not thaw out packed snow and ice near the bottom of the forms. Such a condition in the cellar walls and column piers of a New York Housing Authority project resulted in a honeycomb condition which was beyond correction and the entire section of walls and piers were ordered demolished and rebuilt. (Fig. 10.8). Pile caps poured on frozen ground, in the same project, did thaw out some of the soil and the concrete below the reinforcement mat settled. The result was a horizontal cleavage in the concrete caps at the level of the rods and some of these were also removed and replaced; minor separations were grouted to protect the steel.

The dangers of concrete work in freezing weather are well-

Fig. 10.8 Concrete replacement required by frost and debris in forms.

documented and the necessary frost protection procedures are known, but this knowledge is of no use if the necessary equipment is not assembled and made available before frost strikes. After concrete is frozen, no frost protection methods will cure the trouble—the concrete must not be permitted to thaw partly and refreeze. Any concrete program in the winter where frost is a possibility must be preceded by a complete assembly of all wind breaks, heating devices, and distribution necessary to protect against the coldest day for the largest expected area of work. Only such insurance will protect against weather damage to the work.

10.6 Temperature Change

All structures change dimension with change of temperature. In buildings the better control of internal temperature increases the range of differences at the roof level. Change in size of the roof slab where rigid attachment is made to the walls must result in cracked walls, usually at about the head of the top-story window since the resisting horizontal

wall section is suddenly reduced at this level. Usual expansion of the roof indicates itself in many buildings by such cracking with accompanying leakage, wall staining, and heat losses in the top story. Most older designs called for a projecting brick course at this level to shade the crack and protect the interior against water infiltration.

During the construction of the Mulford Houses, a low-cost project in Yonkers, N.Y., in 1939, the writer suggested that a 1-in. gap be provided between the face of the roof spandrel and the brick covering. This was accomplished by inserting a strip of plywood against the concrete and removing it after the brick facing reached roof level. Although the parapet, 12-in. solid brickwork, covers the gap, the desired freedom of movement seems to have been accomplished; no cracks formed in the brick cover at this project.

Lateral expansion of structures often causes brick covering to push and crack. The cracks never close completely with colder weather; some dust always finds its way into the crevice and subsequent expansions increase the crack size. To control this cracking, in the New York City Housing Authority projects all brick walls covering concrete frames are built with vertical continuous caulked joints at the first window edge on each end of a wall surface. This does not eliminate the expansion, but at least stops the cracking of the brick since the tendency to opening follows a continuous expansion joint.

Roof expansion troubles in concrete structures were analyzed very early in the history of the art. In 1885 a rectangular concrete reservoir for 10 million gal was built in Colombo, Ceylon. Walls were vertical face inside and 5 on 12 batter outside, the floor was 12 in. thick and a concrete slab covered the 30-ft-deep box. When this was filled in 1885 to a depth of 24 ft, cracks were noted near the corners. After repairs and adding another foot of floor concrete, refilling in 1886 to a depth of 28 ft again showed cracks near the top of the walls, which traveled to the ground level in 15 minutes. Sir John Fowler, a Past President of the British Institution of Civil Engineers, was consulted, and analyzed the cause as expansion in the high temperatures of the area. He recommended cutting strips out of the walls and inserting brick piers covered with asphalt. This was done and the reservoir was filled again in 1889. Cracks again appeared at the corners. Measurements were then taken of the roof dimensions and daily variation of 0.01 in. were found. The roof was then covered with earth and no further cracking occurred. This five-year history of repair exemplifies British determination to make a design work.

A similar reservoir in Madrid built of reinforced concrete in 1902 was 22 ft deep and contained 127 million gal. The roof consisted of a series of

parabolic arches resting on 20-in.-deep beams spanning 19 ft. The arches were uniformly 4 in. thick with a rise of 23 in. The beams were supported by 12-in.-square columns 27 ft high, and formed a continuous length of 700 ft. When the first of the four boxes was completed and the roof covered with 31 in. of sand, the girders were noted to be deflected and the columns were being pushed by the expansion of the total length. The deformations were maximum at about 3 p.m. and decreased at night. In about a week some columns broke and the roof collapsed in 1905.

In the February 4, 1960 issue of *Engineering News-Record* an editorial covers this subject completely, noting the lack of definite information concerning thermal strain protection.

"Thermal strains have been the nemesis of many a structure in the past, and occasionally of some recent ones, too. When adequate provision is made in design for the effect of temperature change, however, it need be of no more concern than dead or liveload effects. But what is adequate provision?

"Safety is, of course, a prime consideration, but good engineering also calls for provisions that preclude structural damage at a minimum of cost and that also are no more harmful than the damage they seek to prevent. While many techniques in use today keep thermal strains under control, they do so at the expense of 'overdesign' and sometimes they do more harm than good. In some cases, it may be found to be safer and less expensive to ignore thermal effects or to design a structure to resist the strains than to make provisions for free movement.

"In the building field, there is occasional evidence of the costly result of overlooking temperature effects. There are roofs that expand and push out exterior walls. There are facades that crack because they are prevented from moving, and yet were not designed to resist thermal strains. More attention to thermal effects could eliminate these difficulties.

"In concrete highway slabs, transverse expansion joints are required at frequent intervals by many states, while others have many miles of highway without such joints. Which design is best is plainly an unresolved question, but the possibilities of reduced maintenance and better riding qualities with a jointless pavement argue in favor of further study and trial.

"A study of thermal strains in bridges also would prove profitable. In some foreign countries, for example, bridges are designed for nonuniform temperature distribution, which is associated with smaller movements than those resulting from an assumption of constant temperature. Hence, less elaborate provision need be made for expansion and contraction.

"Furthermore, in some countries, continuous girder spans twice the length of any in the U.S. have been built without suspended spans and interior expansion joints, and have been in service for many years without trouble. This is in sharp contrast with designs that use suspended spans to break continuity and include elaborate provision for expansion and contraction at the ends of those spans.

"Undoubtedly more test data are needed to shed additional light on thermal behavior of structures so that they need not be under or overdesigned for temperature effects. But there also appears to be evidence that all engineers are not yet putting to use all the information now available."

Masonry-covered concrete bridge railings when fitted tight to rigid bridge abutments will bow out of shape with resulting loosening of the mortar joints under sun exposure. One such case is illustrated in the September 5, 1956 *New York Times*, the coping stone being at least 12 in. out of line on the Northern Boulevard overpass at Corona. Temperature expansion distress is usually found near the end of the summer when net heat absorption, in the daytime gain less the night loss, reaches an accumulated maximum.

Normal temperatures permit large structures to absorb enough heat so that surfaces spall, plane areas become curved, and even structural defects occur. Athletic stadia are a good example of the necessity for providing free space for expansion. This is evident in all stadia whether steel or concrete frame. During a detailed investigation of the Polo Grounds in New York in 1940, the magnitude and intensity of the movements was pointed out by the loud and continuous creaking of the structure in the early mornings as the sun started to warm it up. It is not necessary to list the stadia that are under continuous repair programs; an inspection of almost any of them shows a shortage of open joints to permit expansion and often such freedom must be three-dimensional if excessive and expensive maintenance is to be avoided.

Thermal growth of concrete exposed to the sun accumulates over the years and can be the cause of ultimate failure. The curtain walls of the Texas State Fair exhibition building in Dallas were constructed in 1948 by shotcreting mortar on a pressed fiberboard backup to a 2-in. thickness. The layer was reinforced with 4 × 4 welded wire mesh and braced by 2-in. double-channel studs at 4-ft centers. Panels were 24 ft wide by 14½ ft high. Three such panels on the south wall collapsed in September 1966 with little warning.

Some of the delayed distress may be explained by the difference in the coefficients of thermal expansion and of contraction, which apparently are not exactly equal. The *Deutsche Ausschuss fur Eisenbeton* Bulletin

No. 23 reports tests on dry concrete specimens 90 days old, with average coefficients of expansion 0.0000069 and for contraction 0.0000056. Based on these values, a 200-ft length of concrete exposed to a single cycle of warming 50° F and then cooling by the same amount will grow in length by 0.15 in. Such changes will soon close up expansion joints, if any, and where freedom of growth is restrained cracking and spalling must result.

The average coefficient of expansion for concrete is 0.000006 per degree F change. With an elastic modulus of 5,000,000 psi (lower for weaker concretes), each degree change induces a stress of 30 psi. Normally such increase in compressive stress can be easily taken by the concrete, but if no reinforcement is provided to resist shrinkage, only a small amount of cooling must result in cracks when the low tensile strength of the concrete is overcome. The usual requirement of 0.25 percent of the cross-section area in the horizontal direction, and 0.15 percent vertically, lesser amount because of the compression from the weight of the concrete itself, merely extends the range of temperature which will not cause cracking. Experience in temperate zones indicates that 0.65 percent is necessary to resist 100° temperature drop, and it is therefore not surprising to find so few uncracked concrete exposures.

With the accepted architectural use of exposed concrete columns in multistory structures, a new problem has arisen. With interior temperatures fairly constant, the exterior columns change in length with change in season. As a result a differential level develops in the upper floors between the exterior and the interior supports. Floors are actually found to change in level, partitions crack because they cannot absorb the vertical warping, and slabs crack near the interior supports. Exposed concrete shear walls are equally vulnerable with columns, and higher differentials are found in the sun exposure positions. As would be expected, the effects are much greater in northern climates, where such distress has been seen in the upper 10 levels of 25-story buildings. In four 32-story buildings with exposed concrete columns in the end wings, plaster cracking was found above the twentieth floor in the wings only and with greater incidence in the higher floors. To control the high maintenance costs from such action, the Federal Housing Administration in 1964 refused to accept any design with exposed concrete columns taller than 70 ft. Upon proof that the restriction was too drastic and unwarranted, especially that it disregarded variation in climatic exposure, the regulation was revised to permit local FHA offices to accept designs for taller buildings when the owner assures that the designs take into consideration the stresses from any changes in length or height resulting from temperature changes.

In a tall apartment building in Montreal all exterior floor slabs above

the fifteenth floor were designed with an articulated support at the interior end, using a seat on a girder bracket with a caulked joint, and all partitions ending at wall lines were set in channel frames with a separation at the ceilings. Measurements in winter indicate definite rotation of the slabs about the free joint. Where the opening at the top of partitions was not completely provided, the plaster has cracked in a diagonal pattern on the vertical faces. Much has still to be learned about the effect of the vertical reinforcement in the exposed walls and columns with unequal thermal coefficients of concrete and steel, especially when lightweight aggregates are used, and also of the thermal gradient in the concrete thickness. The same problem applies to tall steel column exposures. In the proposed 110-story World Trade Center towers, exterior steel columns will be encased in metal sheaths through which hot air can circulate to keep the exterior and interior columns at the same temperature.

10.7 Deformation and Cracking

Immediate elastic and delayed creep changes in dimension and in shape of concrete structures cause deformations which bring expensive maintenance and repairs although they may not seriously affect the strength of the structure.

Connections of structural members introduce secondary stresses not normally computed. Under the general heading of secondary stresses are those cases where direct stress computations from normal loading would indicate no reason for cracking but the resulting strains induce stresses along other axes which will give trouble. Such conditions always exist but the intensity in usual designs is not high enough to exceed the available, but usually disregarded tensile strength.

Structures like flat slabs transmit loadings to supports by the easiest way which may not be the pattern assumed by the design. Cracks then result as the frame forms hinges and tries to conform to the assumed pattern. Slab loads are assumed to travel to the columns by bending of the allocated bands. Actually, because of the unbalanced bending moments transmitted to them, the spandrels rotate and carry the reactions to the columns partly by torsion. Diagonal torsion cracks then appear in the face of the spandrel section, especially if a deep beam is used. Such rotation has been known to push the masonry facing out of position and form horizontal cracks in the mortar jointing with subsequent rain infiltration into the building. Spandrels must be designed to resist such torsional strains.

Incompatibility of deformation caused some cracks in precast pre-

stressed louver slabs set into grooves of concrete piers in a unique and economical retaining wall in New York City. The imbedded portions of the precast units were roughened to get better bond in the grouted grooves. A rigid connection resulted, and when the wall was backfilled the slabs deflected but the ends were held rigid and the face of the piers in front of the grooves cracked vertically. The pier deflection was small but at right angles to that of the slabs. Reconstruction with a flexible filler corrected the trouble.

Similar torsional action causes corners of roof slabs to curl and distort roof flashing and to crack exterior faces of exterior columns. Storage bins designed for wall tension as a resistance to internal grain or liquid pressure must be free to expand at the contact with rigid base slabs or foundation. As Mensch pointed out, the bottoms of tanks are very vulnerable to failures even if the shell is prestressed.

Trouble must be anticipated at the junction between adjacent bins, whether circular or rectangular, as is described by Vandergrift. A similar action explained the diagonal cracking in a Cuban prestressed concrete pressure main near the joints that were grouted into precast concrete sleeves. The expansion of the pipe as pressure was built up was restrained near the ends by the rigid sleeves. The trouble disappeared when the joints were caulked with elastic material in place of the cement grout.

Concrete floor slabs with monolithic finish, especially in long-span designs, are seldom constructed to satisfactorily level tolerances. The customary procedure of screeding the concrete from midspan to column lines pulls the denser materials toward the columns and on hardening and drying the surface is dished. Built-in camber in the formwork is lost; actually the only resistance to future deformation is the compression of the form supports from the dead weight of the concrete. To neutralize deflection of the slab after the form supports are removed, the top surface should be finished as a slight dome. Even then the later plastic flow of the finished concrete plus any creep from loadings is not compensated and a dished floor results.

In the garage floor below grade in the New York Coliseum a waffle flat plate design with panels 24 × 30 ft, the finish level was satisfactory when first opened to use, but in three years the creep of the concrete, with private car parking imposing at the most not half the design load of 100 psf, the spans had dished to the extent that small weep holes had to be cut in the center of several panels to eliminate the birdbaths. These were connected to the drainage system by small lateral lines. A 1 percent slope on a garage floor will maintain satisfactory drainage for at least 10 years use, with the usual spans. A slope of about 2 percent is a wise precaution and does not hurt the use of a garage floor.

The cause and prevention of cracking in concrete was well covered in an editorial so headed which appeared in the *Indian Concrete Journal* February 15, 1953. It is quoted substantially in full as a complete coverage of this item.

"Cracking is one of those problems in concrete construction which evokes frequent discussion among engineers. One attitude to this problem that seems to be largely prevalent is that cracking is a natural occurrence about which little can be done. And yet it is possible to investigate into the relationship and inter-connection of different types of cracks, to detect their causes, and to apply remedial measures for their prevention.

"Cracks can be broadly classified into those which occur before hardening and those which occur after hardening. The period before hardening is considered to extend from the time of placing the concrete up to the time when it has attained sufficient strength to resist alteration in form, in practice 2 to 8 hours after placing under normal conditions.

"Subgrade movement due to moisture changes and movement of forms due to inadequate design or construction are two frequent causes of cracking in the pre-hardening period which can be prevented by adequate compaction and control of the subgrade and by careful attention to formwork. Settlement shrinkage is another cause of cracking in this group. When concrete is placed in position and before it sets the solid particles of aggregate settle and stabilize themselves. Continued settlement within the skeleton of these large particles can result in mortar settlement and ultimately in the accumulation of water at the upper surface or water gain. Now if the conditions are such that the upper surface of the concrete becomes partially set whilst the interior is still settling, and if the interior contains some rigid obstacles such as reinforcing bars, then the interior will settle around these bars and cause cracks in the partially set concrete above the bars. The remedy lies in the use of dense plastic mixes with well-graded aggregate, low water content, and adequate compaction.

"Then there are those cracks which occur very soon after the placing and finishing, sometimes within a matter of minutes, even under a film of water. These are ascribed to plastic shrinkage caused by the expulsion of free water from the freshly formed silica gel or by a false set. The remedy here appears to be delayed finishing—pressing the plastic concrete to close the cracks. It is found that when this is done the cracks do not reappear.

"Most engineers are familiar with the type of cracking caused by drying shrinkage. This can be particularly observed when freshly poured concrete pavements are exposed to a high wind, low humidity,

or hot sun with temperature differential between the mass and the surrounding air. Prevention lies in covering the concrete with damp sand, hessian, jute bags, or straw as soon as there is no danger of their leaving unsightly marks on the concrete. Drying shrinkage, of course, continues after hardening. Reduction in the cement and water content of the mix with adequate curing to permit complete hydration minimizes the extent of this shrinkage.

"The ramifications of the other group of cracks—those which occur after hardening—are considerable. Moisture movement here again plays a part as subsequent wetting of concrete after drying presents a reversible action which is comparable in magnitude with the original shrinkage. Therefore, structures which are subject to alternate wetting and drying must have provision for such movement or be able in some way to withstand the stresses which are set up, otherwise cracking may be expected.

"The chemical causes of cracking include not only reactions within the constituents of concrete, but also from the presence of foreign bodies. Incomplete combination of the lime with the other raw constituents of cement, and excess of magnesium oxide and gypsum can cause cracking. Reactive aggregates present another source of cracking to which we have made reference in the past. Modern Portland cement specifications are framed in such a way as to prevent these types of cracking. An example of cracking due to the presence of foreign bodies is afforded by the oxidation or rusting of mild steel reinforcing bars. This failure can be prevented by the use of dense concrete and adequate cover to bars.

"Much work has been done to prevent cracks resulting from the differential temperature caused through the heat of hydration of cement, particularly in large mass concrete construction works such as dams. The precautions now employed include limitations on the heights of pours and periods between successive lifts, the use of low-heat cement, the addition of cracked ice to the mix, and elaborated cooling water systems including the use of refrigeration. External temperature variations also cause cracking in structures which can be prevented by the provision of adequate expansion joints.

"Stress concentrations are another source of cracking. Stress concentrations can be caused by stress transfer in reinforcement or by structural form. An example of the former is when a bent bar is placed at the junction of the stairway and landing slabs with insufficient depth of concrete to withstand the stress concentration at the change in direction. Square openings or re-entrant angles such as door and window openings or openings for manholes in pavements cause considerable concentrations of tensile stress which must be resisted by suitably

placed reinforcement if cracks are to be avoided. Foundation settlement and accidental happenings such as overloading, excess vibration, fire, storm, and earthquake are other sources of cracking in concrete structures.

"It is apparent from this survey of the numerous causes of cracking in concrete that the subject is of considerable importance in concrete construction; also that cracking is by no means inevitable and much can be done in design and construction to prevent it."

All of this of course is well-known and included in the same or better detail in the various standards and recommendations of the ACI and the PCA (see, for example, "Prevention of Plastic Cracking in Concrete," *PCA Concrete Information,* Structural and Railways Bureau, 1955) but still the large incidence of cracking exists.

10.8 Precast Assembly

Manufacture of precast elements under controlled conditions permits delivery to the site of relatively dimensionally exact pieces for field assembly. The small dimensions of the elements eliminate thermal change difficulties; better concrete control and steam curing reduces delayed shrinkages and creep deformations. The difficulties are then mostly in the details of the assembly, although fabrication troubles are not eliminated. One such trouble is the shrinkage during setting and curing of precast concrete items made in steel forms that do not give, and cracks develop at the junction of flanges and webs of beam sections and at the contact between tread and riser in L-shaped castings. Casting accuracy is an economical necessity. In a large two-story housing project at Great Lakes Training Station in 1952 the lack of tolerance control on the precast wall units caused a greater labor expenditure in the correction of edges and modification of connection details and welding than the total labor cost of manufacture and delivery. Several cases of prestressed concrete work are described in Chapter 11.

After about 1½ years use the San Diego Transit Shed No. 2 was exposed to a series of earth tremors. The structure is 20 ft high with a single line of columns in the length of the 133 × 880-ft area. Girders are 40-ft-long precast units resting on precast columns and supporting 66-ft-long precast beams framed into wall pilasters. All the girders shifted at the column support, some as much as ¼ in., and the tops of the columns spalled to expose the reinforcement. Corrective work in 1956 added corbels to the tops of the columns and tied the ends of the girders together to provide continuity in the framework.

In 1963 the British House of Commons ordered a technical inquiry into the collapse during construction of an officers' mess building at Aldershot. The Ministry of Public Buildings and Works, owing to a severe shortage of building labor, had contracted for four similar buildings, chiefly of precast concrete, on a design-and-construct basis. The four-story buildings had precast columns, beams, floor, and exterior wall panels. The joints at columns and beams were made by in-place concrete, and the center core area of the 60 × 60-ft building was also cast in place. The framing was a 20 × 20-ft grid.

Columns were cast in story-height with bars projecting from the top and a hole at the bottom to receive the splice steel from the lower columns. Beams were cast full length with recesses for the column steel and formed part of the column when the recesses were concreted. Floor plates were ribbed sections of precast 1½-in. slab and were doweled to the beams by bond splices between projecting rods of the beams and exposed rib bars in recesses of the casting. Bearing of the beams on the columns was noted as 1½ in. but actually was found to be less due to shrinkage in the castings. The ends of the beams were supposed to be rough to provide better bond to the filling concrete, but actually were manufactured smooth. Bent-up dowel bars were shown from the ends of the beams into the columns, but these were often omitted since the bar came close to the form face and could not be concreted readily (Fig. 10.9).

Frames had been erected to full height for all four buildings when one collapsed due to its inherent instability. The investigation recommended that the other three be demolished, but before action was taken another one collapsed. Collapse started by failure at a column joint. With no lateral stiffness the entire assembly collapsed because there was no continuity between beams and no lateral bracing. The design had made no provision for wind loading or eccentric loading on columns. The conclusions of lessons to be learned from the long list of features in the design and construction that contributed to the collapse are an important contribution of the warnings of why structures fail.

"(a) Where a system of building using prefabricated structural components is extended by use in a new building type, a fundamental reexamination of the system design is necessary. This must include a reconsideration of all design assumptions and, if necessary, a recalculation of the structural design from first principles. This is especially true where, as in this case, a building frame which has normally been used with cross-wall stiffeners is used with full-length glazing and open plan. The removal of the stiffening effect of panels is particularly important if,

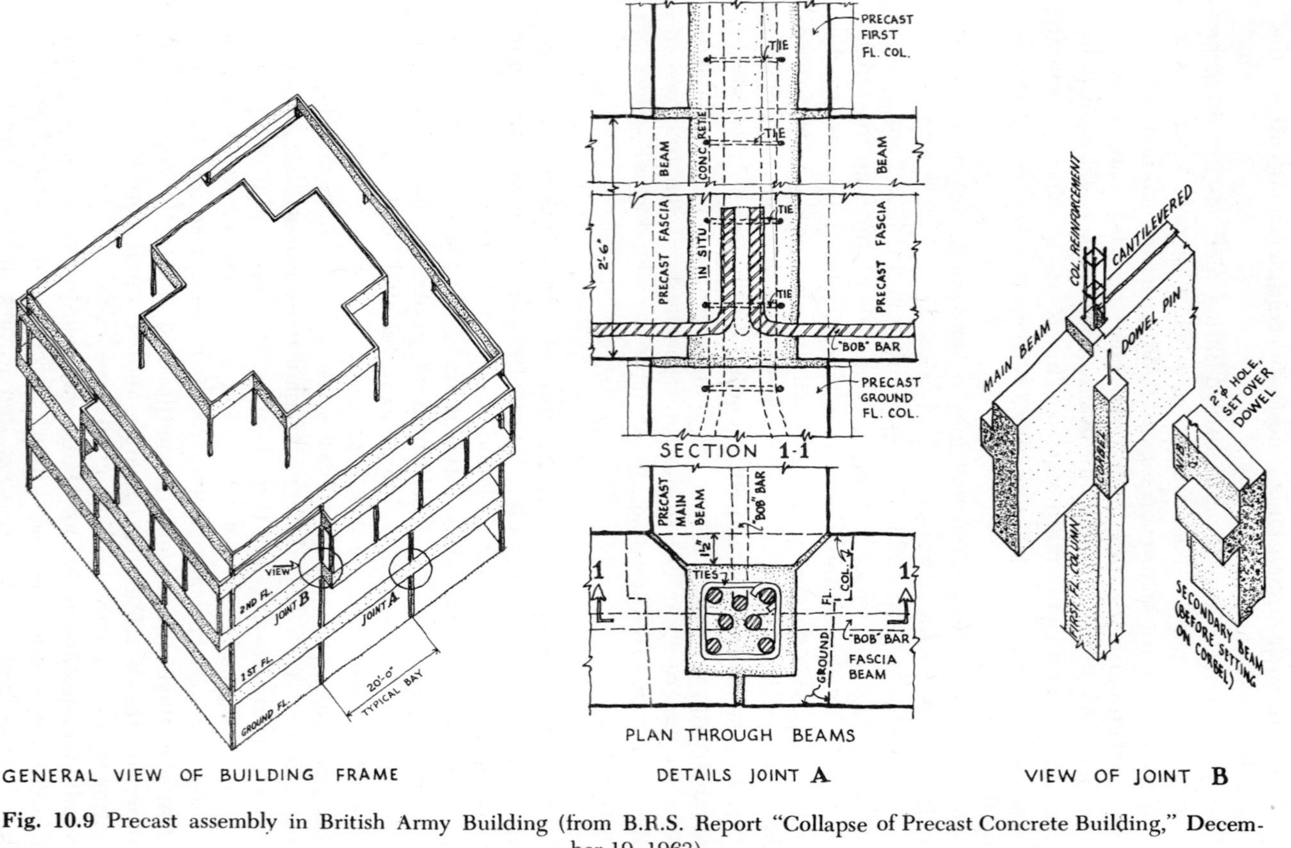

Fig. 10.9 Precast assembly in British Army Building (from B.R.S. Report "Collapse of Precast Concrete Building," December 19, 1963).

as is likely in such circumstances, aesthetic considerations call for the columns to be as slender as possible.

"(b) When novel, or relatively novel building methods are used the thorough and systematic communication of the designer's intentions to the operative is more than ever essential. When traditional methods are used the familiarity of most operatives with the work may allow an occasional uncertainty or ambiguity in drawings and specifications to be corrected by intelligent interpretation. This is not so when structural techniques are used in which operatives have only a limited familiarity with the work.

"In particular, it is essential that fully dimensioned working drawings should be prepared for all joints.

"The designer is responsible for communicating his intentions, particularly with regard to those factors and dimensions which are of critical importance, in distinction from those where some tolerance can be allowed. It is also the designer's responsibility that he does not assume higher standards of accuracy and workmanship that can be attained.

"(c) In systems of construction depending on the assembly of prefabricated structural components the erection procedure is an essential part of the engineering design. It must be specified by the designer in sufficient detail to ensure that the structure is sound at all stages of construction."

During the erection of the precast concrete framework for the Community Hospital, Roanoke Valley, Va., in 1965, one-story columns being set on top of two-story precast columns did not match for the welding exposed ends of the vertical bars. It was found that the reinforcing cages in the columns had not been held in position and the bars lacked the necessary 1½ in. cover. The columns were rejected, after some 59 had been erected before the error was discovered.

The erection of three-story precast screens for the Bellevue Hospital garage, in New York City, disregarded plan requirements that neoprene pads be placed at each of the three bearing lugs provided in the vertical mullions so that inequality of fabrication as well as the irregularity of floor levels would be compensated. No pads were used and the entire weight was then imposed on the cantilevered ribs at the top level. A number of the bearing lugs sheared and required corrective epoxy grouting to distribute the loadings at all three levels.

Small elements assembled into large walls must be detailed to provide sufficient tolerance for imbedded items. Carefully cut hexagonal panes of colored glass showed much distress after a short time exposure to outside heat when edges were too tight in the concrete frames (Fig. 10.10).

Fig. 10.10 Cast concrete elements caused glass failure.

A partial collapse of a seven-level parking garage during erection of the precast elements at the University of Michigan campus in 1966 was blamed on wind conditions. The collapsed section included nine columns, 70 ft high, and 14 63-ft single tee beams. Four of the columns had seven tiers of beams connected and five were standing free except for four anchor bolts to the foundations. About one-third of the 203 × 246-ft garage was erected full height at the time of failure, but only the items erected that day were involved in the collapse. Failure came after all work had stopped for the day and completely wrecked the erection crane. One column was broken about three stories above the ground and one remained intact, but tilted 10 deg.

Precast concrete erection problems also exist in Russia. An article by A. F. Mikhailov in a Moscow technical journal presents the investigations of numerous reinforced structures which were erected with considerable deviation from the allowable tolerances. An 80-ft concrete truss was placed 3⅜ in. eccentrically on columns instead of the ⅜ in. permitted deviation. Cracks discovered in the side faces of a precast truss were correlated with uneven bearing on the columns. The paper concludes that theoretical results and visual inspections show that, un-

less erection tolerances are adhered to, the consequences are often extremely detrimental to the structure.

The domed roof over a Methodist chapel in Atlanta, Ga., collapsed in 1966. The 40-ft-diameter roof was covered by 12 precast radial sections, each supported by two precast wall columns and a center precast concrete compression ring. A weld in the outer steel tension ring gave way, and this warned all people out of the area. Failure came 10 hr later with a crash of 120 tons of precast concrete.

Some 500 tons of precast concrete for a bridge under construction on the extension of the London-Yorkshire Motorway fell into the River Calder in 1967. Two precast deck sections of slab and box girders fell when the common scaffolding collapsed. The bridge design had shared a £5000 first prize in a Ministry of Transport competition in 1964.

To replace an old 800-ft blimp hangar at the South Weymouth Naval Air Station in Massachusetts, a precast concrete design was being erected in 1967, when a 90-ft roof unit being erected by four cranes crashed 70 ft to the floor. The design of the 275-×-260-ft hangar included 15 precast concrete arch ribs. Two half arches were erected into position and pinned at the crown. Three arches were in place and as the fourth was set up, some bits of concrete fell down and the arch members fell, crushing two cranes. The Navy investigation blamed the absence of tie-bars and inadequate vertical shoring during the erection phase.

10.9 Tanks and Silos

Storage structures are exposed to changing pressures as content volumes fluctuate and, in the case of granular storage, may be affected by dynamic loading during rapid drawdown. Reservoirs and swimming pools are prone to fail by cracking of the floor when sudden emptying brings hydrostatic pressure under the floor. Well-built swimming pools will float up if weight or relief valves are not provided to neutralize the ground water buoyancy. Near the rushed completion of a Miami Beach hotel the owner was almost irrational when the tiled pool rose above the tiled terraces after the ground water pumping was stopped. It had been on his express orders, without consultation with the designers, that the pool and adjacent lockers had been lowered 2 ft from plan grades, to get better headroom.

In soils that do not drain readily special leaching provisions are necessary to relieve the hydrostatic pressures under the floor when the pool is emptied. In clay soils mud will enter the pool at all cracks and expansion joint separations and dirty the water. In several such installa-

tions a cure was to excavate pits along the walls to act as sumps below floor levels, jack lengths of perforated drain pipe under the slab, and pump ground water from these sumps whenever the pool was to be emptied. Similar action on floors of reservoirs has caused damage at the Portland, Ore., East Liverpool, Ohio, and the Queen Lane, Philadelphia installations, the former two structures had sloping concrete walls.

Wall failures in reservoirs, where pressure loading is fixed and simply determinable, are usually explainable by the disregard of the fixity at bottom and side edges, monolithically constructed with the adjacent slabs and walls. In 1924 the concrete reservoir at Cudahy, Wis., lost 1¼ million gal of storage when the four sidewalls fell out. Just before being placed into service, a 16-ft high-water settling tank wall fell off in the water treatment plant at Palm Beach Gardens, Fla., in 1967. The reinforced concrete wall, 65 ft long and 12 ft thick, fell over in one piece, but broke up on impact. It was not only the tank wall, but also the exterior wall of the treatment plant.

The failure of the 400,000-gal concrete tank in 1963 at Oakland, Calif., destroyed the folded precast plate design when the walls separated from the induced bending stresses at the breaks in continuity.

Failures of concrete storage bins have been well-documented. The Duluth grain elevator consisting of 15 circular bins 33½ ft diameter by 104 ft high with 6 ft roughly square interstice bins had the distinction of failing twice, first on December 17, 1900, and again after reconstruction on April 16, 1903. Both failures occurred when an intermediate bin was being filled with the circular bins empty, the walls not having sufficient resistance to act as arches. The walls of the circular bins split vertically nearly the full height.

The tilting of the Canadian Pacific Railroad bins in 1914 to an angle of 27 deg and the successful righting of the structure was a classic of engineering talent. Similar operations at the Lincoln, Nebr., grain elevator remedied uneven foundation resistance after the 5-million-bushel elevator settled 8½ in. at one end of its 273-ft length in 1955.

At Fargo, N.D., in June 1955 a new installation of 20 silos, 19 ft in diameter by 120 ft high, failed completely without warning and with no observers. The lower part of the silos broke into about 15 ft lengths and the upper part shattered completely. Theories of failure are either an explosion or a shift of the foundation.

A true explosion failure occurred in the Philadelphia grain storage elevator in 1956. Although large damage was caused to windows and structures, most of the concrete silos filled with grain suffered only superficial damage.

In 1959 one-third of a 100 × 375-ft elevator structure, 115 ft high, in

Port Arthur, Ontario, collapsed. Although 30 years old and founded on fill contained by a Wakefield sheet pile enclosure, the first repairs were found necessary in 1958 when some 500 yd^3 of lost fill was replaced and 250 ft of dockwall steel sheetpiling was added on the water side. The added protection was not sufficient to stop further lateral soil movement and the collapse.

Possibly the study by L. E. Vandegrift in *ACI Journal* 1954 showing that lateral wall failures can result from filling the interstices with main bins empty and the necessity for reinforced fillets at bin wall intersections, may provide some clues to the many bin failures. L. J. Mensch in the November 1955 *Civil Engineering* also points out the necessity for special design features at the bottom of tanks and bins where ring deformations are not compatible with the restraint of the floor.

Necessary rules and regulations for proper design of bins to provide security against lateral unbalanced pressures were formulated in the report of Committee 714 (renumbered 313) of the American Concrete Institute.

Concrete storage bins for cement and for grain have been known to crack vertically and recent techniques of epoxy intrusion are used to repair the walls. In a grain bin at Long Beach Harbor, Calif., 132 ft high and 18 ft in diameter, with 7-in. walls, three vertical cracks from 30 to 68 ft in length were healed by pressure injection of a fast-setting epoxy in 1966. The cracks were first covered with a sealer, leaving open ports on 8- to 10-in. intervals. The epoxy was injected into these ports successively by a hand-held gun. The cracks filled were as small as 0.002-in. wide and carried through the full wall thickness. A total of 204 lin ft of cracks were sealed in 3½ days from hanging scaffold. These cracks are an indication of unequal pressures within the bin combined with thermal distribution.

Cement bins in England were carefully investigated in 1957 to determine cause of cracking in the circular walls. Cracking seemed to be concentrated in the vertical section in contact with the cement being extracted, mostly at a height of 15 ft above the silo floor. Indications were that a temperature differential in the walls and variations in both the vertical and horizontal pressure components while the cement was being drawn out were the causes of the cracking.

10.10 Space Frames

The inherent strength of space frames, usually in excess of the results from simplified computations, once they are erected, protects against

failure. There are troubles, however, resulting from dimensional distortion due to thermal and shrinkage changes as well as a few actual collapses. The older generation of such structures in the form of groined arches and vaulted floors were likewise quite stable with a few exceptions, such as the groined arches in Baltimore in 1913 and at the filtration plant in Philadelphia in 1914. Also a vaulted floor of 22-ft span failed in Columbus, Ohio, in 1918.

Within an hour, two separate quarters of a hyperbolic-paraboloid roof over a gas station in Dallas, Tex., broke off and fell in 1960. The roof was completed five months before the failure came with no load except its own weight. Some warning came in the form of "pinging, cracking, and popping" of the concrete slab which had no tensile reinforcement except a layer of very light mesh.

A folded plate roof over a monumental church in northern Minnesota, a design that received detailed coverage in the architectural magazines in 1961, deflected excessively after completion and was stiffened by anchors attached to the top surface and an additional layer of structural concrete. A folded plate entrance canopy for a college building in the Bronx, N.Y., was refused a building permit without an approval by an outside consulting structural engineer. The design check indicated weakness in shear at the columns and tendency to excessive deflection. The suggested corrective measures were to thicken the slab at the troughs and around the columns, plan changes that were incorporated into the design in 1960. The structure has performed successfully. The incident of a folded plate failure during a load test at Fort Sill, Okla., is described in Chapter 5.

The trouble that comes with three-dimensional space changes is illustrated by the experience with the Kresge Auditorium at M.I.T., Cambridge, Mass., a three-cornered dome completed in 1954. The roof is an equilateral spherical triangle with apices at the ground, supported by pinned shoes. The shell is 3½ in. thick at the crown, increased to 11 in. at the edges and 20 in. at the corner supports. Temperature and plastic movements were first discovered when the suspended lights refused to remain in a plane or at a fixed distance above the floor. Tension members were added to hold the foundations in a fixed configuration and by making the mullions part of the support, some support was provided at the center areas of the three curved edges to restrict roof shape change. The original design was to cover the roof with lead-coated copper as diamond-shaped plates on top of a 2-in. cinder concrete protection of 2-in. glass fiberboard insulation. Mechanical problems in making the joints forced abandonment of the covering material and the use of an acrylic polymer containing fine limestone chips for texture. Thermal changes

cracked this protective layer and water entered the lightweight concrete. During hot sun exposures large bubbles formed and further distressed the covering by freeing it from the structural slab. In 1963, after some two years research on how to solve the problem, the concrete topping layer was scored by an abrasive saw into a grid of 4-ft-square units. Each unit was anchored by 3⅜-in. bolts at the corners and at the center with 4-in.-square stainless steel washes as hold-downs. Lead shingle sheets, about 24 in. square were then laid down, using 6-lb lead, and held by a grid of ¼-in. stainless steel wires fastened to the anchor bolts. Overlap of the lead sheets covered the wires. A vent system at the top of the dome and drip lines equipped with electric resistance heaters at the three corners were provided for control of any future moisture accumulation.

11

Prestressed concrete

Awareness of Failure; Design Error; Concrete Mix; Casting; Bearing; Camber; Erection; Shrinkage and Deformation; Tanks.

The prestressed concrete industry is now a grown-up child in the field of engineering application and must guard against a repetition of the mistakes made in childhood. It is therefore quite proper to describe what has not worked in the rapid development of this technique. It is good that several experienced workers in the field, sometimes in spite of considerable resistance, are publicly presenting summaries of their investigations of troubles, nonsuccesses and costly maintenance details.

11.1 Awareness of Failure

Some papers have been presented as lectures at the Seattle meeting of the American Concrete Institute by Morris Shupack on "Prestressed Concrete Tank Failures" September 1962; by Ross H. Bryan at the American Society of Civil Engineers on "Performance and Problems of Precast Member Connections" (San Francisco), October 1963; and by I. Cornet at the Society of Corrosion Engineers on "Prestressed Concrete Tanks," 1963. The author's discussion at the 1962 International Prestressed Concrete Congress (Rome) on incidents that were not complete successes received so much attention that the magazine *Stahl und Eisenbeton* used 2½ pages of the six-page report of the conference for a

full translation of the discussion on failures. The Congress in Rome included, for the first time, considerable contributions on the subject of failures and the necessity for a new look at accepted procedures and details.

In Theme I on Design, Davydov (U.S.S.R.) in summarizing the papers stated that an unsolved problem is the explanation of cracking. Mikhailov (U.S.S.R.) advised that they were studying the causes and cures of electrocorrosion as well as the effect of cracks on the dynamic rigidity of structures. He also suggested that all members of the Congress join in a consolidated program of research and exchange opinions in an attempt to solve the troublesome questions in this field. Tsamis (Greece) said that corrosion due to strain must be empirically studied from reports of actual cases and cited the troubles from that source at the Gavia Bridge in Brazil and at a syphon in France. His study shows that most trouble occurs when the metal is anodic, since stressing makes it more anodic and strength is then seriously affected by impurities in the steel.

Theme II, on Production, was summarized by Vandepite (Belgium) and he stressed the important items as the following: thicker webs are advisable; wire must not be coiled too tightly on small reels; cold drawn wire is less susceptible to corrosion than the oil-quenched steel; chlorides in the mix combined with steam curing is a very serious combination that is prohibited in Belgium; some breakages in anchorages must be expected and safeguards provided; many beams have hairline cracks following the line of tendons, which the Japanese claim is caused by delayed shrinkage in the grout; in the long winters of the U.S.S.R. large expansion results when the component joints are welded in cold weather; further study is needed of the causes of corrosion and of the effect of grouting. He was especially critical of engineers who persist in designing with unreasonably thin webs so that it is impossible to compact the concrete properly in the bottom flanges; such troubles were reported by representatives of Australia and Portugal. Morujao (Portugal) also reported several cases of concrete failure from crushing behind the anchorages during tensioning. Lippold (Germany) reported on an inspection of the Aue Bridge over several railroad tracks built 25 years ago. The anchor nuts were rusted to a depth of over one millimeter, frequent spalling of concrete resulted from sulfur dioxide engine fumes and the cube strength of concrete in the structure showed reduction by 10 percent from the original tests during construction. Hill and Love (Great Britain) reported some failures of high-tensile prestressed bars when the anchor plates were not truly perpendicular to the axis of the rod. They found that cut threads were worse than rolled threads.

Reports were also received of stress corrosion failure in sewage tanks located in exposed marine industrial areas. Japan reported on spontaneous fracture in 7-mm wire in coils 1.5 m diameter, more in coils covered by sea water spray than those in covered storage, but the same wire in 2.0 m coils even in outside storage showed no failures. There was general agreement that tempered wires were more susceptible to corrosion than cold drawn wires and that chlorides must be kept away from the steel. Anchorage failures were reported from Great Britain, U.S.S.R., Portugal, and Czechoslovakia. The various mishaps included split barrels in single wire anchorages, slip of conical wedges, slip of single wires, undue draw in male cones; stripped threads and breaks of flame-cut anchor plates. Norman (Great Britain) investigated 10 sites where there was doubt about the quality of the grouting and found that blockages occur more often in grouting on hot days; of the units opened up for study 20 percent were poor and 10 percent were bad, with one tendon in 20 showing deterioration and one example of extreme corrosion at three years of age.

As to frost damage in grouted tendons, Czechoslovakian experience showed cracks more frequent along vertical and sloping cables than when horizontal, and more frequent occurrence with metal ducts. In one study in Japan of 453 bridge beams, 203 had longitudinal cracks due to frost or differential shrinkage between concrete and grout, the latter having a higher modulus of elasticity. Norman also added a statement that hair cracks in the sides of top flanges in long beams were traced to low lateral rigidity and oscillation during transportation.

In 1955 at the Second Prestressed Concrete Congress in Amsterdam A. W. Hill reported on some grouting difficulties after the assembly of precast units for an airport building. After the grout had been pressure injected, it refused to harden until the pressure was released, apparently a phenomenon caused by the restraint given by the cast concrete.

Rhodes (Great Britain) reported that of 30,000 anchorages supplied to industry for large sized strands, he had only 12 failures in either the strand wedge or the wire. Shimoni (Israel) stated that in assembling reinforcement for pressure pipes he uses extra wires to allow for some wire failures in the stressing operation. Lowe (Great Britain) warned that all designs must allow space for concrete movements from shrinkage and from deflections.

Theme IV, on Structures also included two comments on failures. Dyrbye (Denmark) reported on a 1000-m^3 tank of precast wall units 1 m square that was prestressed horizontally and vertically but resulted in a very leaky structure. Paduart (Belgium) reported cracking in prestressed

beams cast on long beds, with straight bottom steel and some steel deflected by metal pieces set in the form. Vertical cracks showed up in the webs in line with metal pegs used to hold the deflected wires.

Of course the history of prestressed concrete in the United States is younger than in Europe, but there have been recorded sufficient difficulties to warrant a summary of what not to do, usually in details of design and construction. It makes little sense to apply the prestressed techniques to designs which will not perform properly, where future mainte-

Fig. 11.1 Continuous mat tendon pattern.

nance costs will more than balance any original economies. A great disservice to the industry was the vaunted claims that prestressed concrete will not crack because the concrete is always in compression. Proper analysis shows the fallacy of that statement in many designs. The incidents listed below only cover those personally investigated and others publicly announced. Conversations with men active in the field indicate that at least an equal number of cases have not been published. In the writer's opinion such lack of professional responsibility, often caused by the owners' insistence, is a serious situation and can only

ultimately retard complete acceptance of prestressed techniques by the public.

11.2 Design Error

There have been a few mistakes in design, details, and specifications that should not be repeated. In 1957 at Omaha, Nebr., one of five 100-ft posttensioned beams that had been erected with partial stress and fully

Fig. 11.2 Failure at tensioning.

tensioned in position but not grouted failed when the center buckled upwards 12 in. Bars used in the beams were 80 ft long and were spliced with seven of the 14 splices at the same point where failure occurred. To make room for the splices, 3-in.-diameter sheathes were used, which took out a large part of the concrete section. Such lack of space for concrete is not permitted in reinforced concrete and must be avoided.

A concrete foundation mat for a multistory building was being constructed in 1966 in accordance with a design using tendons draped downward under each column to conform to the bending moment dia-

grams. When posttensioning was applied, the concrete at the column area lifted and broke free of the mat. The design had been based on loaded columns; when the mat was tensioned no columns had yet been constructed and there was insufficient weight to resist the upward component of the curved tendons. The prestressing force was released, and each column was surrounded by heavy anchors drilled and grouted into the soil to engage enough temporary dead load, and the mat posttensioned again, successfully this time. This was a case where erection forces required compensation until the structure was sufficiently in place to provide the necessary resistance (Figs. 11.1 and 11.2).

In 1960 reports appeared both in Sweden and in the U.S.S.R. showing that surface electrolytic corrosion is expedited by the presence of calcium chloride, but it may be inhibited by other admixtures, such as sodium nitrate. Corrosion studies made after defects in the steel at the San Mateo–Hayward Bridge in California were discovered, also indicate that galvanic action may come from the differential concentration of oxygen in the included moistures. Work in South Africa indicates that similar corrosion may come from sodium chloride solution in vapor intrusion. This was definitely found in the concrete at the Key West, Fla., buildings where steel conduit had almost completely corroded and even electric circuits were broken by the internal rust formation.

The failure of a prestressed concrete Vierendeel truss in Lodi, Calif., in 1953 was openly discussed because of the contractor's complaint and attempt to revoke the engineer's license. The failure had no witnesses although four of the five identical trusses collapsed under partly loaded conditions. The truss span was 80 ft with top and bottom chords prestressed with bonded wires at 120,000-psi maximum load. The webs formed seven open areas, there being no vertical at midspan. Due to some difficulty in erection, additional wires had been placed on the exterior of the bottom chords and diagonal steel straps were added in the middle bay of the trusses. The redesign added a vertical member at midspan, and replaced the prestressing wires with high bond bars. The original design contained up to eight #6 bars in both lower and upper chords together with 10 and four ¼-in. diameter high-strength wires respectively for prestressing and for reducing the tension in the unloaded precast members.

11.3 Concrete Mix

Prestressed work requires concrete of high and uniform strengths. Lightweight aggregates save so small a part of the actual weight that the

greater uncertainty of obtaining proper strength should prohibit its use in prestressed work. Weight reduction was the reason for changing to lightweight concrete in manufacturing the girders for the Kenai River Bridge in Alaska. The girders were cast in Oregon in 1955. Design called for 4500-psi concrete, and about one-third of the end blocks cracked at stressing under the 6 × 6-in. plates holding Freyssinet anchors. When the end diaphragms were tensioned, some cracks and spalls appeared at the bottom of two girder ends. By 1958 the cracks had increased from 3 to 5 ft in length and some beam corners were loose. By 1962 complete replacement was ordered, even though samples of concrete then tested 4780 psi, a 5 percent increase in seven years.

11.4 Casting

In the production of precast work some methods must be modified to prevent cracks and costly replacements. In 1958 in a large application of both pretensioned and posttensioned factory-produced I beams, manufactured during the winter of 1957–1958 in a plant specially built for the project, a few beams were rejected before shipment but inspection in the field showed considerably more difficulty. There were then 148 posttensioned beams of 60- to 100-ft span, and 204 pretensioned beams of 30- to 50-ft span to be used for 18 highway bridges in connection with a major traffic artery in Virginia. The total order was for 572 beams. Some of the bridges were already complete with poured concrete deck in place, others were in process of erection, and many of the beams were in storage both at the bridge site and in the plant. Cracks were found, of intermittent length, outlining the curved metal sheathes in the web and along the straight lines in bottom flanges. Of the 148 long beams, 101 had serious cracks, some at more than one location. Cracking was noted in 33 webs, 89 bottom flanges, and 4 top flanges. Of the 208 short beams, seven showed cracks in the webs and in 16 the bottom reinforcement was exposed, mostly in poor concrete covering.

The latter beams were corrected by patching and some epoxy-cement filling of cracks and honeycomb pockets. Most of the longer beams were too badly cracked for repair and in some of them moisture leaked out from the cracks.

First analysis put the blame on the freezing of the grout and many of the beams were rejected and new ones were ordered. When, however, similar difficulty was found in the beams made in early summer, a new investigation was started. The cause was then determined to be the unequal shrinkage of the different concrete thicknesses when the steam

curing was shut off. Typical cycle of steam application consisted of 5-hr normal setting at about 60° F room temperature, from 1½ to 2 hr steam added slowly to raise the temperature to 165° F, and holding at 165° F for 22 hours. Then the steam was turned off and covers removed.

Certainly the metal conduit acted as a radiator to cool the thin concrete cover more rapidly and some fine separation cracks developed. The grouting under pressure, after the posttensioning was completed, opened up the cracks. It seems that the combination of metal sheath inside beams being steam cured is an uncertain or at least an unadvisable one. The literature indicates that similar difficulties have been encountered elsewhere, and in some countries the use of metal sheaths is prohibited when steam-curing is to be used.

Similar type of beam cracking was noted at the second Cooper River Bridge at Charleston, S.C., in 1964.

Shrinkage of the concrete during setting and curing of items cast in steel forms has been known to pull the flanges and webs apart since the forms do not give. Precast concrete joists often show fine cracks at the web junction with the flanges, but when reinforced with diagonal steel wires or mesh, full-scale load tests reported by the author in 1945 indicate no reduction in strength from such cracking.

11.5 Bearing

The highest stress concentration is under the bearing plate of a posttensioned anchorage. Here the highest concrete strength is required, with some hoop or spiral reinforcement to resist lateral expansion under the high stress. The combination of bearing plate, often with attached anchors, cable, or rod sheath, stirrups, and hoops is not conducive to getting a dense concrete. Usually there is no room for a vibrator and densification of the concrete mix depends on manual rodding and tamping "as far as possible." Sometimes the result is not sufficient.

In one very large prestressed concrete viaduct project in New York the bearing plates set in the web face of the outer girders for cross-tensioning the roadway width, crushed the web when the jacking was applied. The concrete in back of the plate was found quite porous; there was too much steel in back of the plate, as described before, to permit a proper filling of the web with concrete.

In a fine monumental building in Philadelphia the main floor consists of circular sectors with field posttensioning required to join the pretensioned castings. The narrow end of the sector consisted of two separate bearing areas, 7 in. × 14 in. each, with a total of four stressed rods. The

bearing plates were located at the top of the casting with little clear distance along three edges. With each rod stressed to 130 kip, the average unit pressure under the bearing plates was 4000 psi for a design concrete mix of 5000 psi. With only the upper rod stressed to full value, the edge pressure was higher, depending on the closeness of contact. Some localized shear failures occurred in back of the bearing plates. The result was a complete stoppage of work, with a tentative order to demolish part or all of the structure built at that time. This order was later rescinded, but only after extensive tests, inspection, and arguments. In the broken end bearings it was found that two small steel stirrups had been omitted in the plant manufacture, "to provide space for the concrete."

Additional concrete strength was provided by filling the gap between the two separate bearing areas, a gap that had been left for possible vertical pipes in the wall. The project is now completed and a credit to both the architect and the engineer, but a great deal of trouble and unfavorable publicity could have been avoided by providing greater bearing areas with more easily accomplished details. After the spalled and cracked ends had been rebuilt, all future posttensioning was performed in a two-step procedure, each rod being stressed to 50 percent value, then the other 50 percent added in sequence. Since the rods were too closely located to provide clearance for jacks, this required four operations, but it reduced the high local eccentric stress concentrations during the stressing.

In 1954 Belche reported on some prestressed arches in Belgium where 7-mm wires curved to a 20-cm radius snapped when bent around the end of the arch. He blamed corrosion, superficial defects, and stresses due to curvature. Later in this chapter the subject is covered in connection with tank and pipe incidents.

Polivka has warned that account must be taken of temperature effect on long prestressed casting beds. A change of 30° F in 500 ft, not unusual in many plant locations, changes the wire or cable length by 1¼ in. and this must cause some loss in bond values. Similarly, placing heated concrete on cold steel will relax the stress and the casting will not have the expected strength.

11.6 Camber

Several long precast beams, up to 113 ft for a skew bridge crossing, had been detailed with steel bearing plates to be anchored to the abutment seats. The posttensioning induced camber in the beams, according to

design, with straight profiles to be attained under full dead load of the stringers, deck slab, and paving. But the beams were erected in the abutments in the "naked" condition and the weight alone was insufficient to reduce the camber. Great care had been taken to set the bearing plates on the abutment seats in perfect position with the anchor bolts exactly to fit the templates of the concrete beam ends. Of course the ends did not fit. Not only were the beams apparently short, but the shoe plates were slightly rotated. Repairs had to be made with 80-ton beams held in the air by two cranes, an expensive and time-consuming operation. Shoe and bearing details must be designed to provide for the shape and length changes as a prestressed beam is loaded. Direct copy of satisfactory details from structural steel or normal concrete design is not advisable.

The end of a posttensioned, cast in place girder carrying a building column over the garage passage was designed to rest on an exterior concrete wall for the full thickness of the wall. The girder was cast short to leave room for masonry covering at the face of the wall. After the girder was tensioned and two stories of column load were applied, the bearing was insufficient and the girder tore away from the slab and out of the wall, forming a 90 deg bend in its alignment. Fortunately there were no people near the immediate area and only replacement cost was lost by the accident.

11.7 Erection

Erection failures have occurred and can be guarded against by proper care in lifting, transporting, guying, and stressing until all permanent bracing is in place and connected. Lack of such care can seriously delay the job at considerable cost and wipe out all the economies of a prestressed design. In a bridge realignment job in Yonkers, where 32 identical prestressed beams were ordered and shipment by rail synchronized with the erection program, the last beam was dropped when it just did not clear the side of the car. Replacement cost was not too serious, but the bridge construction program was upset, and the completion time could have been bettered by using steel girders. Hauling long beams through congested areas requires special traffic permits and can only be done in overtime work periods; it has resulted in girder failures like the case in Providence when a 100 ft beam turned over as the trailer made a sharp curve (Fig. 11.3).

The long haulage over some 300 miles of lightweight precast T's 80 ft long for a school gymnasium roof in Arizona may have released the bond

Fig. 11.3 Haulage failure.

of the pretensioned strands. Although lifted from the bed and placed on the trailer and erected on the walls without incident, the beams developed a sag of 6 to 8 in. When a heavy rainfall caused further deflection the roof was rejected. The combination of dynamic loading during transportation and the lower bond developed in lightweight concrete seems to be the reasonable explanation.

A more drastic failure was reported in 1955 in New Zealand where 80 beams of 105-ft span for the Hutt Bridge near Wellington, were posttensioned each with 96 wires of 0.267 in. diameter and grouted before shipment. A beam weighing 40 tons tipped over in shipment and exploded so violently that the shock wave touched off failure of a second beam, still in storage, which also broke into fragments.

In 1962 a 100-ft-long precast beam transported in Seattle, Wash., on a low, two-unit trailer broke its moorings and slid down to the pavement, breaking into separated segments. In 1957 a 100-ft long posttensioned T beam at Omaha, Nebr., buckled as it was unloaded by a somewhat off-center hitch.

Prefabricated roof single and double T's must be laid out on the plans with sufficient tolerance for the usual deviations in line and camber of the component units making up the roof or floor area. In 1961, in the

first use in New York City of precast tensioned T's as a roof over a lecture hall, the deviations from straight lines at the contacts of the flanges required wider joints than expected and the total length of the roof had to be made up by cutting several inches from the last unit. To compensate for the variation in camber, units were selected from the delivery to match as closely as possible, the variation when the 40-ft lengths were set down on the supports were over an inch, and steel insert plates were welded to force the castings to work as an assembly.

Precast elements have little reserve strength until connected to the supporting frames. During erection of the roof over a one-story plant in Bermuda Hundred, Va., in 1963, some 11,000 ft^2 of 40-ft-long double T's fell to the ground.

Jacking apparatus must be in good condition and sufficient stressing capability. In 1960 in Montreal a 35-ton beam collapsed from the shock of sudden stress reversal when a jack failed during posttensioning. At a viaduct construction in Paris in 1963, a 206-ft girder with 13 of the 14 cables stressed, failed as if eccentrically loaded.

Failure of a 5-million-gal tank at Littleton, Colo., was caused by wind load on one side of an almost completed job. The 100-ft-diameter water-storage tank was made of precast vertical staves 6 × 28 ft set with edges tight. After more than half the height had been covered with die-stressed wound wires and completely grouted, six staves fell inward, releasing all wire tension. All the wires were then removed. While the discussion of responsibility was going on an identical tank with all staves in position but with no wire yet installed also collapsed from lateral wind pressure.

A monumental two-span, 90 and 60 ft, folded-plate roof near Chicago collapsed in 1959 when the temporary shores under the 90-ft span were removed before completing the welding of steel and concrete over the center support. The added cost of making each span sufficiently strong to carry its own weight would seem to be a good investment. The notes on a drawing are not sufficient warning that the precast elements need continuity for safety (Fig. 11.4).

Some prestressed pile failures have been reported during the driving. Zollman at the Baltimore Bridge noted that three out of 44 16-in. square piles failed during driving at 31 to 49 ft from the top. Donovan Lee commented on that report that failure can occur when the pile is being driven too hard into soft ground, and such combination can set up tensile stresses in the pile. Dean in 1957 stated that piles with 400-psi prestress when driven in soft soils cannot take the rebound effect; it will destroy the pile. An initial stress of 800 psi is needed in such cases.

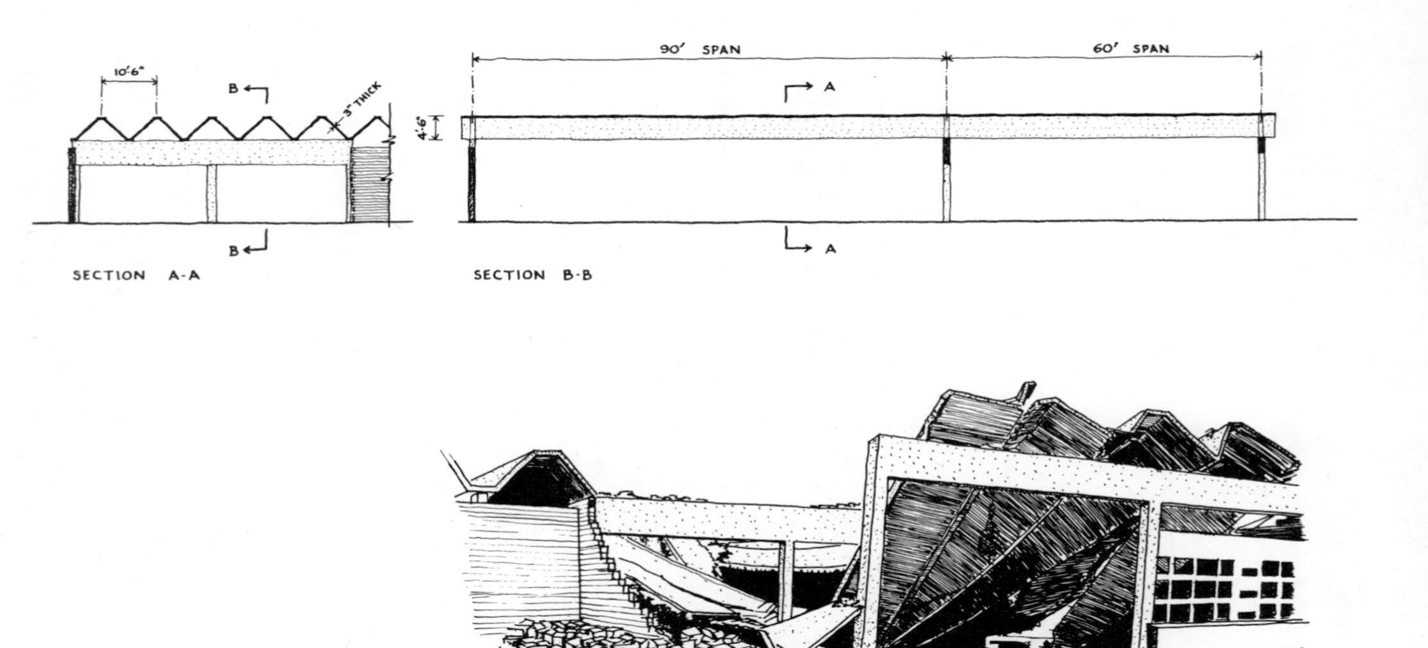

Fig. 11.4 Combined span not completed at removal of supporting shoring. (Drawing based on editorial article in *Engineering News Record*, March 5, 1959).

Some prestressed piles at the Fort Randall Dam in South Dakota also developed cracks during installation.

11.8 Shrinkage and Deformation

In a large industrial building on the West Coast the main roof (used for automobile parking) is framed with precast posttensioned beams and girders, supported by precast columns. The bays are 40 × 44 ft, columns are 2 × 2 ft in section, the girders 44 ft long, pinned to the columns and made continuous by adding heavy negative moment steel over the supports in the cast-in-situ roof slab, 4.5 in. thick. Beams are located at column lines and at the third points of the girder spans, sitting on brackets cast with the girder webs. The beam support is free, with an asbestos packing seat layer and concrete filling in the gap between beam ends and girder web. To reduce total construction thickness, the end blocks of the beams were notched, 12 in. above the bottom. Theoretical bearing areas are 12 × 5 in.; somewhat smaller areas are actually found where shrinkage of the precast concrete resulted in beams of shorter than designed length. Such shrinkages reduced the bearing area by as much as 20 percent in some locations.

The concrete design mix was 4000 psi in 3 days and 5000 psi in 28 days. Test cylinders of the concrete produced, containing satisfactory glacial sand and gravel, w/c ratio 0.435, Type III cement with 1½ to 2 percent calcium chloride, after steam curing showed strengths of 6600 psi at 28 days and 6000 psi in three days for beams and 5565 for girders. Beams were cast within steel forms, stripped at 3-hr age, including 2-hr steam and then steam cured for 20 to 24 hr at 100–140°F. Girders were stripped at 5-hr age, including 4-hr steam and then steam-cured for 30 hr at 90 to 100°F. Posttensioning of the high-strength steel rods came at three-days when the concrete strength indicated by a Schmidt hammer was at least 4000 psi. Of the 606 units manufactured, only four required rejection. Beams and girders had stirrup loops extended for composite action with the slab concrete.

Shortly after the completion of the roof slab, failures of the beam ends were noted at the expansion joints and a more positive sliding seat detail was installed (Fig. 11.5). After that repair over a period of six years, progressive distress of end blocks in both beams and girders developed. Typical cracks in beams were vertical shear separations above the bearing area and diagonal tension cracks originating at the reentrant corner of the seat recess. It was evident that the posttensioning stress did not provide a uniform compressive stress across the full section of the beam.

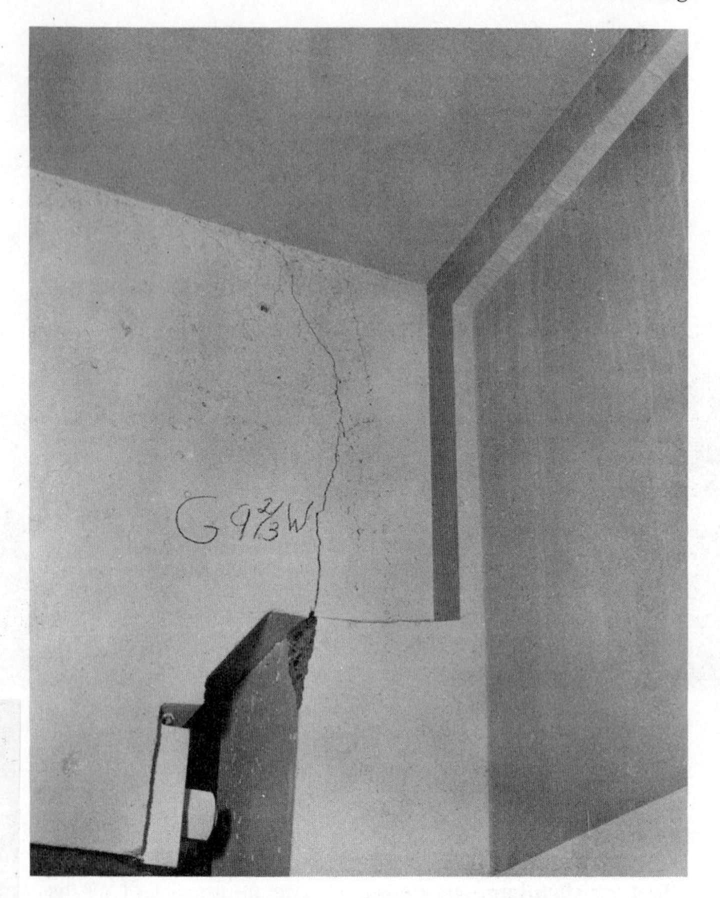

Fig. 11.5 Precast roof cracking from shrinkage.

Actually the analysis of the beam just above the bearing, as a free body, indicates tensile stresses at the edge of the seat. The beam acting as a simple span, partly restrained by the roof slab, must have freedom to rotate when loaded. The sloping end will modify the bearing reaction and had caused the faces of the girder brackets to spall. Similarly the shrinkage of the girder lengths had caused a number of columns to crack vertically.

Corrective work consisted of through-bolt anchored steel bridles added at each beam end to transfer the reaction without reliance on the girder seats and with pickup at the bottom of the full depth of the beam. Columns that showed distress were encased at the top with tensioned

Fig. 11.6 Saddle supports for beams to neutralize cracking.

strap iron, installed as a bolted yoke, and gunited for fireproofing (Fig. 11.6).

The effect of shrinkage in changing the geometry of an assembly of members must be considered in the design and in the details of jointing. High early strength concrete, obtained by special cements, calcium chloride and steam curing, is liable to greater shrinkages and over longer periods of time than are normal mix concretes. Some recent data on concrete shrinkage appeared in articles by T. C. Powers and M. Mamillan, *Revue des Materiaux* No. 545 (Paris), February 1961.

In all operations of hoop stressing the diameter of the vessel is reduced and freedom of movements must be provided at contacts with the unstressed members. Such incompatibility of deformation caused serious diagonal shear cracks to develop in a concrete penstock pipe in Cuba in 1958. The joints were covered by a sleeve that was tightly grouted. When pressure was applied the greater stiffness of the sleeve resisted expansion of the pipe ends and the shear cracks opened. The cure was to dig out the grout and caulk the joint with a plastic filler.

Ooykaas in the March 1952 *Magazine of Concrete Research* reported similar difficulties where pipe had not been prestressed for the full length to provide space for a sleeve joint. He warns that pipe must be prestressed to the extreme ends to avoid secondary bending and shear stresses.

In 1960 the failure of almost half the roof of the factory at Harrisonburg, Va., resulted in some frank discussion on tolerances to be permitted in precast work. The precast beams came with seat angles 2 × 2 × 5/16 in., 14 in. long and were specified to be field-welded as required for erection only. The tops of the columns had bearing angles 1½ × 1½ × 5/16 in. by 14 in. long. Eventually the top steel was to be welded over the columns and concrete added to provide full continuity. Beams were 34-in.-deep I's spanning 40 ft and carrying 50-ft-long double T's. Shrinkage and fabrication tolerance reduced the required bearing area and increased the bearing stress above the expected 3460 psi (assuming support uniform with no load taken by the steel anchors). The edges of the concrete columns sheared off and 24,000 ft^2 of the roof fell with considerable loss of life. Too many designers detail and dimension precast work as if they were dealing with components for cabinetwork or even watches. Calling for "exact" dimensions does not guarantee the performance of the unrealistic (Fig. 11.7).

Dean in 1957 reported on condition and correction of the prestressed beams in the Lower Tampa Bay Bridge. Cracks had developed in the bottom of the webs and camber changed with age. The deflection could not be predicted to the same degree of accuracy as in steel beams. In the later designs he changed the I-beam section by using a heavier flange bulb, 6-in. webs in place of 4 in. in 40-in. deep I's, and including shear reinforcement in all webs.

Placing long spans on slender walls is an invitation to failure. In 1959 three precast concrete arches, spanning 49 ft on top of 46-ft-high walls over a steel mill near Lisbon, Portugal, fell 60 days after erection.

In 1956 a prestressed concrete roof, 56-ft span, over the cafeteria of a school in Orlando, Fla., was erected on 8-in.-thick concrete block walls. About a week later eight of the 5-ton double-T pretensioned beams, 14 in. deep, fell to the floor, bringing down with them one wall and a roof section 32 ft wide. Smaller span roofs of similar design were not affected. The walls were quite flexible, especially as there were three 7-ft-wide window openings in the 32-ft length of wall that failed. An almost identical failure occurred in 1958 when the precast hollow slab roof fell and brought the top of the hollow block wall down with it in the erection of a one-story commercial building in Waltham, Mass. The roof consisted of 30-ft-long hollow precast units, 9 in. thick by 16 in. wide,

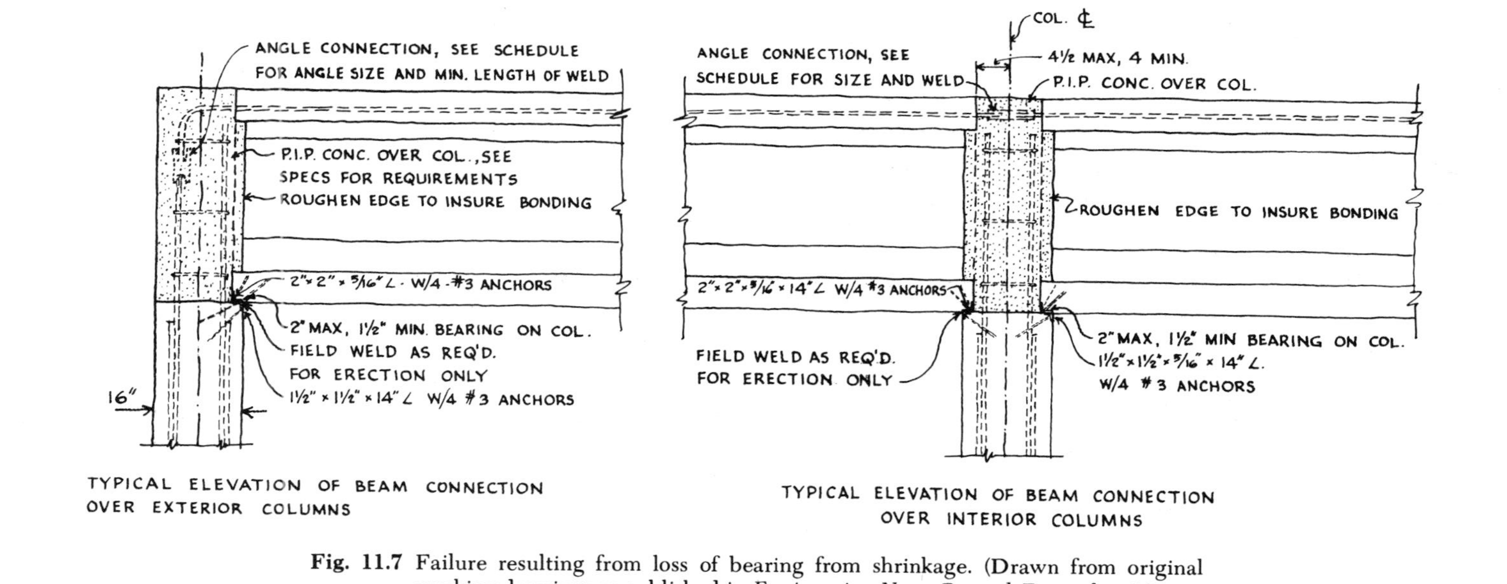

Fig. 11.7 Failure resulting from loss of bearing from shrinkage. (Drawn from original working drawings as published in *Engineering News Record*, December 16, 1960).

resting on an 8-in. hollow block wall that also was pierced by window openings.

11.9 Tanks

Although there had been some experience with trouble from the use of die-stressed wire prestressing, it took a spectacular failure and sufficient professional responsibility on the part of a municipal agency to make everyone aware of the possible dangers.

The collapse of a 2.5-million-gal sludge tank in 1961 in New York City, fortunately without any damage to life and little to property, brought attention to the internal disintegration of highly stressed die-strained wires. Although the wires on the outer face of the tank, which had been brick covered and indirectly subject to exterior rain or temperature, were the immediate cause of the collapse, and were considerably eroded and corroded, the major portion of the tank perimeter was inside the workroom where climatic conditions were favorably controlled. The wires removed from this part of the wall were also found, with a few exceptions, severely rusted and with brittle cone ends and longitudinal splits. Some of the wires, which for unknown reasons showed no outside rusting, when bent were also found to have corroded longitudinal splits in the interior. Identical straight wires, used for vertical posttensioning and located on the inner face of the walls, were found in perfect condition. These wires were not die-stressed, but were jack-tensioned and covered with about 1-in. cement mortar, as part of the interior surface of the tank.

The failure at the Owls' Head Plant was of one primary digestion tank in a battery of eight identical units built in 1950–1951. Records of the construction show that walls were concreted between April 20, 1950, and January 5, 1951, using normal 3000-psi concrete and with all cylinder tests satisfactory. The concrete mix had no admixtures and no calcium chloride was used. Eight similar tanks were constructed at the Hunts Point Sewage Disposal Plant (also in New York City) in 1951–1952. Complete investigation of all tanks showed that the earliest indication of trouble was in August 1958 when some gunite protection on Tank #7 at Hunts Point spalled off, exposing corroded horizontal wires. This area was within the plant operating room formed by the space bounded by the eight tanks and not exposed to outside temperature or rain. When the builder explained this localized failure as caused by electrolysis, a careful survey was made and no stray electrical currents

were found. In a number of areas on most of the tanks, especially, but not limited to locations near metal sleeves through which pipes had been led into the tanks, gunite protection spalled off (1958–1961) and the wires were cleaned, painted with red lead, and the mortar replaced.

Failure of Tank P-4 was a complete separation and bodily displacement of the northeast quadrant for a height of 15 ft, from the exterior brick shelf where the concrete wall was 23 in. thick to the level where the roof steel bars, bent into the wall, were terminated (Fig. 11.8). This blowout of the wall was followed by a vertical collapse of the upper part

Fig. 11.8 Collapse of wire wound concrete tank.

and of the roof structure. About 1.6 million gal of liquid came out as a "tidal wave," overturning three railroad freight cars from a nearby track, but fortunately without injury to any persons. This section of the tank had been covered by a brick facing sitting on the concrete shelf and tied to slotted galvanized steel anchors imbedded in the gunite covering the prestressed wires. Exposure was the worst possible with the driving northeast rains saturating the brickwork. In addition the roof flashing had broken and rain water leaked through the base of the parapet wall.

The concrete parapet wall was not prestressed and the stressing of the tank wall had caused a horizontal separation at this level when the tank diameter was reduced by the circumferential tensioning.

A careful inspection was then made of the remaining tanks in both locations. Numerous areas of loose and cracked gunite covering were found on the walls within the operating room areas. When the loose gunite was removed two conditions were observed. In some of the locations an aluminum connector used at the ends of wires when a complete reel had been used up were found completely disintegrated.

Fig. 11.9 Typical wire failures.

Chemical analysis of the white powder found in the cavity showed it to be almost pure aluminum oxide. In other locations, where the wires exposed were found to be considerably rusted, longitudinally split, and of greatly reduced section, some wires were fractured with typical conical brittle failure (Fig. 11.9).

Where the brick facing covered the walls there generally was a horizontal crack at the level at which the bent-down roof bars end. Considerable leaching of lime deposits were noted, a condition observed for

several years before 1961, and some frost damage of the masonry with brick actually pushed out.

At the time of failure the sludge level in the tank was 3.5 ft above the level of the ceiling slab at the wall contact. The tank was designed for pressure in excess of full depth to allow for the effect of gas or foam trapped by the domed roof. A number of times since completion the tank had been emptied for servicing and inspection and then refilled. The tank had been empty from late 1960 to 10 days before failure. There was no report of any unusual structural conditions during this period. Immediately after filling some leakage was observed at the exterior weep holes in the brick facing. The first tests of this seepage indicated presence of sludge but later tests showed no sludge content in the seepage liquid and the seepage soon decreased.

An inspection of the debris brought out the following facts:

1. The concrete was sound and dense; the sludge had not penetrated or discolored the interior but had left a paper-thick black skin on the inner surface.
2. The mild steel reinforcement in the roof and in the upper part of the walls was not corroded and seemed well bonded to the concrete.
3. The vertical poststressing wires were in perfect shape, free of rust or corrosion and with splice connectors intact.
4. The horizontal prestressing wires were completely rusted, corroded, and not a single clean piece of wire could be found. A careful search also failed to locate a single wire connector. End of wires showed typical reduced sections of cone breaks at about half diameter, thin skin shells with the interior of the wire consumed and split ends with laminar stripping.
5. Gunite covering seemed dense and firm with many areas of rust stain outlining the wire positions. The rust stain indicated that the gunite was not totally impervious.
6. The metal slot anchors for holding metal ties imbedded in the brick mortar joints had been located as vertical strips within the top layer of gunite covering. Space between the wires and the slots varied from zero to about ½ in. Generally the metal slots were badly corroded, chiefly from roof water seepage in the air gap in back of the brickwork, resulting from the fault in the coping and flashing at roof level. Wires in immediate contact with the slots were always rusted, but were not the only rusted wires.

Samples of horizontal and vertical wires were tested with the following results: (Wires taken from different areas in the failed walls).

Percent Included	S	P	C	Mn	Si
Horiz. wire	a 0.044	0.023	0.82	0.70	0.28
Horiz. wire	b 0.035	0.019	0.84	0.65	0.26
Horiz. wire	c 0.038	0.018	0.82	0.65	0.24
Vert. wire	e 0.034	0.018	0.72	0.84	0.22
Vert. wire	f 0.038	0.014	0.76	0.88	0.24
Vert. wire	g 0.030	0.015	0.74	0.74	0.20
Specification	0.05 max.	0.05 max.	0.60–0.70	0.76–1.00	—
Clean wire	d 0.017	0.014	0.60	0.79	0.24

The clean wire was found in the later demolition at the base of the tank. The physical tests on three samples of each wire, using wire in the best condition as found in the debris, showed on basis of ASTM A421-58T procedure:

	Diameter in.	Yield kip/in²	Ultimate kip/in²	Percentage Elongation in 10 in.
Vert. Wire	0.1916–0.192	183–194	240–244	5–6
Horiz. Wire	0.140–0.1405	220–223	232–241	1.25 (one sample)

Failure of all vertical wires and two horizontal wires was ½-cup fracture and the last wire failed by split fracture.

In the demolition of the tank debris all of the wire was removed from the wall, most of which had adhered to the tank with the gunite attached. This area (about two-thirds of the perimeter) was within an

operating room, not exposed to outside weather or to the leakage from the roof parapet. The interior operating room is kept at working temperatures and normal humidity; the exposed iron piping and electrical conduit shows no corrosion. Of all the horizontal wires exposed, about 10 percent were clean and shiny, about 10 percent were corroded and fractured, the rest were all rust covered and in various stages of corrosion. In one localized area near the outside wall there were a number of aluminum wire connectors in place and unaltered. Nowhere else was a connector found. Wires showed laminar erosion patterns and splitting, exactly like those in the wall section that failed. The exterior exposure does not seem a necessary condition for the wire disintegration.

The conclusions were therefore reached that the tanks were not structurally sound and must be strengthened by inserting steel tanks within them to take full hydrostatic load. The actual direct cause of failure was the reduction in sectional area of the horizontal wires to the point where there was insufficient hoop tension resistance to hold the liquid pressure.

A reduction in sectional area of the wires could result from one or a combination of several factors, here separately enumerated and discussed as to probability and intensity.

1. Surface corrosion from outside sources:
   (a) Sludge within the tank.
   (b) Seepage from the roof.
   (c) Seepage through the wall surface.
2. Surface erosion of metal by electrolytic action:
   (a) Stray electrical currents at pipe sleeves.
   (b) Chemical cells at ends of wires at connectors.
   (c) Chemical cells at crossing of wall slots.
3. Stress corrosion internally:
   (a) Dissimilar metallurgy of fibers.
   (b) Separation of fibers due to die-stressing.

1. Surface corrosion from the sludge did not occur as is proven by the condition of the vertical wires which were located much nearer the liquid, and although of similar metallurgy, were entirely unaffected. Seepage from the roof and/or from the wall surface would only affect the exterior parts of the tanks. Such seepage, if reactive, would form a coating of rust, which by the expansive effect would tend to seal the porosity of the gunite covering, would stain the gunite, but would be nonexistent in the wires within the operating room areas. It is true that the outer sections are much more corroded than the operating room

sections, but the latter also had some surface corrosion. Seepage had some effect on decreasing the wire sectional areas but not to the extent where failure would occur. In other words, without any surface corrosion from roof and wall seepage, failure might have been delayed, possibly by a year, but would have occurred. The chemical analysis of the deposit on the exterior face of the brick covering showed it to be 96 percent calcium carbonate. This material is noncorrosive to steel. It is only the solution of salt air vapor in the presence of oxygen that would give serious surface chemical corrosion.

2. Surface erosion of the metal by electrolytic action did occur and was quite evident at locations of the aluminum connectors and in the vicinity of pipe-sleeve crossings. At these places the chemical change in the metal caused large expansion of volume with resulting popping off the mortar protective coat.

The chemical analysis of one pocket of white powder under such a blister in the mortar covering showed it as aluminum oxide, apparently the remains of an aluminum wire connector. Electric currents are generated where dissimilar metals, like steel and aluminum, are in contact in the presence of moisture. Stray currents will always find their way along pipelines and are often generated by the flow of liquids within such lines. All such currents will continually rob the metal surfaces and move the material to other positions. The thinning of the wires in many places indicated the existence of such losses and consequent reduction in sectional area of the wires. If the aluminum couplings were the only source of such trouble, the separations in continuity of the wire would be randomly positioned and the bonding of the gunite should be sufficient to transfer the tension loss. The losses at or near pipe crossings would only affect the wires within the operating room areas.

Similarly the electrolytic action where galvanized slotted anchors crossed the wires would affect the wall areas covered by brickwork and would only be serious where the gunite was lacking in thickness or in porosity. It is difficult to prove or disprove the actual contact of wires and slots. The corrosion of the galvanized slots from wall and roof seepage could occur without affecting the wires. The flow of metal would be from zinc to steel, rather than the reverse, and not sufficient steel loss in the wires can be explained to permit the complete failure. The condition of the gunite covering with the impress of the wires does not indicate serious exterior changes in the wires. Surface erosion was a more serious factor than surface corrosion, but not sufficient by itself to cause failure.

3. Internal stress corrosion caused by electrochemical action when the internal fibers are disarranged during die-stressing, then unequally

pulled by tensioning around a curve, is probably the largest factor in cross-section reduction. Stress-corrosion only occurs with tensile stresses. Internal microcracks or flaws open up great areas for internal corrosion. Fibers of nonmetallic slags and of nonuniform metallurgy all become areas of stress concentration with resulting internal flaw cracks and corrosion. The examination of the failed wires indicates considerable loss of internal metal with surfaces rusted but intact. There is no known corrective measure that can inhibit or counteract this internal disintegration of the wires while under service.

There are several intensive research studies in progress to explain the failure in metals called "stress corrosion." Most of the reports now available indicate that exposure to some corrosive liquid or gas is necessary. Some agree that the nonhomogeneity of the metal in a wire is, however, sufficient to start internal changes and corrosion. Recent work in the field of solid physics on the internal structure of metals, especially in a disturbed state along the interlocking contact of the atomic structure (known as the Fermi surface), seems to explain the internal disintegration which is loosely called stress corrosion. Such study is going on at the General Electric Company Research Laboratory, Schenectady, N.Y. (Dr. Walter A. Harrison), at the Honeywell Research Center, Hopkins, Minn. (Dr. Charles J. Speerschneider), and at the Corrosion Research Laboratory of Ohio State University, Columbus, Ohio (Dr. David K. Priest, now at Pfaudler Research Laboratory, Rochester, N.Y.), and probably at many other laboratories in the United States and elsewhere.

The important lesson is that highly stressed wires must not be mistreated and it seems that die-straining with the consequent heat generation is a dangerous procedure. Possibly the method can be modified by some procedure which will dissipate the heat as soon as generated, and thereby eliminate the danger of metal damage.

Several other tank failures have been reported, especially by M. Schupack and by I. Cornet; the latter is actively involved at the University of California in the subject of stress corrosion.

In 1955 prestressed wires in the concrete wall of a 12-million-gal reservoir at Richmond, Calif., failed as brittle-end factures, minute longitudinal cracks as well as surface erosion. The failure was charged to high local combined stress, but the indications suggest stress corrosion as the cause. This tank was 200 ft diameter and 43 ft high, stressed by 360 parallel wire cables, each extending over 90-deg arc, each made of 18 high-strength 0.196-in. wires, encased in a metal tube for later grouting. Cable anchorages were at eight concrete pilasters. After tensioning but before grouting, so many wires were found parted that all of the cables were replaced.

Failure of highly tensioned wires has been studied by Van derHeiden who in 1958 computed the speed of failure as 85 fps at the instant that the deformation energy is converted into kinetic energy. In failures on circular tanks, however, he notes that the speed of breaking wire has been found to be many times greater than expected, and the snapped wires have been known to cut through 1-in. wood planks.

Several cases of failure of slurry and sludge holders have been reported. In 1954 at Buffalo, N.Y., a 21-ft-diameter, 45-ft-high clay slurry tank had a section about 4 ft × 3 ft blow out near the bottom manhole. It was found that the circular tension hoops were not continuous over the height of the manhole frame. In 1955 at San Antonio, Tex., eight 75-ft sludge tanks with prestressed concrete dome roofs required repair of all roofs after serious wire corrosion was noted at two-years of age; one dome almost failed.

In 1955 at Menlo Park, Calif., two 500,000-gal die-stress tensioned tanks for sludge were found at four years to have considerable areas of spalled gunite and corroded wires, both at anodic and cathodic areas. And in 1959 an investigation of three tanks, 55 ft diameter by 57 ft high containing an alkaline slurry weighing 100 lb/ft^3, showed serious corrosion of steel and one tank failed at two years. The tanks were located in the Los Angeles Harbor with humid saline exposure.

Also located along the West Coast are several tanks 40 to 320 ft diameter by 40 to 20 ft high for sea water holders, where prestressed wires have broken from corrosion and a 200-ft-diameter tank for magnesium hydroxide slurry required new circular bands that were asphalt coated, after 25 years use. Other water holders where corrosion of wires required repairs are the 1.5-million-gal tank south of Honolulu, Hawaii, which was completely rewound and gunited in 1960 after five years and the 130-ft-diameter Rodeo Reservoir in Richmond, Calif., built in 1952, where cathodic protection has been installed to control wire corrosion first noted at one year.

There are a number of leaking prestressed sludge holders in which there is evidence of wire corrosion, but the tanks are kept in service by backfilling the outside almost to roof level and disregarding the seepage percolating through the fill.

Domed roofs over prestressed tanks must be provided with freedom of differential movements for full and empty conditions. One of the early designs of this type, 160 ft diameter and 25 ft high at Halifax, Nova Scotia, required considerable repair of the walls and roof, after the concrete showed signs of disintegration from such differential movement.

A serious corrosion failure of 36-in. precast pipe at Regina, Saskatche-

wan, in 1952 caused the complete abandonment of the finished work and substitution with steel-lined pipe. The pipe had been die-stressed with 0.142-in. wire that was covered with Kalicrete, a high sulfate-resistant cement. Of 84 pieces cast without admixtures, 29 were inspected and found to be sound. Of 58 pieces in which calcium chloride had been added to the mix, all showed wire failures. The report by C. R. Young stated that the wires had been heated to 280° F during the die-stressing, which may add 50,000-psi stress in the wire as it cools to normal temperature. The complete report by Robert Legget in the *Proc. Inst. C.E. (London)*, **22,** 11–20 (1962), states that the corrosion was seen only on the inner faces of the wires next to the concrete core and no corrosion was found in the longitudinal straight tensioned wires. Reports of stressed concrete pipes in Belgium, France, and Brazil were reported in an article in *Wire Products*, **30,** p. 316 (1955).

The largest failures come from improper small details. Joints must allow for differential thermal, creep, and shrinkage, all of which are time-dependent changes. Bearing stresses may be increased by axial forces, with serious bracket stresses and often splitting of the concrete seat on the beam. Prestressed members will continue to change camber with time and some provision for rotation at the support must be provided.

Finally, there have been at least two failures of prestressed work during load testing. At a prestressed folded plate roof built at Fort Sill, Okla., a water load test seemed satisfactory until exposed to an air shock during some ballistic firing tests, which was enough to cause complete failure. Investigation indicated omission of some of the reinforcement at the columns. At a Detroit motel job, load tests of a questioned precast plank floor to 80 lb/ft^2 caused the unfortunate death of two inspectors who were trapped by the fall of the floor. One can never be too careful in such tests and certainly the fail-safe concept should prevail in the shoring during all load testing.

12

Formwork and Temporary Structures

Formwork Failure; Lift Slab Manufacture; Forming Problems; Temporary Grandstands.

12.1 Formwork Failure

A system of formwork for the reception of wet concrete at a height above the previously constructed floor is not the most stable structure. The weight is almost entirely at the top and is supported by an array of posts which are not rigidly connected to the form at the top or to the floor at the bottom. Considerable resistance against lateral sway can be provided if the columns are poured at least a day ahead of the floor. The unit labor cost of placing the concrete may be somewhat increased, but the small expenditure is cheap insurance. Actually, in multistory flat-plate designs many contractors are convinced that if the columns are poured as soon as the deck is installed and before any slab reinforcement is placed, a saving will result in the reduction of labor of pouring columns through the concentrated maze of rods at the columns and in the easier control of bleeding that reduces finishing time and costs. At the same time the columns are permitted to take their shrinkage, a procedure usually specified but often not obeyed.

Posts supported on a lower completed floor can be assumed to have equal and uniform bearing. However forms for the lowest slab level are often supported on "mud-sills" that are not on solid ground, usually on backfills recently placed with great probability of softening from flow of

Fig. 12.1 High shore form supports.

water, either natural runoff or wash water from forms or from truck mixers. Unequal settlement of the sills seriously disarranges the designed equality of post reactions with good possibility of overloading the posts which do not settle from load carried over as the exterior line of posts and sills settle. High posts without lateral support will buckle and break apart (Figs. 12.1 and 12.2).

The pressure of liquefied concrete during and after vibration is pretty well established by the recommendations in the report of "Pressure of Concrete" published by ACI Committee 347 (formerly 622). The lateral force from a vibrator striking against a beam side is not known or considered in form design. The often-seen bulge or drift of exterior faces on spandrel beams can well be attributed to the careless use of a

Fig. 12.2 Failure shapes of shores.

vibrator against the outside form in the attempt to get a dense exposed concrete face. The lateral force of a power buggy stopping to unload is another unknown which needs evaluation. But the common experience of vibration and shiver when a fully loaded buggy stops "on a dime" convinces one of the magnitude of such lateral force. The transfer of these loads through the legs of the runway panels, often wedged against the reinforcement introduces a rotating couple into the formwork system.

So many formwork failures are reported that space permits only a summary of the types of failure. Unfortunately only seldom is the real cause divulged. It does seem, however, that the pattern is fairly uniform: accidents occur when the concrete is being placed too fast, often during the stage of construction when forms have been used several times and work is routine. Few failures are reported at the beginning of the jobs; and few on small operations.

Examples of Form Difficulties

Stairwell openings in a steel frame building were found to be too narrow. In this job forms were suspended by tie wire from the steel

beams. To reduce time and labor in stripping, the wood joists were hung by wires bent over the top beam flanges all on one side of the beam. The weight of the wet concrete caused a torsional pull at one edge of the steel beams. The concrete hardened around the twisted steel beams and fixed them. Exposure of the steel showed that top flanges at midspan were from ½ to ¾ in. out of level and the faces of the concrete fireproofing had drifted into the clear width required for the stairways.

The frequent necessity for chopping concrete in all shaftways has been taken as a normal expectation. This error in unbalanced loadings on steel beams was then explained at a large office building in New York, and the contractor measured the edge elevations of the steel beams during concreting and found that all the beams were twisting. Upon publicizing the results, orders were issued by the New York City Building Department that tie supports must be alternated and the difficulty has been eliminated. Of course there is an added cost to stripping, since one-half of the wires must be cut before the joist can be rolled out and removed. But the cost of chipping and other construction adjustment is eliminated and that must be a net saving in overall cost.

Engineering News, **48,** 478 (1902), reports failure of fourth floor forms in a Chicago apartment building. Forms were hung from the 10-in. I beams but were also supported by diagonal timber frames wedged to the completed floors. The concrete was 21 days old and was being loaded with a 6-in. layer of cinder concrete filling when a workman removed some of the timber supports. About 40 ft in length of a 13-ft span fell through all four floors to the basement, since the unbraced timbers were not strong enough to carry the weight before the concrete had reached sufficient strength.

AREA Bulletin, **20,** (1903), reports failure of the formwork for the roof of a nine-story apartment in Pittsburgh. Spans were 12 ft-9 in. and reinforced with expanded metal. The timber used for shores was defective and when the load was applied, the shores broke and the concrete slab fell right through to the basement. *Engineering News*, December 17, 1903, reports the same accident and states that the shores were three stories high since the spans corresponding to the failed area in the two tiers immediately below had not been put in, so that the mass of concrete fell three stories before striking a floor.

Engineering News, **49,** 328 (1903), reports failure during a Milwaukee factory construction. A form area 55 × 18 on the fourth floor gave way during concreting and the mass fell through the completed floors. In the same issue on page 324, the editor warns that "some degree of skilled labor should be employed, . . . In the carpentry work for forms and falsework, especially, ..."

Similar form failures are reported continuously in almost every year with little definite explanation of causes and as jobs got larger, so did the failures.

Among the more notable examples were the fall of the roof form at the University of California Mechanical Engineering building at Berkeley in 1931; the roof over the filtration plant at Toronto, Canada, in 1934, where the shores were 20 ft high; the roof over the reservoir at Allentown, Pa., in 1937, where the last of 18 sections, an area 20 × 60 ft, collapsed under the load of 75 cy when a timber stringer failed; the roof over the Villa Scholastica Chapel in Duluth, Minn., in 1937, when an area 28 × 30 ft did not have adequate falsework support; and the roof over the Vegetable Market at Los Angeles, Calif., in 1940, where the main beam, 18 in. wide by 16 ft deep and spanning 100 ft collapsed when 60 cy of concrete had filled two-thirds of the volume of the beam.

Formwork for the roof over a Washington, D.C., reservoir failed in 1941 during concreting. Shores consisted of double-tier wood posts, supporting a domed rib slab, and steel column forms had been used. Investigation reports were not made available, but the failure was officially charged to sabotage of the war effort.

Construction Methods, **48,** (May 1949), reports failure of a 40 × 85-ft area of roof deck over the Santa Monica, Calif., reservoir. Columns in steel forms some 20 ft high, were being poured together with the 5-in. deck. Forms were supported on typical extended jacks. The failure photographs show that duck boards were used as runways for rubber-tired buggies. The columns remained erect after the floor area collapsed. Jack shoring of the adjacent forms were kicked out of plumb and there is no indication of any diagonal bracing.

Engineering News-Record **36,** (June 23, 1949), reports failure of a 36 × 76-ft area in the fifth floor of an art gallery in New York. The area which collapsed was a clear span between girders on four sides; the form was 22 ft above the completed floor. Shoring consisted of 8-ft high adjustable shores supported on 4 × 4 timber studs 14 ft long, at the junction braced horizontally in two directions. Concrete had been placed in adjacent panels on short posting in the morning and the failed area was being poured after lunch. Failure was localized to the area being poured and the mass was successfully carried by the completed floor below. Photographs indicate that concrete was being placed by large wheel buggies running on duck boards. The failure area was limited almost exactly to the entire form area on high shores (Fig. 12.3).

Wall-form failures were not common until the availability of faster and larger mixers so that rate of pouring was increased beyond normal practice. Such an incident came in 1952 at the Dodge Engine Plant in

Fig. 12.3 Typical form failure of double-tier shores.

Chicago, Ill., when the form of a 20-ft-deep engine pit, 16 ft long, blew in from excess liquid pressure.

A tank roof area about 90 × 168 ft covering half of a fuel storage facility in Everett, Mass., failed in 1953. Some 400 cy had been placed from 7 a.m. to 12:05 p.m., when collapse occurred within a few minutes of pour completion. The form was supported by 36 round columns spaced on 21 × 18-ft centers and 30 ft high, and by 4 × 4 spliced timber posts set between columns and with some diagonal bracing. Similar form supports in adjacent tanks had been successfully used. The entire form collapsed leaving the columns, previously concreted, standing.

The failure of about 10,000 ft^2 of formwork at the New York Coliseum construction of the main exhibition floor caused much public comment. The discussions were summarized in an editorial by *Construction Methods*, 53 (June 1955), forseeing that such failures will cause new laws controlling methods as well as results, and this will be bad for the industry. The design was a waffle flat slab, the immediate area of failure also had several deep girders framing out the main stairway and escalator opening. Form shoring was of two tiers with maximum height 22 ft. The bottom tier consisted of wood posts spaced on 2.5 × 4-ft grid with a line of horizontal bracing in two directions and some diagonal bracing.

The upper tier sat on a cap sill and consisted of adjustable pipe shores with flat cap plates to support the timber joists.

Similar form design had been used on this project for some 60 pours of similar extent and some of these had been also 22 ft high. Some 700 cy of a scheduled 1000-cy pour had been placed by 2 p.m. when failure occurred without any warning. Two inspectors were on the deck as it failed. The watch crew under the form however was almost entirely away at the time. Placing was by power buggies running at a reported 12 mph on timber duck boards, each carrying 12 ft^3 of mix.

An immediate exhaustive investigation failed to uncover the exact cause of failure, but later work was performed with slower speeds on the buggies, rubber cushions under the duck boards where they rested on the completed concrete floor edges, channel shaped top caps on the shores to permit horizontal nailing into the joists, and diagonal ties for the pipe shores. Power buggies at high speeds, if in synchronism, will impose definite lateral thrust into the formed deck and the shoring must be properly designed and detailed to transfer such horizontal force to the supporting concrete deck below.

The probable case of failure was not determined until all rubbish had been removed. The deep girders framing out the escalator were directly over the pit framed out in the street floor level. The 6 × 6 post form supports rested on horizontal timbers spanning the pit walls. The horizontal forces transmitted from the deck had caused two such posts to walk off the timbers. Under the weight of the concrete, they fell to the 4-in. concrete floor of the pit and punched through the slab. The lower ends at these posts were visible and were clean cross sections of timber.

In 1967 a small area of similar form support at the Juilliard School in New York, with diagonal plank bracing all in one direction, with the posts resting on 12 × 12 timbers spanning from a concrete wall across excavated rock to concrete footing piers failed from the same action. The uncovered conditions showed where the posts had rested on the timbers and moved laterally until they dropped into the excavation. Only 12 cy had been placed by power buggies.

Failure of some 1000 ft^2 of formwork at the Louisville, Ky., Fairgrounds Stadium also came in 1955. Forms were supported on adjustable length open-web steel joists resting on 2-in. pipe shores 16–18 ft high, at 30-in. spacing. Only 50 cy of concrete had been placed at the time of failure. Photographs indicate that the columns remained in place but apparently the joists had deflected enough to pull out their ends from the supporting timber sills.

Failure of a 20 × 30-ft bay in a five-story garage under construction at

Jackson, Miss., was reported in 1956. The failure was in the second floor form, supported by pipe shores 30 ft high, braced horizontally only at midlength. Concrete was being vibrated at the instant of form collapse.

In a three-tier concrete school building in Brooklyn, N.Y., the long span slabs were found dished in smaller amounts from first floor to roof but in sympathetic shapes. Investigation in 1956 showed that the lowest level had been supported on mud sills which settled, and the upper layers were poured in formwork with precut posts set on the deflected slabs. No check of floor level had been made, the pours were merely uniform thicknesses of concrete on the respective forms.

A 16 × 100-ft section of a 150-ft-diameter dome tank roof 30 ft high, in Melrose Park, Ill., failed in 1957. One 8-ft-wide band had been completed, and failure occurred during the placing of the next 8-ft band. Similar form design had successfully been used in many previous projects.

In a large garage construction in Yonkers, N.Y., where a single bay of forms on the first floor collapsed and investigation indicated possible poor soil bearing for the mud-sills, a casual remark by a worker explained the failure more properly. A 7½-cy truckload of ready-mix had backed up to the form and the entire load was poured in the center of a 15 × 30 end panel which was already almost filled with its design capacity of 12 cy, when the load was dumped.

A bridge deck form failed in 1958 in Denver, Colo., where a 60-ft section dropped 15 ft to the ground after some 39 cy were placed. Form support was a system of tubular steel-braced towers and concrete was being placed by bucket and crane. Pipe shores rested on mud sills.

In a Miami, Fla., hotel job, also in 1958, use of steel open-web joists for form support was accompanied by rotation of the joists as concrete was linearly loaded on the form. This action is similar to the unbalanced loading on steel beams from which forms are hung, previously described. Emergency bridging was added to avoid collapse. The slim joists had insufficient lateral strength to resist one-sided loadings.

Flat-plate forms supported on expanded-steel joists stringers, cambered for bending deflection, were used at a Washington, D.C., apartment in 1959. This did not take care of the load compression at the column form, which was more rigid than the slab shoring. The result was a lip in the concrete ceiling surfaces, which required expensive grinding and plastering to provide the desired smooth ceiling.

The year 1959 had several form failures. At a Los Angeles bridge over the Pacific Electric Railroad, timber form shores collapsed during concreting. The *New York World-Telegram* pictured the accident under

the heading: "Bridge Came Tumbling Down," instead of noting the incident as a local form-support failure. At the Oakland, Calif., Hall of Justice, a section of 14-in.-thick first floor slab 20 × 40 ft on adjustable steel shores supporting a horizontal steel scaffold beam, fell 22 ft to the basement level. In Atlanta, Ga., two successive failures of 21 × 27-ft bays occurred on the second floor which was being concreted by push buggy from a hopper. The forms were metal pans 14 in. deep with 4-in. topping, supported on plywood deck spanning 4 ft between metal pipe scaffolding. The legs of the scaffolds were extended above and below the frames to make the 26-ft height. The top pipe extensions were bent out of shape. At the end of a 750 yd^3 pour on the roof deck of the Johnson & Johnson plant in Montreal, Canada, a section 120 × 160 ft collapsed. Form supports were placed in two directions and posts rested on 6 × 8-in. mud-sills set in run of crusher compacted base layer. The bays were 30 × 32 ft for a 9½-in. flat-slab roof, 20 ft high. Concrete was being placed from two truck cranes. Pipe scaffolding was used with one level of horizontal bracing. The pour was the fourth of five sections for the entire roof; the first three were installed without incident. Witnesses testified to checking all supports of the form and blamed a 45-mph wind acting against tarpaulins which had been slung to cover the open sides of the building. The Montreal Coroner's jury reported it "An Act of God."

Three failures were reported in 1960. In a seven-story wall-bearing apartment house in Arlington, Va., a 24 × 30-ft section of the first floor gave way during the concreting. The slab was reinforced with paper-backed mesh which also acted as the form and was tied to open-web steel joists. Rotation of the joist supports was easy since no ties had been installed to connect joists together and provide resistance to overturning of the end joist under unbalanced load. At Newark, N.J., about 120 yd^3 of concrete in the roof of a three-story garage fell as the men were screeding the completed surface of an 18-in. slab designed to carry 4 ft of earth. The fall onto the second floor slab, then two weeks old, also broke out the same area, 2000 ft^2, and then cracked the slab below (Fig. 12.4).

Concrete had been ordered with high early strength cement and both floors had been reshored with steel pipe screwjacks. Investigation after the failure proved that the concrete delivered contained Type I cement and the strength of the concrete floor carrying the formwork was far lower than indicated by the laboratory cured cylinders. The low strength of the concrete had caused unexpected high loadings at the shores surrounding the columns and failure apparently started there. The

steel forms for the circular columns and conical heads, still in place on the lower story, were pulled over as if, relatively, the column had punched upward.

While finishing a completed section of the first floor for the Smithsonian Institute Museum in Washington, an area 40 × 60 fell about 25 ft into the basement. Reports blamed the supporting timbers, possibly a displaced wedge, as starting the chain reaction of failure.

The year 1961 had three major form failures. At Beverly Hills, Calif., timber falsework for a five-story garage was undermined by a water

Fig. 12.4 Failure of concrete at Newark garage.

main break and collapsed (Fig. 12.5). At the Toronto subway construction, roof forms collapsed when a timber bulkhead on a deep pour moved outward. At Cleveland, Ohio, falsework for a bridge was supported by 18-ft-high shores resting on the pavement of the lower road. Under the weight of the concrete, the pavement support gave way and the forms collapsed.

Four major form failures are on record for 1962. At Rapid City, S. Dak., double-tier timber falsework for a 57-ft span bridge collapsed. At the rocket test silo in Tullahoma, Tenn., a 150-ft. diameter deck was

placed on a ring of steel pie-shaped form sections braced and supported by brackets fixed against the completed wall. Some 450 cy fell 50 ft when brackets gave way. At Hillsboro, Ore., the roof of a 150-ft diameter tank, 65 ft high was supported by timber shoring which gave way and dropped 119 cy of concrete mix. A sloping tunnel shaft being concreted for the Rapid Butte Dam in Oregon lost part of one pour when the pin supports for the inclined formwork sheared.

While no formwork failures on buildings were noted in 1962, there were two in the following year. At an apartment building in Washing-

Fig. 12.5 Failure of form with undermined supports.

ton, D.C., a 7-in. slab supported on pipe shores which rested on a sloping floor, fell when the shores slipped. A roof over a gymnasium in St. Louis, Mo., a 50 × 57-ft area on 20-ft high timber posts failed 45 minutes after the concreting was complete. The failure was charged to poor lumber used for the supports.

Four major form failures were reported in 1964 with various causes. At Merritt Island, Fla., a deck form collapsed under an overload of reinforcing bars. At Nashville, Tenn., 350 cy of concrete in a 32-ft wide bridge of 64-ft span fell when the timber falsework gave way. A bridge at

Stockholm, Sweden, with steel formwork-truss spanning the water, lost 130 cy of concrete when the steel gave way. The second floor of an addition to the National Press Building in Washington, D.C., supported on double-tier shores lost 950 ft^2 of slab when the shores buckled.

In 1965 again four major form failures were reported. At Jacksonville, Fla., the forms at the First Baptist Church failed during concrete placement with conventional timber supports. At Concord, N.H., 4000 ft^2 of the second floor of the General Services Administration contract dropped 200 tons of concrete when the steel shores gave way. At Montreal, Canada, the form for a railroad tunnel arch 9 ft thick collapsed, apparently from insufficient lateral resistance for the pressure of such depth of wet concrete. At Platteville, Wis., a waffle slab roof pour of a university building collapsed the high-shore supports.

Failures of building construction formwork were fewer in 1966, but there were five major incidents, three outside of the United States. Ottawa, Canada, had two failures, the formwork for a highway bridge over the Heron Road, and the fourth floor of a 12-story building. At Athens, Greece, 4500-ft^2 formwork in a two-story building collapsed. At the second Delaware River Bridge near New Castle, the forms for the anchorage mass gave way, and at the Ohio State University nuclear accelerator, 100 tons of concrete were lost.

With concrete placing rates of some 100 cy/hr and the use of power buggies with good brakes stopping a 2500-lb load "on a button," formwork design must take into consideration the horizontal and possible synchronous impacts from such equipment. Forms are supported on continuous framework and, like all continuously supported structures, must have relatively fixed reactions. As the plastic concrete covers the forms, reactions on the posts change and there is good possibility of uplift at posts beyond the area covered by the concrete. Careful nailing or bolting for uplift resistance must be provided. Wedging posts to level up forms are operations which must be done under careful supervision, to avoid leaving posts disconnected from the deck. This would provide a possible beginning of failure if the reaction thereon became negative. How to obtain sufficient and capable direction for formwork erectors to avoid all possible trouble is the present problem. No amount of inspection can replace that direction.

The concentration of formwork failures in 1959–1960 was noted in an editorial in *Engineering News-Record* of April 17, 1960 (after the failure at the Newark garage) which read:

"A recent formwork collapse, costing thousands of dollars, though it miraculously took no lives, originated with careless reshoring practices, according to a building department official of the city concerned.

"Construction of this building, like so many others, placed temporary loads on the reshored slabs far greater than their permanent design loads. If this is a practice that can't be avoided, then reshoring must be counted on to take the extra load.

"To prevent overstressing when the concrete is least able to take it, reshoring must reduce spans and bending moments and distribute loads between two or more floors. Usual practice calls for reshoring two floors below the floor being cast.

"The building official's inspection after the collapse disclosed careless reshoring in many areas of the building. Some of the pipe screw-jack reshores appeared to have been bent before installation. Some were not carefully plumbed. These two defects produce eccentric loading, which can lead to buckling failure at a fraction of the reshore's load capacity under concentric loading.

"Plumbness of reshores should be carefully checked, perhaps with a carpenter's level or a plumb bob frame. Defective reshores should be removed and replaced. When timber reshores are used, tight wedging at the bottom and cap plates at the top are necessary to insure concentric bearing. Providing sufficient bearing area and rigidity in all timber mudsills is essential for all types of reshores.

"Contractors, engineers, building inspectors and others concerned with construction safety must not let the fast pace of construction blind them to these important safety details. One or two defective reshores—slipping, settling or buckling under load—can trigger a serious failure."

1. The weakness in the designs is in details rather than the main structural members, since similar designs have been successfully used.
2. High shoring is more susceptible to failure especially when not diagonally braced.
3. Shock and vibration from the use of duck board runways must be controlled.
4. Lateral force from power buggies must be provided for in the design and details.
5. Details that are difficult to perform, such as the overhand driving of nails to connect flat-cap plates of a shore to an unloaded wood joist, will not be properly performed and may start a failure.
6. Forms are continuously supported structures and must be provided with uniform bearing at each support, otherwise settled mudsills or shrinkage at timber post splices will completely upset the computed reactions, with possible overloading of some posts.
7. Wedging of posts to counterbalance load compression must be done under proper supervision so that a previously properly assembled form support is not disrupted.

8. A repetition of an editorial warning in 1903 will cover the most important item: "some degree of skilled labor should be employed, or that at least the entire work should be under strict and constant supervision by skilled foremen, architects or engineers."

12.2 Lift Slab Manufacture

There have been some poor results from the improper separation of slabs poured on each other for later vertical lifting. In a building at Utica, N.Y., in 1956 separation was forced by eccentric jacking and large areas of the ceiling face were ripped off, exposing the reinforcing and requiring extensive repair. Similar difficulty in a lift slab operation in eastern Pennsylvania was corrected by drilling through the top slab and jarring the contact surface with black powder. The experience of local coal miners was used to good advantage.

In 1960 at Salisbury, Md., a roof slab lift took with it the finished mesh reinforced grade floor slab, nearly 8000 ft^2. After unsuccessful attempts to pry the slab loose, small holes were drilled on a 1 × 1.5-ft grid and 3500 charges of 1 oz of 40 percent gelatin, shot off four holes at a time, were used to demolish the unwanted form slab from the ceiling.

The great number of successful separations in lift slab operations indicates that a small amount of proper care in the application of a proper bond prevention medium would eliminate such difficulties.

12.3 Forming Problems

Inflated Forms

Use of inflated rubber surfaced flexible forms has produced a number of economical concrete domed structures. A few of such designs have failed during construction. In 1942 at Los Angeles, Calif., a 100-ft-diameter, 32-ft-high dome concreted on an inflated balloon failed because the ½-oz. of internal pressure was too low to sustain the weight of the wet concrete. The second attempt was also a failure, when one of three blowers operated to maintain pressure went out and the two units were insufficient to balance the air leakage. In 1962 a plastic sheet dome form in Cincinnati, Ohio, failed when the joints opened up from lack of bonding seal. In 1964 at Dalles, Ore., the plastic dome for the test laboratory of the Bonneville Power Authority, 200 × 100 ft in elliptical plan and 60 ft high, collapsed during concrete placing.

Premature Form Removal

Formwork is a mold for casting a wet mix that will harden and become self-supporting after the passage of time, and the form support must not be removed until the concrete is self-sustaining. Even after hardening, plastic flow and creep can seriously modify the shape and support must be provided long enough to permit the concrete to obtain its permanent shape. Failure of concrete caused by too early form removal was very common before 1920 but better control and testing have practically eliminated this trouble.

The need for reshoring slabs too young to support the formwork of a newly poured slab was carefully discussed in the literature after the failure of a roof slab for a five-story building at Long Beach, Calif., in 1906, when no shores were used below the fifth floor. In the same year the Eastman Kodak plant being constructed in Rochester, N.Y., lost part of the roof slab when the forms were stripped too early, but here there were contributory defects as well. The rods were too short for proper anchorage in the columns and wood shavings were left in the beam forms. In 1907 the shores at a three-story factory in Philadelphia were removed by error in 5½ days after concrete placement and the slab fell. Similarly, in 1910 at Saybrook, Conn., the forms for 37-ft span beams for the carbarn roof, were removed after 10 days in fairly cold weather, and the beams fell. Also in Cincinnati, Ohio, in 1912, beam bottom forms for a movie house were stripped too soon in December and the beams collapsed. The complete collapse of the Stark-Lyman building at Cedar Rapids, Iowa, in 1913 came when shores were removed below the fifth floor of the seven-story frame; the concrete had not hardened enough. At a Rockford, Ill., factory forms for flat-slab floor poured on November 18, 1915, were stripped the next day, with complete collapse. And in 1921 the roof slab of the Masonic Temple in Salina, Kan., rested on the bottom chord of the trusses 38 ft above the floor level. All shores had been placed on mudsills which gave way and three stories of construction failed.

Formwork Fires

Commenting on reports of fire damage to forms and to the contained concrete, a technical journal headed the editorial "Forms were not for burning." Fire damage is an expensive consequence of lack of care and protection to the heating equipment used in winter concreting. In 1962 the United States Embassy building in Mexico City had a fire which destroyed part of the first floor. Among several formwork fires in the winter of 1962–1963 in New York was the thirty-first floor of a 35-story apartment building where propane containers were the fuel. The next winter the top of the service core of the 21-story IBM building in

Philadelphia, Pa., and a hospital building in New York had similar incidents. Although a fire watch is usually on duty full time with some portable fire extinguishers available for emergency use, the open structures are affected by dangerous drafts especially in the high stories and the necessary access and high-pressure water lines are normally absent for fire fighting personnel. Here is a condition where a little expenditure in protection is worthwhile insurance against large loss, both in completed work and in time of performance.

12.4 Temporary Grandstands

The framework for temporary seating stands used to increase peak capacity of permanent stadia, for reviews and spectators of parades and demonstrations and for assembly of crowds for religious, political, and cultural programs is similar to that used as formwork support. Their temporary nature is often taken as an excuse to degrade the factor of safety from that required in normal structural design. Yet the temporary nature of the seating usually results in higher loadings than is permitted in the permanent stands and the light dead weight does not provide the mass of a permanent structure to absorb shock and vibration from an audience interested in the spectacle. Public news is made by these failures since human life is lost or injured. Three typical examples of such failures are cited to show the various reasons for structural insufficiency.

In 1926 a stand erected for the Tournament of Roses in Pasadena, Calif., failed under a load of 900 people. The stand was 68 ft long, 41 ft wide, and 20 ft high with 22 rows of seats. The static live load was 62 psf. Failure was caused by the use of poor lumber with no bracing for the posts. In 1947 a bleacher stand at Lafayette, Ind., collapsed and 1000 spectators were injured. There were 3600 seats in a stand 125 ft long, 81 ft wide, and containing 42 rows. The frame was assembled of wood with steel joint connectors, and was carrying a static load of about 53 psf. Failure was at the joints when steel hook bolts and stirrup hangers gave way. A circus bleacher in 1948 at Redwood City, Calif., containing 1000 seats in eight rows had posts resting on adobe soil. After a rain the posts squeezed into the softened soil and the stand collapsed.

Structures carrying load are susceptible to failure whether they are temporary or relatively permanent in nature and no distinction can be made in design and detail procedures, in the quality of materials, or in the control of assembly.

13

Ignorance and Negligence—Responsibility

Ignorance or Incompetence; Negligence; Control and Supervision; Responsibility; Tolerances; Legal Responsibility.

13.1 *Ignorance or Incompetence*

All structures pass through a series of critical stages during the assembly of the components. Only after overcoming such crises is the completed work a stable and successful performance of a design. Each stage may involve "operational shock," requiring special control and watching for any sign of misbehavior. In a well-planned program of construction each critical phase is provided with a backup resistance that automatically comes into play if the primary assembly fails, similar to the fail-safe procedures characteristic of space vehicle design. Examples of "operational shock" are common in such work as tunnel excavation, bridge erection, underpinning, and concrete falsework.

In the public press failures from unexpected movements of the temporary devices used in construction are not distinguished from those coming as a result of design deficiency. The forces that cause failure during construction are not of the same magnitude or even in the same direction as the loads used for the structural design. At the same time the strength of fresh concrete, unbraced steel, or blasted rock faces and roofs is lower than the resisting values desired by the completed design.

Construction sequences require some technical evaluation as to safety and sufficiency. There is little difference in this context, whether the structure fails during erection because of ignorance or incompetence on the part of the responsible person. Ignorance, the state of not being informed of what is required, always entails incompetence, the inability to do what is required whether one knows or does not know. There is little doubt that experience is the best teacher for the experienced construction manager and an extra sense of prewarning of imminent danger is a valuable asset. Yet such simple errors as leaving an unbraced one-story wall standing for the wind to blow over, or the eccentric loading of such walls by hanging scaffolds, are so common that both ignorance and incompetence need eradication. Prequalification and registration of construction superintendents may eliminate some of the trouble. Safety inspectors provided by the insurance carriers have done much to reduce construction accidents on large jobs, but the same service is much needed in the many more smaller jobs. In the smaller operations safety is often neglected because of the protection against financial loss, which is assured by the various insurance policies carried by the builder and usually insisted on by the state and financial interests. More chances are now taken than in the past, when every failure meant financial loss or ruin to the builder.

Failures in completed structures, although still too frequent, are much less common than when steel bridges were being sold complete with design by nontechnical salesmen on open price competition, or when less was known about concrete than the much-too-little now commonly known. With very few exceptions, failures in completed sturctures are caused by dishonest performance or by noncompliance with accepted practice. Ignorance is more often at fault than inadequate design. Sometimes, as traffic gets denser, loads increase at the same time that joints work loose and members corrode. This was the likely cause of the collapse of the 40-year-old I-bar suspension bridge over the Ohio River on a cold night in December 1967. Suddenly everyone realizes that structures are not everlasting.

A report by the L. B. Foster Company at the Construction Section of the 1967 National Safety Congress rated soil cave-in as the Number 1 cause of construction worker deaths. In the preceding two years, trench cave-ins had been responsible for more than 125 deaths, where the causes were usually unshored trenches, improper shoring, and backfill placed too close to the edge of the excavation. In half of the trenches at accident sites no shoring had been placed. The New York State Industrial Code lists specific requirements for minimal protection of trenches based on depth and type of soil. Similar legal requirements are common in most jurisdictions, so that ignorance is not an excuse. The incompetence of the control needs correction.

Building collapses in 1967 in Rio de Janeiro caused furor in the press about laxity of control, incompetence of design, and ignorance of the proper limit of safety. The collapse of the Rua Rosario apartment building in February, killing nine persons, immediately following a fatal crash of a reinforced concrete marquee when rainwater accumulated in the cellular shaped cantilever, brought a harsh article in the magazine *O Cruzeiro*. Listed were 20 buildings that collapsed in Rio in the past 10 years and photographs of several buildings leaning 8 to 12 in. located on Rio's main avenues. Although none of these structures had been condemned by the city, the municipal government had abandoned use of space in one of them. The evident separations between adjacent buildings were hastily covered up. Reports that the settling buildings were relatively safe were hurriedly prepared but little constructive corrective measures were required.

Probably the major cause is the very light penalty in Brazil for construction failures. For instance, in a legal proceeding in Sao Paulo where a concrete building collapsed, killing six workers, apparently because almost five of the seven bags of cement per cubic meter of concrete had been diverted into the black market (1948), the "Engineer-Responsevel" of the project, who acted both as the designer and contractor, was penalized by a revocation of his license for six weeks.

Elimination of ignorance and incompetence would go a long way in the elimination of construction failures.

13.2 Negligence

One is negligent if he is likely to omit what ought to be done. Disregarding the obvious benefit of hindsight when omissions become obvious, the difficulty arises in defining what should have been provided in the original design and in the construction. When ignorance or incompetence cannot be proven, the cause of failure is explained by negligence. The failure of shoring that caused the collapse of the precast bridge members over the River Calder near Wakefield, England, in 1967 was charged to the use of incorrect steel beams as grillages. The correct beams, double the strength of those actually used, were available at the site, but no identifying marks had been placed to distinguish the lighter and heavier sections. Other safety precautions, such as web stiffeners on beams and diagonal bracing for the supporting piles, had also been omitted but the negligence in using the wrong beams was considered the prime cause of failure.

Recent legal determinations that liability for damage to neighboring structures resulting from blasting vibration effect is not dependent on proof of negligence, brings up a new sequence of responsibilities. In effect, it becomes negligence when any procedure is used, with the usual care and precautions, where the natural consequences produce forces that cause damage or nuisance. Such was the decision in a New York court in 1966.

Negligence was charged against an engineer for the design of a windmill constructed on top of a 100-ft-diameter circular barn in Indiana. The barn had a conical roof with a 3-ft air shaft in the center framed in heavy timbers. A steel tower windmill weighing 2000 lb was added above the shaft. The windmill collapsed and the design was claimed to be deficient in that reasonable care had not been taken to determine whether the existing structure could sustain the new loading.

Negligence often is found in the preparation of contract documents, especially in the lack of coordination between plans and site conditions and in the discrepancy between plans and specifications. The voluminous record of litigation stemming from failures chargeable to such causes, as well as the many claims arising from such negligence, is summarized as to legal implication in the last part of this chapter.

Proved negligence on the part of a professional usually results in the revocation or temporary suspension of his license to practice. A license is revoked not as a punishment but in the exercise of the state's discretion under its police power as to whether the person holding the license is properly qualified to continue in his profession.

13.3 Control and Supervision

To control is defined as to exercise a direction, restraining, or governing influence over the production. Just as many centuries of tradition have taught that a ship can be well run only when the captain is in sole command, so is it also in construction. There can only be a single control, and that control of production can only rest in the hands of the contractor.

The meaning of supervision in the common language differs somewhat from the accepted meaning in construction. When supervision is defined as having a general oversight of the work, or as inspection of the work, there is no conflict or disagreement in meaning. When defined as equivalent to superintendence, however, which is commonly defined as having charge and direction of the work or as manager, there is a distinct conflict. Having charge or managing the work entails a type of control that would not be agreed to by the contractor or accepted by the

professional inspector. To avoid the possible conflict of authority, whether intended or not, it would be wise to eliminate the use of supervision and call the service by its correct name, inspection. There is then no possible conflict in the necessary single control of production, and supervision is then the province and duty of the contractor. Inspection is the province and duty of the designer, usually delegated to a representative, as a service to protect the economic rights of the owner. He is on the job to see that the owner receives what he has bought and, because most of the work is eventually covered from view, inspection must be provided at each step of the production.

There is another facet of field service that can best be defined as technical coordination, or even technical control. Included is the always present information seeking, the explanation of meanings, the fitting together of details involved in separately specified items, sometimes performed by different contractors. The allocation of space for the various services, such as plumbing lines, ventilation ducts, electrical feeders, in the modern requirements of buildings entails a major expenditure of technical help.

In one medical laboratory building recently completed in New York City the concentration of such services at the cellar ceiling was not properly coordinated in the field, and an expensive removal and rearrangement was necessary. The first subcontractor on the job took the prime space, up against the ceiling. By the time the electrician started to install the bus-ducts, all plumbing, heating and ductwork had been hung and the high tension bus-ducts ended up within reach of anyone walking on the cellar floor. This position was quite properly rejected as a hazard by the municipal inspector and a reassembly of the maze was forced on the contractors, with sizable monetary claims and a considerable delay in completion.

Many of the examples of structural failure described in the previous chapters could have been avoided if proper field control were provided. Many more could have been similarly eliminated if knowledgeable technical field inspection were available to bring to attention a misfit or an omission in detail. Unfortunately, such inspection service is usually not provided in jobs where the need is greatest, that is, where the contractor control is minimal or even nonexistent. In work based on plans with no full explanation of details, again in the type of project where minimum planning fees are combined with lack of sufficient field control, the stage is always set for something to go wrong.

Reliance by many owners on the inspection service of local building inspectors, usually on the argument that such service has been paid for by the building permit fee, is of questionable value. In some jurisdic-

tions, the District of Columbia being a noteworthy example, the field inspection is competent and thorough; but in many areas it is nonexistent, even to the extent of (an actual occurrence) disregard of the fact that an additional story had been built on a hotel building, with no change in the plans approved as a basis for the construction permit issued by a city department.

The problem of obtaining proper inspection leads to many and diverse solutions. Some public agencies have their construction departments to take care of inspection of the designs prepared both by their own design sections and by private consultants. Misinterpretation of the latter's intent is often found. Some private corporations with large building programs similarly separate design and inspection of performance. The high cost of maintaining a full-time inspection staff has brought the concept of a separately hired entity as construction coordinator or project manager to take over all inspection and sometimes even act as the owner's agent in both the technical and financial controls of the performance. The cry in the technical press is for the designer to inspect his own work and such cry is loudest after a concentration of construction failures. Many building codes are gradually adding requirements, as more complicated designs are permitted, for certification by the designer that the production complies with the intent of the approved drawings. If honestly provided, such inspection can significantly reduce the number of construction failures.

Inspection, or the lack of it, has never caused a failure. It can only serve, by warning or even halting the work, to prevent the failure caused by some error or omission for which others are responsible. Competent control, on every level of responsibility, is the best insurance against mishaps. The shortage of such personnel, especially at the foreman level, directly in charge of labor, is a critical situation, which must be corrected by concerted efforts of training and selection. Labor organizations must cooperate in this program and avoid interference with promotions and with wage increases to men who qualify for foreman status and duties.

13.4 Responsibility

Every incident where there is damage to life or property is followed by an attempt to pin down the responsibility. Unless the cause of failure is first determined, the search for the responsible person is a fruitless effort, accompanied by incriminations and unethical claims and counterclaims. Even with the cause determined, there is often complete disagreement about assigning responsibility.

Standard forms of agreement, in which the limits of responsibility are fixed, separately prepared by associations of architects, engineers, contractors, subcontractors, and vendors, are seldom sufficiently coordinated to eliminate gaps or overlaps. In the last 10 years efforts have been made to reach industry-wide agreement on the division of responsibility. In the many joint committee meetings, conferences, symposia, and round-table discussions, there is general agreement that a clear-cut division of the total responsibility to the owner and to the public is desirable and as yet nonexistent. Some unilateral attempts to solve the problem by requiring the contractor to provide "hold harmless" protection for the designers was counteracted by the passage of a New York State law prohibiting such broad protection which might exonerate the errors and omissions of the designer.

In a symposium on this general subject the writer offered a formula for division of responsibility among the facets of a construction operation, to provide proper performance and result.

Four factors, all of which must be provided, are required:

1. A proper design with competent specifications to provide safety and suitable to the desired use of the structure. A proper design is not necessarily a scientific solution of the problem, but a technical determination of requirements based on the geographical and climatological conditions.
2. The design in all its phases of architectural and engineering detail must be buildable, not for watchmaking but for construction, by available materials, labor, equipment, and local experience.
3. The contractor, and all his subcontractors and vendors, must have the capability of reading the plans, and must read the specifications before they take on the work, and must have the desire to follow the contract requirements and the desire to do a good job.
4. An inspector or somebody to watch that the contractor performs what he is contracted to do, is essential, but the inspector cannot be expected to do the impossible. He cannot correct for the incompetence or the lack of interest of others who make up the whole team.

If each of the team members does his job well and avoids trying to do the work of anyone else, the performance will be successful and the product will be satisfactory. Attempts to bypass this division of responsibility, by design-construct contracts, or inspection by contractor formulas, have not eliminated construction failures, quite the contrary is indicated by the record.

The personal control of a design project by an individual with a small staff of assistants and the traditional responsibility for controlling and coordinating the building process by a general contractor whose pride is

the quality of the completed project are no longer always available in producing the tremendous volume of needed construction. Responsibility is assigned, subdivided, and endlessly fragmented. The possibility of oversights, gaps, and errors becomes much greater. Unless the industry develops a workable arrangement of control and responsibility, the incidence of failure will increase and greater governmental control will result. This will not solve the problem. It will merely interfere with the work of each member of the construction team and increase the costs of production. Shifting responsibility, at least for monetary loss, to insurance carriers similarly does not solve the problem and has the same result as from governmental control.

13.5 Tolerances

Whether manually or mechanically produced, all items of identical design and material specifications are not mathematically equal. Some are "less equal" than others. The most minute quality control and inspection has not eliminated the inequalities in component production of the most expensive electronic hardware with some really expensive failures in the space-probe program. Most of us have learned from bitter experience that all machines and equipment of the same model and vintage, in spite of the carefully documented advertising, include some "lemons." In the comparatively roughly produced and inadequately supervised construction industry, how much deviation from the desired level of accuracy can be permitted?

If one could guarantee that the deviations from normal would be in random fashion, above and below the desired level, some statistical procedure would be warranted for determination of acceptability. However, actual experience indicates that when one phase of the work is deficient other deficiencies also exist. From the necessity of safety, it seems wise to so design and plan the operation that each individual phase is sufficient after allowing the normal deviations for each step. This, of course, is very wasteful in the properly controlled jobs and becomes a penalty on good construction. However, until the various phases of the industry can develop proper control to indicate that a closer tolerance is warranted, the facts of life must be admitted and trouble should be avoided, at costs that are a small increase over the ideal. The increase in cost should be considered an insurance against the much higher costs of job stoppage, investigations, load tests, partial or total demolition, and reconstruction. The value of deleting unfavor-

able publicity is hard to fix, but in the long run it may be the most important factor in deciding on permissible tolerances.

When concrete operations were small and daily volumes produced at the site were made of the ingredients locally stocked, a fairly good control of uniformity was easily obtainable. With most of the concrete coming from distant ready-mix plants, supplying many jobs with various types, sizes, and gradation of aggregates and with limited number of bins for storing the various different items, a variation from the samples used for preparing a design mix must be expected. It is too common to find gravel delivered in a stone mix, sudden change in color because the cement bin was filled with a different brand, continuous blending of different lightweight aggregates, change in gradation of the course materials as the crane bucket is set on the top of the storage pile and the larger sizes are separated by gravity and accumulate at the base of the pile, continuous fluctuation of moisture content and in some plants, even change in the water used. Yet we naively accept the laboratory report based on a small sample of each ingredient. There is a possibility, quite remote, that the materials used in the mix are superior to the combination of the samples used in the laboratory. One would be extremely foolish to rely on such chance. The limitations in possible economical production of the concrete mix must be recognized and the reasonable tolerance permitted.

Factory produced construction materials, such as steel, timber, glass, and masonry units are all controlled by standard specifications. Variations in dimension and physical properties must be expected and designs must tolerate such variations. Use of standard specifications without reference to the actual grade or type desired is a worthless requirement.

Compatibility of tolerances for different adjoining materials must be studied and resolved. One such error was disclosed in a monumental precast concrete facade with colored glass units. The variations in dimension of the castings caused complete misfit of the carefully cut glass units, with subsequent breakages. In a large bus terminal with odd-shaped window openings cast in the walls the prefabricated frames required complete reconstruction to fit in the openings as built with normal tolerances for such work. Tolerances in fabrication of concrete units differ from those that must be provided for prestressed units. Tolerances in rolling large jumbo section steel columns differ considerably from the control of dimension and shape of thinner steel sections. The accuracy of building production becomes more critical with the greater use of prefabricated and subassembly components. Fitting and adjustment of such members in the field requires special skills and

unless properly done will result in local and possibly total failures. The concept of tolerances permitted for construction must be reevaluated, since the present practice of specified tolerances for individual items must be made part of a total or system approach to the problem.

LEGAL RESPONSIBILITY

by Nathaniel Rothstein, Member, New York Bar

Duties and Standards

In early days there were no contractors or engineers. The architect was all. He was the "architekton" or "master builder." He planned and supervised the construction from beginning to end.

With the passage of time and introduction of new construction materials, new construction techniques, and new technological developments, the architect was no longer able to maintain the position of both designer and builder. And so was born the general contractor who assumed the task of hiring the workmen and subcontractors, and supervising their work; and the general contractor became responsible for the methods and procedures used to erect the building, albeit in conformity with the architect's plans and specifications. So, too, engineers were employed by the architect who farmed out to these specialists the detail work of various portions of his design.

Today the architect is primarily concerned with the visual concept of the proposed structure, and in portraying graphically the spatial relationships of the structures. He employs separate engineers for the structural design, for the mechanical design, for the electrical design, for site-planning, etc. and he then coordinates the several engineering designs with his own overall plan. The architect has become the master planner-administrator who must be familiar with good construction techniques and adequate engineering designs.

Although the architect-engineer (the term is used to describe both members of the design profession) generally participates in the construction phase of a project, he now rarely supervises the work to the extent of exercising direction and control of the manner of performance of the contractors' work. Rather, it is more usual for him to visit the jobsite two or three times a week to inspect the work of the contractors and make sure that the plans and specifications are being followed.

Thus have the duties of the "master builder" been divided and subdivided. Presently, the general contractor is responsible for construction methods, techniques, practices, and procedures. He hires the various subcontractors, each a specialist in his own field, to perform the various

portions of the construction work, and he assumes (by contract) legal responsibility for the work of these subcontractors.

The design professional, be he architect or engineer, is required to possess and use the same degree of skill, knowledge, and ability possessed and used by other members of his profession [1], and he is charged with the exercise of ordinary care and his best judgment in carrying out his assignment. However, he does not guarantee a perfect set of plans nor does he warrant a satisfactory result [2]. In his papers on "Professional Negligence of Architects and Engineers" [3], Professor George M. Bell states:

"An architect or engineer does not warrant the perfection of his plans nor the safety or durability of the structure any more than a physician or surgeon warrants a cure or a lawyer guarantees the winning of a case. All that is expected is the exercise of ordinary skill and care in the light of the current knowledge in these professions. When an architect or engineer possesses the requisite skill and knowledge common to his profession and exercises that skill and knowledge in a reasonable manner, he has done all that the law requires. He is held to that degree of care and skill and that judgment which is common to the profession."

In *Bayne* v. *Everham* [4] the rule is thus stated:

"This court has held that the responsibility of an architect does not differ from that of a lawyer or physician. When he possesses the requisite skill and knowledge, and in the exercise thereof has used his best judgment, he has done all that the law requires. The architect is not a warrantor of his plans and specifications. The result may show a mistake or defect, although he may have exercised the reasonable skill required."

We must be mindful that an architect or engineer is neither a manufacturer nor vendor of any product. He is a professional man, and, like the doctor or accountant or lawyer, he renders purely professional services. Hence, unless he expressly guarantees a specific result, his liability in negligence (or malpractice) [5] is akin to that of other professionals; he is not legally responsible for an unfortunate result if he has followed the normal, accepted practices used by other members of his profession. It is only when he deviates from good practice, resulting in property damage or personal injuries, that the design professional subjects himself to liability in damages.

Liability during the Construction Phase

Because of the gradual division of duties of the "architekton" it is not surprising to find overlapping areas of responsibility involving architects, engineers, and contractors. However, their respective responsibilities are often modified by their contractual arrangements. Note how sharply limited has become the architect-engineer's duty of overseeing the work performed by the contractor. What was formerly termed "supervision" in the owner-architect contract has now been reduced to "observation." The 1967 A.I.A. contract between owner and architect makes quite clear that the owner does not intend that the architect be responsible for the construction methods, or even for faulty work, by the general contractor. Paragraph 1.1.14 of this agreement provides:

"The Architect shall make periodic visits to the site to familiarize himself generally with the progress and quality of the Work and to determine in general if the Work is proceeding in accordance with the Contract Documents. On the basis of his on-site observations as an Architect, he shall endeavor to guard the Owner against defects and deficiencies in the Work of the Contractor. The Architect shall not be required to make exhaustive or continuous on-site inspections to check the quality or quantity of the Work. The Architect shall not be responsible for construction means, methods, techniques, sequences or procedures, or for safety precautions and programs in connection with the Work, and he shall not be responsible for the Contractor's failure to carry out the Work in accordance with the Contract Documents."

Paragraph 1.1.21 states:

"The Architect shall not be responsible for the acts or omissions of the Contractor, or any Subcontractors, or any of the Contractor's or Subcontractors' agents or employees, or any other persons performing any of the Work."

These provisions do much to define and delimit the architect's responsibility and tend to clarify what has been a gray area. The architect assumes the task of visiting the jobsite one or two or three times a week to check the progress of the work and to determine whether the contractor is doing the work in general conformity with the architect's design. However, many contracts refer to these jobsite visits as "supervision" when nothing more than "inspection" is intended. The use of the term

"supervision" has led many courts to impose an unjustifiably heavier burden of responsibility on the architect-engineer, and for this reason "supervision" has been completely eliminated from the current A.I.A. contract form.

In the recent case of *Pastorelli* v. *Associated Engineers, Inc.* [6] an engineer was required to "supervise the contractor's work throughout the job." The plans called for heating ducts to be suspended from the ceiling by hangers. The actual installation of the ducts was supervised by the general contractor and was performed by a subcontractor, who, with an amazing display of bad construction technique, nailed the hangers to a ⅞ in. thick sheathing instead of causing the hangers to be screwed to joists above the ceiling. About 15 months after the system was completely installed, a 20-ft section of the ductwork, weighing close to 500 lb, fell and injured the plaintiff.

The engineer testified that he had made two or three visits a week to inspect the work but was not aware of the way these particular ducts had been installed. He had never actually observed the workmen installing the particular ductwork that fell, nor had he ever climbed a ladder to ascertain exactly how the ducts were affixed to the structure. The engineer's attorney raised the plea of "no privity of contract." This defense was quickly brushed aside by the court and the engineer was held jointly liable with the contractor and subcontractor.

One would hardly expect the engineer to be *in pari delicto* and share liability jointly with a subcontractor clearly guilty of such gross negligence. Under the doctrine of active versus passive wrongdoing and the weighing of comparative culpabilities [7], the engineer should have obtained indemnity from the subcontractor and the contractor. Perhaps the engineer's attorney simply neglected to crossclaim over against his codefendants. At any rate, the subcontractor was clearly the primary, active tortfeasor on whom ultimate responsibility should have fallen.

Compare the Pastorelli case with the Louisiana case of *Day* v. *National–U.S. Radiator Corp.* [8], where the architect designed a hot-water heating system that called for the installation of a pressure-relief valve on the boiler. The O-A contract called for the architect to give "adequate supervision of the execution of the work to insure strict conformity with the working drawings," and so on. The contractor completed the installation and decided to test the heating system. However, he failed to share this decision with the architect, who was not aware that the system was ready for final inspection or testing. While the boiler was being tested it exploded, killing a workman.

The plaintiff charged the architect with faulty supervision in failing to discover the absence of the safety valve on the boiler during the super-

vision of the work. The lower courts found against the architect, but Louisiana's highest court reversed, pointing out that a contract requiring architects to exercise adequate supervision of execution of work did not charge them with the duty to inspect *methods* employed by the contractor or subcontractor. The court held in part:

". . . Under the contract they as architects had no duty to supervise the contractor's method of doing the work. In fact, as architects they had no power or control over the contractor's method of performing his contract, unless such power was provided for in the specifications. Their duty to the owner was to see that before final acceptance of the work the plans and specifications had been complied with, that proper materials had been used, and generally that the owner secured the building it contracted for.

"Thus we do not think that under the contract in the instant case the architects were charged with the duty or obligation to inspect the methods employed by the contractor or the subcontractor in following the contract or the subcontract."

When an attorney represents an architect charged with faulty supervision, it behooves him to "educate" the court as to the true function of the architect and the limits of his responsibility in this so-called supervisory capacity. The architect does not tell the contractor how to do the work. The method and manner of performance are properly the responsibility of the contractor, unless the owner-architect agreement provides to the contrary. As noted above, today's A.I.A. contract specifically states that the architect shall *not* be responsible for the construction methods employed by the general contractor. What, then, is the architect's responsibility in the construction phase? It is to see that the *result* of the contractor's work brings about a building that is in compliance with the architect's plans and specifications.

In the *Day* case the general contractor and the subcontractor who performed the work would be liable to the estate of the deceased workman. But if the decedent were in the direct employ of the general contractor, he could not sue him because of the exclusive remedy provided under the Workmen's Compensation Laws. Since the amounts recoverable in Workmen's Compensation are relatively small, the decedent's estate, through its lawyer, is always pressured to look about for likely third parties against whom court action can be maintained; and more often than not we find the lawyer attempting to stretch the areas of responsibility so as to encompass a third-party architect or engineer or contractor who is not the employer of the injured party.

This attempt to seek out third parties as prospective defendants leads

to peculiar lawsuits. Not too long ago a woman was walking along the street where construction work was going on when she came upon a pile of debris. She stepped into the roadway to bypass the rubbish, was struck by a passing motor vehicle, and was killed. Her estate sued not only the owner of the car but also the general contractor, the subcontractors, the architects, the engineers, and so on. Thus were design professionals made defendants in a lawsuit although they had no responsibility whatever for this sidewalk condition. Although ultimately they were not held liable for this unfortunate accident, they were obliged to incur substantial legal costs and expenses in defending the case; and these expenses are not recoverable.

The Citadel of Privity

Not too long ago, privity of contract, that is, contractual relations of the parties, sharply proscribed the responsibility of a contractor or architect in a suit by a third party. Thus, in 1891, we find a Pennsylvania Court holding:

"If one who erects a house or builds a bridge . . . owes a duty to the whole world that his work . . . contains no hidden defects, it is difficult to measure the extent of his responsibility and no prudent man would engage in such occupations upon such conditions [9]."

As late as 1926 this strict "privity" doctrine still prevailed in many of our jurisdictions. Thus, when a theater roof collapsed shortly after installation, causing the death of a patron, the court ruled that neither the contractor nor the architect could be held liable in damages to the estate of the deceased patron [10]. The reason: no privity of contract. Only the theater owner, who was in privity with the decedent, could be held liable in such case.

The requirement that an injured party be in privity with a prospective defendant in order to recover damages against him was shattered by Judge Cardozo's famous decision in *MacPherson* v. *Buick* [11], and since then there has been a continual erosion of the privity doctrine. However, remnants of privity still obtain to defeat a third party's claim to damages.

In *Olsen* v. *Chase Manhattan Bank* [12] the plaintiff Olsen was employed by the foundation contractor. While working on the foundation of the bank building he was struck and injured by a drill that fell from a temporary platform some 30 ft above him. He sued the architect and engineer on the ground that their contract called for "complete

supervision of the work of the various contractors." Note the strong language, which here gave the architect *supervision and control* of the work. Indeed, testimony on behalf of the engineer showed he was to supervise the manner of the contractors' work and to review and approve for safety at all times, temporary construction as well as permanent installations.

The jury's verdict was against the architect and engineer. On appeal the judgment was reversed, the appellate court pointing out that there was a difference between the duty owed by the architect to the workman and the duty owed to the owner. The Court stated:

"On the evidence presented, Skidmore & Moran, as architects and engineers, could be charged only with nonfeasance. For that, it is possible they could be liable to the Bank (owner), but not the respondent (workman)."

The *Olsen* case cited the earlier New York case of *Potter* v. *Gilbert* [13] where Potter, a carpenter, was killed when one of the walls of the building under construction, collapsed. The architect was sued by the estate of Potter claiming that the architect supervised the construction of this building and was therefore responsible for the collapse of the wall. The Appellate Court in dismissing the complaint stated:

". . . The architect would be liable to his employer, the owner, for a failure to properly supervise the work; but a failure in this regard amounting to no more than nonfeasance would not give rise to a cause of action in favor of a third party, whose claim would merely be that, if the architect had attended to his duties more diligently, he would have discovered a departure from the plans or specifications by the contractor and might thus have prevented the accident.

"The architect owed the decedent and every one lawfully on or about the premises the duty of preparing plans and specifications under which the building could be constructed with safety, and the decedent's employer owed him the duty of following the plans and specifications; but the architect owed no duty of active vigilance to the decedent to supervise the work of the employer of the decedent, although he may have owed such duty to the owner by whom he was employed."

We thus find New York holding that an architect-engineer owes a greater legal duty to the owner with whom he is in privity than he owes to a third party. To establish liability there must be active wrongdoing, that is, misfeasance, by the architect-engineer; mere nonfeasance will not be sufficient to render the A-E liable to third parties. The *Potter* case was relied on very recently by a Special Term Court in New York

[14] in granting the architect's motion for summary judgment and dismissing, without a trial, the complaint of a third-party workman on the job. The court held in part:

"It appears from the papers submitted on support of and in opposition to the motion made by the defendants that said defendants were architects who owed no duty to the plaintiff. Their sole duty was to see that the general contractor's work conformed to the plans drawn by them. The complaint charges an omission of duty on the part of the architects while acting for the owner constituting nonfeasance for which they may be liable to the owner but not to third parties."

Crossclaims and Third-Party Actions for Indemnity

When a structural failure occurs the owner will often sue the contractor alone, on the theory that the failure resulted from faulty construction technique or from the contractor's neglect to follow the architect's plans and specifications. The contractor will defend by claiming that the failure was the result of the architect's defective design. More often than not the attorney for the contractor will then implead the architect under a third-party complaint charging faulty plans and specifications, and demanding that if judgment is obtained by the owner against the contractor, that the contractor in turn have judgment against the architect for the same amount.

Such third-party complaint, which appears at first blush to be tenable, is dismissable for legal insufficiency. If, as claimed by the contractor, the plans or specifications were faulty, the owner alone has the right to sue the architect. The owner, as plaintiff, may choose which party or parties he wishes to join as defendant. The contractor is not thereby placed at a disadvantage. He can make out a complete defense by proving he used proper construction technique and procedures and by showing that he complied with the architect's plans and specifications. Thus the contractor will not be held liable to the owner if it develops that the structural failure was due to improper design; and, since he is being charged by the owner with being an *active* tortfeasor, he may not claim indemnity from the architect who, at most, might conceivably be a joint tortfeasor. This is the general rule [15].

This type of situation was involved in an unreported case tried recently in the Supreme Court of the State of New York, County of Bronx [16]. There, 12 newly built, attached houses were purchased by the plaintiffs. Soon thereafter these houses began to sink and the exterior

walls opened up in ever-widening cracks. It was believed that this condition resulted from the failure of the piles on which the houses rested for support. When these homeowners were compelled to vacate their newly acquired homes because of the dangerous condition that developed, they brought an action for damages against the owner-builder, the piling contractor, the piling engineer, and the City of New York. The piling engineer then impleaded the architect who had designed the piling foundation which ultimately failed. At the end of the plaintiffs' case this third party complaint was dismissed as a matter of law, the Court holding:

"A plaintiff can choose among joint tortfeasors, joining all or some, but a defendant does not have the same choice to bring in other parties whom the defendant thinks should be liable either in place of or jointly with those the plaintiff has selected."

In that case the piling contractor and the piling engineer were both held liable to these plaintiffs, but the judgment for some $200,000 was not collectible.

Another interesting case is *Earle* v. *The City of New York* [17] where the plaintiff brought an action against the Triboro Bridge and Tunnel Authority when the automobile in which plaintiff was riding struck a guard rail and came in contact with a protruding beam attached to this rail on the Long Island Expressway. The Authority served a third-party complaint on an engineering firm with whom the Authority had contracted to design the Expressway and to provide overall engineering supervision and inspection during its construction.

In its third-party complaint the Authority alleged that if the plaintiff sustained injuries as alleged in the main complaint, such injuries were occasioned by the affirmative negligence of the Engineers in preparing a faulty design and/or in providing faulty supervision of the construction work. However, the Authority, on pretrial examination, admitted it had inspected the guard rail in question before its final acceptance of the work and that it had maintained the guard rail in the same condition for approximately two years before this accident. The Appellate Court affirmed the dismissal of the third-party complaint against the Engineers, stating:

"In the circumstances, the Authority is chargeable with *actual notice*, and may not be heard to say that it is a passive tortfeasor only, or that it is not *in pari delicto* with the third party defendant."

Thus the Authority, having been found to be an *active* wrongdoer (because it had actual notice and knowledge of the claimed defective

condition) could not obtain indemnity from the engineers who designed and supervised the construction work, even though their work might have been faulty.

A landmark case in this field, which will undoubtedly be followed in the majority of jurisdictions in this country, is *Inman* v. *Binghamton Housing Authority* [18]. The *Inman* case was an action against the architect, the builder, and the owner of the house, for personal injuries sustained by a tenant's child. Plaintiff claimed she fell off a one-step stoop at the rear of the house. The building had been completed in 1948, some six years before plaintiff's accident, and the complaint alleged (as did the third-party complaint of the owner against the architect) that the architect's design was faulty in that it created a dangerous condition resulting from (a) the absence of a protective railing; (b) the fact that the door opened outward on the porch of which the stoop was a part, causing anyone on the porch to back up dangerously close to the edge; and (c) the step was in the center of the porch and did not extend its full length. The contractor was charged with negligence in that he followed plans that plaintiff claimed were so obviously defective as to put the contractor on notice that he was thereby creating a dangerous condition that would be likely to cause someone personal injuries.

Normally, a contractor is fully protected in following the architect's plans and specifications, but his protective mantle falls away if the plans are so patently defective that any contractor of average skill and ordinary prudence would not blindly follow such plans. Proper practice calls for the contractor to direct the designer's attention to any portion of the plans that appear to be deficient or inadequate.

In applying the *MacPherson-Buick* doctrine to structures involving real property, the Court of Appeals in *Inman* pointed out that where the prime complaint was not expressly based on a hidden defect or unknown danger, neither the architect nor the contractor could be held liable. The Court dismissed both the prime complaint and the third-party complaint against the architect, and stated on page 146:

"Whatever the defect, it may not be said to have been latent, and, whatever the danger, it certainly was not hidden. That being so, it is evident that the requirements of the *MacPherson-Buick* rule have not been met; the complaint of the Inmans against the architects and the builder is without legal basis and was properly dismissed at Special Term.

"We turn, then, to the third party complaint, and, first to the question of the Authority's right to recover over against the architects and the builder on the ground that those defendants were 'actively negligent,'

while the Authority was only 'passively' so. Since, however, the architects and the builder breached no duty and were, therefore, guilty of no negligence, the Authority's claim to recover over against them (founded, as noted, on their assertedly 'active or primary negligence') may not prevail. There is still another basis for the conclusion. According to the complaint of the Inmans against the Authority, the injuries alleged were owing, as Special Term aptly summarized it, to 'negligence . . . in the construction, maintenance and continuance of an allegedly known defective condition of the premises.' Taking the allegations of the complaint to be true, as we must at this juncture, the Authority is cast 'in the role of an active tortfeasor and as such, is not in a position to compel . . . indemnification . . . absent an agreement so to do.' "

We have already noted the reason why the contractor may not implead [19] the architect-engineer when the contractor alone is sued by the owner. However, the converse is not necessarily true. When the charge is limited to faulty design, of course the architect alone stands liable. If, however, the charge is faulty construction work and the architect merely performed periodic inspections of the contractor's work, then the contractor would be *primarily* liable for the failure and the architect only *secondarily* liable. In such event it is proper that the architect implead the contractor if he has not been made a party defendant [20]; and if the contractor is being sued jointly with the architect, then the latter should crossclaim over against the former.

Indemnification Provisions

It is clearly unfair to compel indemnification of an architect-engineer for errors in *design*, and this is now part of the statutory law of New York that makes such indemnification agreements void as against public policy [21]. However, in view of the architect's purely secondary liability during the construction phase, it is not unfair that the contractor indemnify the architect for any defects occurring in the performance of the work. Here, again, the most recent revision of the A.I.A. General Conditions contains such indemnification agreement. Article 4.18.1 reads as follows:

"The contractor shall indemnify and hold harmless the owner and the architect and their agents and employees from and against all claims, damages, losses and expenses including attorneys' fees arising out of or resulting from the performance of the Work, . . . caused in whole or in part by any negligent act or omission of the contractor, any subcontrac-

tor, anyone directly or indirectly employed by any of them or anyone for whose acts any of them may be liable, regardless of whether or not it is caused in part by a party indemnified hereunder."

However, this is modified by Article 4.18.3, which provides:

"The obligations of the contractor under this paragraph 4.18 shall not extend to the liability of the architect, his agents or employees arising out of (1) the preparation or approval of maps, drawings, opinions, reports, surveys, change orders, designs or specifications, or (2) the giving of *or the failure* to give directions or instructions by the architect, his agents or employees provided such giving *or failure to give* is the primary cause of the injury or damage [22]."

Some courts have gone far indeed in expanding the area of the architect-engineer's legal liability. In Mississippi [23] an architect named Malvaney approved a progress payment to the general contractor near the end of the job, in reliance on the contractor's oral assurance that all subcontractors and materialmen had been paid in full. It later turned out that the general contractor had not fully paid his subcontractors and liens were filed against the job. The surety on the contractor's performance bond paid off the subcontractors and then sued the architect to recover its loss. In vain did counsel for the architect argue that his client owed no duty to the surety who was not privy to the owner-architect agreement; and further, that it was the surety's own principal, that is, the general contractor who deceived the architect, and who was in fact the real villain in this piece. The court, however, held the architect liable to the bonding company, and this case was soon followed by a somewhat similar holding in Minnesota [24].

A problem that frequently arises deals with the responsibility assumed by an architect when he hires competent engineers to prepare the various engineering drawings. If the engineering plans prove to be inadequate, may the owner or a third party hold the architect responsible on the basis of negligent design?

In a simpler age an architect and a contractor constituted the entire team in constructing a building. Today the erection of any multistory structure has become so complex that one cannot expect an architect to possess such detailed engineering knowledge as to be capable of preparing special plans and specifications for the structural work, the electrical work, the heating and ventilating work, etc. These plans must be prepared by specialists in the particular branch of the engineering profession. The architect's function is to prepare the overall design for the building, farm out the engineering work to competent consulting engi-

neers, and then coordinate these engineering designs with the architect's own design so as to produce a building both functional and eye-pleasing.

Realistically, one must accept the fact that an architect is normally not qualified to determine whether every detail in each engineering design is technically correct, because of the great complexity of modern engineering techniques and devices. The architect must therefore rely on his engineer, who is an independent contractor. In the usual case the engineer performs his work independent of direction by the architect, hence there is no principal-agent relationship and the doctrine of *respondeat superior* does not obtain. It follows that when the architect has used the same degree of skill used by other architects, defects in an engineering design, although adopted by the architect, should not render the architect liable unless he has specifically assumed such obligation in the owner-architect agreement. The negligence of his engineer, acting as an independent contractor, should not be imputable to the architect who exercised due care and his best judgment in the selection of such engineer.

It has been argued that the architect can avoid this problem by requesting that the owner hire directly his own engineers (as is sometimes done) to work with the architect. This is normally not feasible because the architect is better qualified than the owner to select the various engineers. The architect will select those engineers whom he believes are competent and with whom he will have a harmonious relationship. "Leadership" by the architect is necessary to accomplish the task of coordination of the engineer's plans with those of the architect, but this should not result in holding the architect as guarantor of his engineer's plans. The duty of reasonable care and best judgment should remain the criterion of the architect's liability to the owner.

Not to be overlooked in this discussion is the role of the vendor or materialman in the construction arena. If material delivered to the job proves to be defective or not in accordance with specifications, the supplier is, of course, liable. So, too, if a vendor or manufacturer warrants that his product is fit for the purpose for which the architect has indicated he will use it, such vendor is liable for any resultant damages if the product fails to measure up to the warranties. However, an architect or engineer may not rely on the manufacturer's or vendor's representations or brochures in specifying a new, untested material. He will be liable for the failure of such material if it is not adequate for the job; but in such event, the architect-engineer may look to the vendor for indemnification. Needless to say, the designer who specifies a new material, and does not test it himself, should require written warranties as to

the sufficiency of the material for the purpose intended, and he should seek to obtain a written indemnification from the vendor in the same writing so as to protect himself in the event of the failure of the material.

In Pennsylvania [25] an architect designed a rayon and nylon mill, and the plans called for the upper portion of this mill to be air-conditioned so as to maintain a constant temperature of 80° F plus a relatively high humidity. The architect provided for a vapor seal to be installed in the roof to prevent leakage of moisture from the outside and also to prevent condensation of moisture on the inside. After the construction work was completed the owner sued the architect claiming that his specifications were faulty in that the glass fiber insulation material specified by the architect was not suitable as a vapor seal.

At the trial the architect testified that he had originally intended to specify foamglass insulation instead of glass fiber, but because of the then existing shortage of foamglass he substituted glass fiber in reliance on the advertisements of the manufacturer that this material was suitable for insulation. The court held the architect liable in damages to the owner, stating:

"Viewed in this light, there is ample evidence in the record to prove that the defendants were aware of the intended use of the building; that the problem of proper insulation was of paramount importance; that they recommended the use of insulating material which proved to be unsatisfactory; that this material would absorb and retain moisture which fact was within their knowledge; that they made no previous tests of this material nor did they have any knowledge, specifically, of where it had been satisfactorily used for similar buildings; that they were aware that, unless the design and construction was such as to create and maintain a hermetically sealed envelope about the insulation material, moisture would infiltrate causing it to lose its insulating efficiency and function; that they did not specify in the plans that such a complete envelope or enclosure be constructed; and, that in fact such was not done. All of these facts contained in the record clearly constituted a basis for the fact finding tribunal to conclude that negligence existed."

We have deliberately avoided discussing the owner's responsibility in construction cases because that is a subject by itself. However, we feel impelled to point out that it is the owner who is often the real but undisclosed villain. This is especially true when the owner is planning a speculative type building so that virtually the entire cost will be paid for out of the mortgage money. If, when bids are received, the cost exceeds the owner's tight budget, there is an immediate squeeze on the architects, the engineers, and the contractors to find a way of reducing the cost. This

price squeeze is often reflected in a structure that represents something less than quality work and quality material. The deficiencies may usually be observed in inadequately performing heating systems, poor mechanical installations, plus high cost of maintaining mechanical equipment

The evils brought about by excessive price cutting are rarely ascribed to the owner. The legal responsibility never shifts from the contractors or the architect or the engineers who undertake what are often difficult if not impossible projects, hoping somehow that they will work out satisfactorily. When such a building fails, causing either property damage or personal injuries, it is the architect-engineer or the contractor, not the owner, who will ultimately bear the legal liability for the failure and who will be required to respond in damages.

Notes

[1] *White* v. *Pallay*, 247 Pac. 316 (Ore., 1926).

[2] *Chapel* v. *Clark*, 76 N.W. 62 (Mich., 1898); *Bayshore Dev. Co.* v. *Bonfoey*, 78 So. 507 (Fla. 1918).

[3] See Roady and Andersen, *Professional Negligence*, Vanderbilt University Press, 1960.

[4] 163 N.W. 1002, 1008 (Mich., 1917). See also 5 Am. Jr. 2d, Architects, sections 8 and 23.

[5] "Malpractice" bears no sinister implications. It means nothing more than negligent conduct by a professional.

[6] 176 F.S. 2d 159 (R.I., 1959).

[7] The weighing of comparative culpabilities or the relative delinquencies of the wrongdoers appears to be the real test for determining who is the active and who is the passive tortfeasor. See *O'Dowd* v. *American Surety Company of New York*, 144 N.E. 2d, 359 (N.Y., 1957).

[8] 128 So. 2d 661 (La., 1961).

[9] *Curtin* v. *Summerset*, 21A.244 (Penn., 1891).

[10] *Ford* v. *Sturgis*, 14 Fed. 2d 253 (D.C., 1926).

[11] 111 N.E. 1050 (N.Y., 1916).

[12] 205 N.Y.S. 2d 60, aff'd 9 N.Y. 2d 829, 175 N.E. 2d 350 (1960).

[13] 115 N.Y.S. 2d 425, aff'd 196 N.Y. 576, 90 N.E. 661 (1909).

[14] *Maltese* v. *Reader-Green et al.*, N.Y. Sup. Ct., Kings Co., N.Y.L.J., March 30, 1967, p. 19, col. 4.

[15] 43 *Corpus Juris Secondum*, 27.

[16] *Acevido* v. *Bella Homes*, Index No. 13, 260/62.

[17] 260 N.Y.S. 2d 670, 24 A.D. 2d 476 (N.Y., 1965).

[18] 3 N.Y. 2d 137, 143 N.E. 2d 895, 59 A.L.R. 2d 1072 (1957).

[19] To implead is to bring into the action as a third-party defendant.

[20] Cf. *Miller* v. *DeWitt*, 208 N.E. 2d 249 (Ill., 1965), where the architect was held to be an active tortfeasor under the Illinois Structural Work Act and therefore not permitted to obtain indemnification from the contractor.

[21] New York General Obligations Law, section 5-324, eff. September 1, 1965.
[22] We would eliminate the portions that we have underscored. Because the architect does not normally direct the contractor as to the manner of his performance or instruct him as to proper construction techniques or procedures, the indemnity agreement should not be qualified with the words "or the failure to give."
[23] *National Surety Company* v. *Malvaney*, 72 So. 2d 424 (Miss., 1954).
[24] *Peerless* v. *Thorshov & Cerney*, 199 F.S. 2d 951 (Minn., 1961).
[25] *Bloomsburg Mills, Inc.* v. *Sordoni*, 164 A. 2d 201 (Penn.).

Index

)5 A

5 A

7 A

3 A

8 A

2